Towards a Smarter Battery Management System: 2nd Edition

Towards a Smarter Battery Management System: 2nd Edition

Guest Editors

Chris Mi

Zhi Cao

Naser Vosoughi Kurdkandi

MDPI

Basel • Beijing • Wuhan • Barcelona • Belgrade • Novi Sad • Cluj • Manchester

Guest Editors

Chris Mi
Department of Electrical and Computer Engineering
San Diego State University
San Diego, CA
USA

Zhi Cao
Department of Electrical and Computer Engineering
San Diego State University
San Diego, CA
USA

Naser Vosoughi Kurdkandi
Department of Electrical and Computer Engineering
San Diego State University
San Diego, CA
USA

Editorial Office

MDPI AG
Grosspeteranlage 5
4052 Basel, Switzerland

This is a reprint of the Special Issue, published open access by the journal *Batteries* (ISSN 2313-0105), freely accessible at: https://www.mdpi.com/journal/agriculture/special_issues/J881G82LXU.

For citation purposes, cite each article independently as indicated on the article page online and as indicated below:

Lastname, A.A.; Lastname, B.B. Article Title. *Journal Name* **Year**, *Volume Number*, Page Range.

ISBN 978-3-7258-4423-4 (Hbk)
ISBN 978-3-7258-4424-1 (PDF)
https://doi.org/10.3390/books978-3-7258-4424-1

Contents

batteries

MDPI

Editorial

Towards a Smarter Battery Management System

Zhi Cao, Naser Vosoughi Kurdkandi and Chris Mi *

Department of Electrical and Computer Engineering, San Diego State University, San Diego, CA 92182, USA; zcao2@sdsu.edu (Z.C.); nvosoughikurdkandi@sdsu.edu (N.V.K.)
* Correspondence: cmi@sdsu.edu

Batteries play a critical role in achieving a sustainable energy future, enabling the integration of renewable energy sources and supporting electrified transportation and smart grids [1–3]. Advanced Battery Management Systems (BMSs) are essential in harnessing the potential of various battery chemistries. BMSs ensure safety, optimize performance, and prolong battery lifespan through advanced monitoring, state estimation, thermal management, and fault diagnostics [4,5]. Recent advancements in algorithms, sensors, and hardware have significantly enhanced the capabilities and intelligence of BMSs, making them increasingly adaptive and efficient.

The first edition of this Special Issue, "Towards a Smarter Battery Management System", gained remarkable success, with 11 high-quality papers published, covering essential topics related to smart BMS solutions. Inspired by these systems' reception and the continuous developments in the field, this second edition expands upon these critical research areas to further highlight the latest research and perspectives. It covers diverse topics, including advanced modeling techniques, state-of-health (SOH) and state-of-charge (SOC) estimation algorithms, battery balancing technologies, battery durability, second-life applications, and emerging chemistries such as sodium-ion batteries. The issue attracted strong interest, receiving 11 high-quality submissions, reflecting ongoing advancements and diverse research efforts within the BMS field.

Research Papers:

1. Advanced Algorithms for State Estimation

Akram et al. [6] developed a novel SOH estimation model that integrates Distribution of Relaxation Time (DRT) parameters with a Long Short-Term Memory (LSTM) neural networks. This hybrid approach enables the capture of both electrochemical dynamics and temporal trends, significantly improving estimation accuracy and adaptability to various cycling conditions.

LeBel et al. [7] conducted a detailed analysis on the impact of entropy change in lithium-ion battery electro-thermal modeling. By incorporating entropy change into thermal prediction frameworks, the study achieved better alignment with experimental temperature profiles and improved the fidelity of battery thermal models.

2. Battery Testing and Durability

Neupert et al. [8] proposed innovative data-driven load cycle generation methods for battery testing. By leveraging Gradient Random Pulse strategies and advanced Generative Adversarial Networks (GANs), they produced synthetic profiles that closely resemble real-world usage, enabling more robust and flexible battery evaluation.

Tian et al. [9] evaluated the degradation behavior of lithium-ion batteries under frequency regulation conditions. Their findings provided critical insights into how high-rate cycling impacts capacity fade and electrochemical stability, which is vital for grid-supporting applications.

Muresanu and Dudescu [10] investigated the structural integrity of cylindrical lithium-ion cells under mechanical compression. Through experimental and simulation-based approaches, they characterized deformation modes and provided recommendations for mechanical protection design.

3. Hardware Innovations

Song et al. [11] introduced a new inductor-based active balancing circuit capable of operating efficiently across a wide voltage range. The proposed hardware architecture reduces balancing time and energy loss, offering practical improvements for high-energy battery packs.

Martínez-López et al. [12] investigated flow dynamics in organic redox flow batteries. By introducing electrode obstacles to guide electrolyte movement, they demonstrated improved concentration distribution and mass transport efficiency, which can boost the performance and durability of flow battery systems.

4. Second-Life Management

Cao et al. [13] assessed the second-life potential of commercial $LiFePO_4$ battery packs retired from electric vehicles. Their study evaluated capacity consistency, balancing challenges, and overall system integration, offering practical strategies for repurposing used batteries in stationary energy storage applications.

Review Papers:

Andrenacci et al. [14] reviewed battery storage developments within European smart mobility contexts, discussing performance, safety, regulatory challenges, and sustainability considerations.

Bača et al. [15] systematically reviewed sodium-ion batteries, presenting a detailed analysis of their properties, advantages, challenges, and suitability for stationary storage applications.

Jose et al. [16] provided an in-depth exploration of artificial intelligence applications in battery recycling processes, emphasizing their role in enhancing recycling efficiency and environmental sustainability.

These contributions significantly advance the field of smart battery management systems, providing essential references for future research and practical applications. Future studies might further address advancements in the software and hardware of intelligent BMSs, the integration of BMSs with emerging battery chemistries, and standardized approaches to managing second-life batteries. We encourage researchers to submit their work to upcoming editions of this series.

Funding: Financial support from the California Energy Commission, grant number EPC-19-053, is gratefully acknowledged.

Acknowledgments: We express our sincere gratitude to all authors, reviewers, and the editorial team at *Batteries* for their invaluable support in realizing this Special Issue.

Conflicts of Interest: The authors declare no conflicts of interest.

References

1. Nyamathulla, S.; Dhanamjayulu, C. A Review of Battery Energy Storage Systems and Advanced Battery Management System for Different Applications: Challenges and Recommendations. *J. Energy Storage* **2024**, *86*, 111179. [CrossRef]
2. Waseem, M.; Ahmad, M.; Parveen, A.; Suhaib, M. Battery Technologies and Functionality of Battery Management System for EVs: Current Status, Key Challenges, and Future Prospectives. *J. Power Sources* **2023**, *580*, 233349. [CrossRef]

3. Tran, M.-K.; Panchal, S.; Khang, T.D.; Panchal, K.; Fraser, R.; Fowler, M. Concept Review of a Cloud-Based Smart Battery Management System for Lithium-Ion Batteries: Feasibility, Logistics, and Functionality. *Batteries* **2022**, *8*, 19. [CrossRef] [PubMed]
4. Krishna, T.N.V.; Kumar, S.V.S.V.P.D.; Srinivasa Rao, S.; Chang, L. Powering the Future: Advanced Battery Management Systems (BMS) for Electric Vehicles. *Energies* **2024**, *17*, 3360. [CrossRef]
5. Cao, Z.; Gao, W.; Fu, Y.; Mi, C. Wireless Battery Management Systems: Innovations, Challenges, and Future Perspectives. *Energies* **2024**, *17*, 3277. [CrossRef]
6. Akram, A.S.; Sohaib, M.; Choi, W. SOH Estimation of Lithium-Ion Batteries Using Distribution of Relaxation Times Parameters and Long Short-Term Memory Model. *Batteries* **2025**, *11*, 183. [CrossRef]
7. LeBel, F.-A.; Messier, P.; Blanchard, M.; Trovão, J.P.F. Evaluating the Role of Entropy Change in Lithium-Ion Battery Electro-Thermal Modelling. *Batteries* **2025**, *11*, 84. [CrossRef]
8. Neupert, S.; Yao, J.; Kowal, J. Advanced Load Cycle Generation for Electrical Energy Storage Systems Using Gradient Random Pulse Method and Information Maximising-Recurrent Conditional Generative Adversarial Networks. *Batteries* **2025**, *11*, 149. [CrossRef]
9. Tian, Y.; Wang, L.; Liao, C.; Yan, G. Comprehensive Investigation of the Durability of Lithium-Ion Batteries Under Frequency Regulation Conditions. *Batteries* **2025**, *11*, 75. [CrossRef]
10. Muresanu, A.D.; Dudescu, M.C. Modelling of a Cylindrical Battery Mechanical Behavior under Compression Load. *Batteries* **2024**, *10*, 353. [CrossRef]
11. Song, H.; Hredzak, B.; Fletcher, J. Inductor-Based Active Balancing Topology with Wide Voltage Range Capability. *Batteries* **2025**, *11*, 77. [CrossRef]
12. Martínez-López, J.; Fernández-Gamiz, U.; Sánchez-Díez, E.; Beloki-Arrondo, A.; Ortega-Fernández, Í. Enhancing Mass Transport in Organic Redox Flow Batteries Through Electrode Obstacle Design. *Batteries* **2025**, *11*, 29. [CrossRef]
13. Cao, Z.; Gao, W.; Fu, Y.; Turchiano, C.; Vosoughi Kurdkandi, N.; Gu, J.; Mi, C. Second-Life Assessment of Commercial LiFePO4 Batteries Retired from EVs. *Batteries* **2024**, *10*, 306. [CrossRef]
14. Andrenacci, N.; Vitiello, F.; Boccaletti, C.; Vellucci, F. Powering the Future Smart Mobility: A European Perspective on Battery Storage. *Batteries* **2025**, *11*, 185. [CrossRef]
15. Bača, P.; Libich, J.; Gazdošová, S.; Polkorab, J. Sodium-Ion Batteries: Applications and Properties. *Batteries* **2025**, *11*, 61. [CrossRef]
16. Antony Jose, S.; Cook, C.A.D.; Palacios, J.; Seo, H.; Torres Ramirez, C.E.; Wu, J.; Menezes, P.L. Recent Advancements in Artificial Intelligence in Battery Recycling. *Batteries* **2024**, *10*, 440. [CrossRef]

batteries

MDPI

Article

Second-Life Assessment of Commercial $LiFePO_4$ Batteries Retired from EVs

Zhi Cao [1], Wei Gao [1], Yuhong Fu [1], Christopher Turchiano [2], Naser Vosoughi Kurdkandi [1], Jing Gu [2] and Chris Mi [1,*]

1 Department of Electrical and Computer Engineering, San Diego State University, San Diego, CA 92182, USA; zcao2@sdsu.edu (Z.C.); wgao2@sdsu.edu (W.G.); yfu@sdsu.edu (Y.F.); nvosoughikurdkandi@sdsu.edu (N.V.K.)

2 Department of Chemistry and Biochemistry, San Diego State University, San Diego, CA 92182, USA; cturchiano1590@sdsu.edu (C.T.); jgu@sdsu.edu (J.G.)

* Correspondence: cmi@sdsu.edu

Abstract: $LiFePO_4$ (LFP) batteries are well known for their long cycle life. However, there are many reports of significant capacity degradation in LFP battery packs after only three to five years of operation. This study assesses the second-life potential of commercial LFP batteries retired from electric vehicles (EVs) by evaluating their aging characteristics at the cell and module levels. Four LFP cells and four modules were subjected to aging tests under various conditions. The results indicate that LFP cells exhibit long life cycles with gradual capacity degradation and a minimal internal resistance increase. Module-level analysis reveals significant balance issues impacting capacity recovery. Incremental capacity analysis (ICA) and post-mortem analysis identify the loss of active materials and lithium inventory as key aging mechanisms. This study provides the optimal working conditions of second-life LFP batteries and suggests that, with proper balancing systems, LFP batteries can achieve extended second-life use in stationary energy storage applications, emphasizing the importance of effective balance management for sustainable battery utilization.

Keywords: LFP battery; second-life battery; echelon utilization; battery balance; battery aging

1. Introduction

The adoption of electric vehicles (EVs) has surged in recent years. In 2023, over 14 million EVs and plug-in hybrid vehicles (PHEVs) were sold worldwide, capturing 16% of the total vehicle market. Predictions indicate that EV and PHEV sales will reach 60 million units by 2035 [1]. Consequently, tens of millions of EV battery packs, totaling hundreds of gigawatt-hours, will be decommissioned annually after serving 8 to 15 years [2–4]. Utilizing these retired batteries for second-life applications is considered the most cost-effective and environmentally sustainable option compared to direct material recycling [5,6].

$LiFePO_4$ (LFP) batteries are widely used for EV and battery energy storage system (BESS) applications due to their high-power capabilities [7], low fire risks [8], low cost, and long cycle life [9]. Compared to Li-NCM batteries, which suffer from aging knee issues [10,11], LFP batteries' capacity fading is generally mild, following an approximatively linear pattern at room temperature [12], and the positive electrodes usually do not exhibit a significant decline under moderate ambient temperatures [13].

Various studies have investigated LFP batteries. In ref. [14], a cycle aging study on a commercial LFP cell investigated the influence of the temperature, C-rate, depth of discharge (DoD), and state of charge (SoC) on the capacity and impedance degradation. The findings highlighted the significant impact of the DoD and SoC in terms of capacity reduction and a resistance increase, while the C-rate had a minimal effect. Some works have studied the aging mechanisms of LFP batteries [15,16]. Jiang et al. [17] used incremental capacity analysis (ICA) and IC peak area analysis to identify the aging modes. They found a consistent aging pattern across all batteries studied, characterized by a loss of lithium inventory and active materials on the negative electrodes. Some post-mortem analyses have been conducted to further verify the aging mechanisms of LFP batteries [18]. In ref. [19], macroscopic and microscopic techniques were used to identify the aging phenomenon occurring in the cell on the positive and negative electrodes. The results showed a stable positive LFP electrode but significant deposits and cracking on the negative electrode. Zhu et al. [20] investigated commercial 100 Ah prismatic graphite/$LiFePO_4$ batteries cycled under 45 °C and found that electrolyte loss due to the enhanced evolution of the solid electrolyte interface (SEI) film resulted in the loss of the lithium inventory. They observed that larger LFP particles are prone to fracturing and that Fe deposition accelerates SEI film formation, which blocks graphite layers and impedes lithium intercalation.

Despite the long life cycles of LFP batteries at the cell level, there have been many reports of LFP battery packs in EVs or BESSs retiring earlier than expected, usually due to balance issues [21,22]. Ramirez-Meyers et al. [22] found significant variations and mismatches in the SoH within the modules of used HEV packs, rendering them unsuitable for reuse. Jiang et al. [23] discovered that LFP battery module parameter dispersion increases with battery aging, and SoC imbalance is the main factor contributing to capacity fading in the battery pack.

To study retired LFP batteries and evaluate their second-life value, a study is carried out at both the cell and module levels. The aging characteristics of LFP battery cells under various working conditions are investigated, and the aging mechanisms are revealed by different methods. The health conditions of LFP battery modules are analyzed in terms of their capacity, impedance, balance state, and consistency. Based on the analysis, we provide a comprehensive evaluation of the second-life use of LFP batteries and clarify the prospects for the second-life utilization of LFP batteries with a proper balance system.

The remainder of this paper is organized as follows. In Section 2, the battery cell and module test are introduced. Section 3 analyzes the battery cell aging characteristics, module health conditions, second-life evaluation, recommended second-life working conditions, etc. In Section 4, the battery post-mortem analysis is demonstrated. Conclusions are drawn in Section 5.

2. Experimental Description

2.1. Description of Experimental Procedures with LFP Battery Cells

This study utilizes LFP battery cells manufactured by China Aviation Lithium Battery Co., Ltd. (CALB, Luoyang, China). The main specifications of the CALB LFP batteries are detailed in Table 1 [24]. Four battery cells were collected for this study, and their initial conditions are summarized in Table 2. Cells 01 and 02 were new. According to initial tests, the actual capacity of the new cells was 107 Ah, which was used as the nominal capacity instead of the labeled 100 Ah. Cells 03 and 04 were retired from electric forklifts after four years of service.

Table 1. Specifications of the CALB LFP batteries.

Parameter	Value
Nominal Capacity	100 Ah @ 0.3 C
Nominal Voltage	3.2 V
Charge (CC-CV)	1 C/3.65 V
Charging Time	Standard: 4 h Quick charging: 1 h
Discharge	2 C/2.5 V
Recommended SoC Window	10–90%
Charge Temperature	0–45 °C
Discharge Temperature	0–55 °C
Internal Resistance	≤0.9 mΩ
Weight	3.4 kg
Cell Packaging Type	Prismatic
Dimensions	142 mm × 67 mm × 219 mm
Energy Density	100 Wh/kg

Table 2. Initial conditions and test dates of CALB LFP battery cells.

Cell Number	Initial Capacity (Ah)	SoC	Initial Capacity Test Date	Cycling Test Start Date
Cell 01	107.2	100.2%	July 2020	September 2020
Cell 02	107.1	100.1%		March 2022
Cell 03	95.0	88.8%		September 2020
Cell 04	94.6	88.4%		September 2020

The battery test system shown in Figure 1 was used to conduct accelerated cycling tests on the LFP batteries. A Chroma 17010 battery cell tester, with 24 channels, was employed for the battery aging experiments. Each channel operated within a voltage range of 0 V to 5 V, with a maximum current capacity of 100 A. For testing under varying temperature conditions, a CSZ Plus 8 chamber was used for high temperatures, while a THERMOTRON thermal chamber was utilized for low-temperature tests.

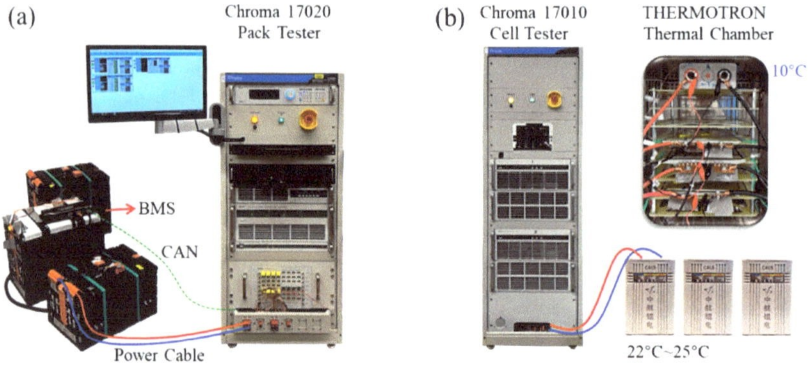

Figure 1. Battery test system: (**a**) battery pack tester; (**b**) The Chinese characters (中航锂电) in the figure stand for CALB.

To study the aging characteristics of the LFP batteries during second-life use, the working voltage range was downscaled from 2.0 V~3.65 V to 2.8 V~3.60 V. The aging tests were divided into different groups, each with different working conditions. In the initial stage, all battery cells were cycled between 3.10 V and 3.45 V. The charge current was 60 A, and the discharge current was 95 to 100 A. In the following stages, the cycling range was adjusted to different levels, such as 2.80 V to 3.45 V, 2.80 V to 3.60 V, etc. The current was also varied in the different stages to study its effect on battery degradation. All battery cells were tested at room temperature (23 °C), except Cell 03, which was tested at a high temperature for some time. The test conditions used in the aging tests were selected based on common operating conditions for LFP batteries in electric vehicles and energy storage systems. The cycling test start dates for all battery cells are given in Table 2. Cell 02 rested for 1.5 years before the cycling aging test.

Figure 2 presents the aging test data for Cell 03 across various stages. Each stage included 55 aging cycles under specific conditions, followed by 5 characteristic test cycles, which were used to assess the battery capacity and impedance. Data during cycling were captured at 5 s intervals. The characteristic tests, performed at room temperature, consisted of a 0.2 C capacity test, two 0.5 C capacity tests, and a hybrid pulse power characterization (HPPC) test to determine the battery impedance. When aging cycles occurred at high temperatures, the thermal chamber was set to 23 °C and allowed to stabilize for eight hours before conducting the characteristic tests. The test conditions remained consistent within each stage to ensure the reliable evaluation of the battery parameters but were adjusted between stages to account for significant changes in capacity and impedance. The

0.2 C capacity from 2.80 V to 3.55 V was considered the cell's capacity, and the HPPC test data were recorded at 1 s intervals for the battery parameter analysis.

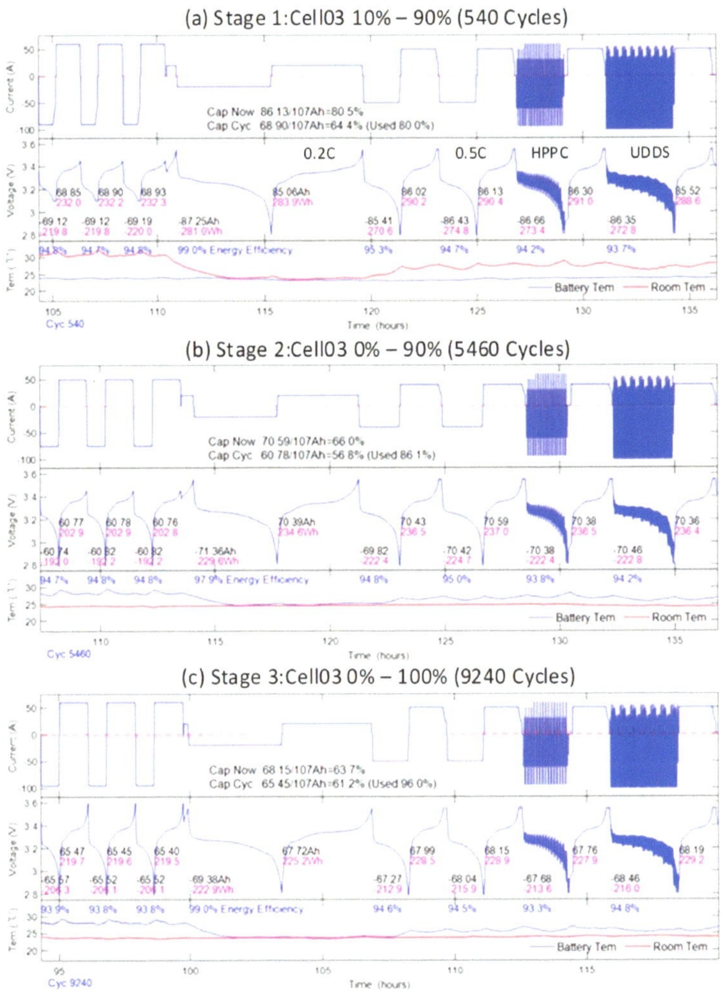

Figure 2. CALB battery cell aging test data. (**a**) Cell 03 after 540 cycles; (**b**) Cell 03 after 5460 cycles; (**c**) Cell 03 after 9240 cycles.

2.2. Description of Experimental Procedures with LFP Battery Modules

The LFP battery modules used in this study were sourced from BYD. Four retired battery modules were collected for the study. After around four to five years of service in California, USA, these battery modules were retired from BYD K9M electric buses. Modules 01 and 02 came from the same vehicle, with a configuration of 1P12S. Modules 03 and 04 were from another vehicle, with a configuration of 1P8S. The main specifications of the BYD LFP batteries are detailed in Table 3.

Table 3. Specifications of the BYD LFP batteries.

Parameter	Value
Nominal Capacity	270 Ah
Voltage Range	2.8–3.8 V (3.2 V Nominal)
Module Weight	59.5 kg
Cell Packaging Type	Prismatic
Dimensions	415 mm × 60 mm × 145 mm
Energy Density	116 Wh/kg
Charge Current	Standard:100 A; Max: 200 A @25 °C
Discharge Current	Standard: 200 A

To evaluate the SoH of the retired BYD battery modules in terms of the module capacity, cell capacity distribution, and internal resistance, the following tests were designed.

(1) A battery module test system was used to test the battery modules and obtain their capacities. The internal resistances of the cells were calculated afterward.
(2) A battery cell tester was used to obtain the cells' capacities within the module. Then, all cells were fully charged to have the modules achieve a top-balanced state.
(3) The modules' capacities were tested again after balancing.

The battery cell test system was the same as the one used for the CALB cell testing. The module test system, shown in Figure 1, consisted of a Chroma 62024P-80-60 DC power supply, a Chroma 63203 DC electric load, a dSpace AutoBox as the main controller, and a BMS to measure the battery cell voltage and temperatures.

2.3. Battery Impedance Calculation

The 1-RC equivalent circuit model was applied to calculate the battery parameters [25,26]. This model comprises a voltage source represented by an open circuit voltage (OCV)–Ah curve, along with an ohmic resistance R_0 and an RC network that includes diffusion resistance R_1 and capacitor C_1. The battery impedance is characterized by the parameters R_0 and R_1. The 1-RC model is structured as follows:

$$\begin{cases} \dot{U}_1 = -\frac{1}{R_1C_1}U_1 + \frac{1}{C_1}I_t \\ U_t = U_{OC} - U_1 - I_tR_0 \end{cases} \quad (1)$$

where U_1 represents the voltage across the RC network, I_t denotes the load current, U_{OC} is the OCV, and U_t refers to the terminal voltage. Its discrete-time format can be expressed as

$$\begin{cases} U_{1,k+1} = \exp(-\Delta t/R_1C_1) \times U_{1,k} + [1 - \exp(-\Delta t/R_1C_1)] \times I_{t,k}R_1 \\ U_t = U_{OC} - U_1 - I_tR_0 \\ D_1 = \exp(-\Delta t/R_1C_1) \end{cases} \quad (2)$$

where k indicates the step, Δt represents the time interval (1 s in this study), $U_{1,k+1}$ is the voltage U_1 at time step $k+1$, $I_{t,k}$ is the current I_t at time step k, and D_1 is the time constant.

The model input is I_t while the output is U_t. Both the input and output are measurable in the physical system. The parameters U_{OC}, R_0, R_1, and D_1 are unknown. The particle swarm optimization (PSO) algorithm is employed to estimate these parameters using the HPPC data. The specifics of the algorithm are detailed in references [10,27] and not repeated here.

3. Results and Discussion

3.1. CALB Battery Cell Aging Test Results

3.1.1. Aging Speed Analysis

Figure 3a–d present the aging trajectories of Cells 01 to 04, including the capacity and internal resistance versus test cycles and full equivalent cycles (FECs). The FEC is defined as the total discharge throughput capacity scaled by the battery's nominal capacity. Retired

Cells 03 and 04 have completed over 11,000 test cycles under varying conditions, with average aging rates of 3.41% and 2.56% per 1000 cycles, respectively. This is equivalent to 32 years of actual operation at one cycle a day. In terms of the FECs, Cells 03 and 04 could complete 7000 cycles. New Cell 01 lost 33.3% of its capacity after 10,390 cycles (8000 FECs), with an average aging rate of 3.275% per 1000 cycles, while Cell 02 lost 19.83% of its capacity within 4510 cycles, with the highest aging rate of 4.397% per 1000 cycles. The initial and final capacities of the four cells are summarized in Table 4. It is evident that LFP battery cells have very long life cycles.

Table 4. Battery cell statistics at the start and end of the aging tests.

Cell Number	01	02	03	04
Initial capacity (Ah)	107.1 (100%)	104.48 (97.64%)	95.0 (88.79%)	94.6 (88.41%)
Final capacity (Ah)	70.69 (66.7%)	83.26 (77.81%)	51.45 (48.08%)	62.53 (58.44%)
Testing cycles	10,390	4510	11,930	11,710
Average aging speed per 1000 cycles	3.275%	4.397%	3.412%	2.560%

Battery Cells 03 and 04 operated for four years before retiring. The estimated first-life cycle count was 1200. Based on their capacity at retirement, the estimated aging rate during first-life use was 9.58% per 1000 cycles. In the aging test, new Cells 01 and 02 were cycled under conditions that mimicked the first-life use of Cells 03 and 04. Their capacity degraded to the initial capacity of Cells 03 and 04 after approximately 1400 cycles. The first-life aging rate in the laboratory test was similar to that in real-world conditions.

Figure 4 presents the working conditions and aging speeds during the aging cycling tests. It is observed that the aging speed is faster at the beginning of the test and then slows down gradually. For example, the aging speed of Cell 01 decreases significantly after 3000 cycles. Even though the discharge cutoff voltage decreases from 3.10 V to 2.80 V during this period, the aging speed does not change. However, a lower discharge cutoff voltage, such as 2.6 V between 5100 and 5700 cycles and 2.5 V between 8400 and 8760 cycles, causes a slight increase in the aging speed. However, increasing the charge cutoff voltage within 3.60 V does not contribute to an accelerated aging speed. Similar aging behaviors are observed in Cells 03 and 04. After 3840 cycles, although the discharge cutoff voltage of Cell 3 decreases from 3.10 V to 2.80 V, the aging speed remains near zero. However, a lower discharge cutoff voltage, such as 2.60 V from 6420 to 7080 cycles and 2.0 V from 10,030 to 10,570 cycles, raises the aging speed. The effect of the charge and discharge current on the aging speed is less evident from the testing results. The testing temperature for Cells 01, 02, and 04 was always 23 °C, while the temperature of Cell 03 was raised from 45 °C to 55 °C between 9680 and 10,160 cycles. The high temperature caused a increase in the aging speed (about 15.5% per 1000 cycles), after which an aging knee occurred. Cells 03 and 04 degraded faster when their capacity dropped to 65% of the nominal capacity after about 10,000 aging cycles. Even with the cycling current reduced to 30 A, the trend did not change.

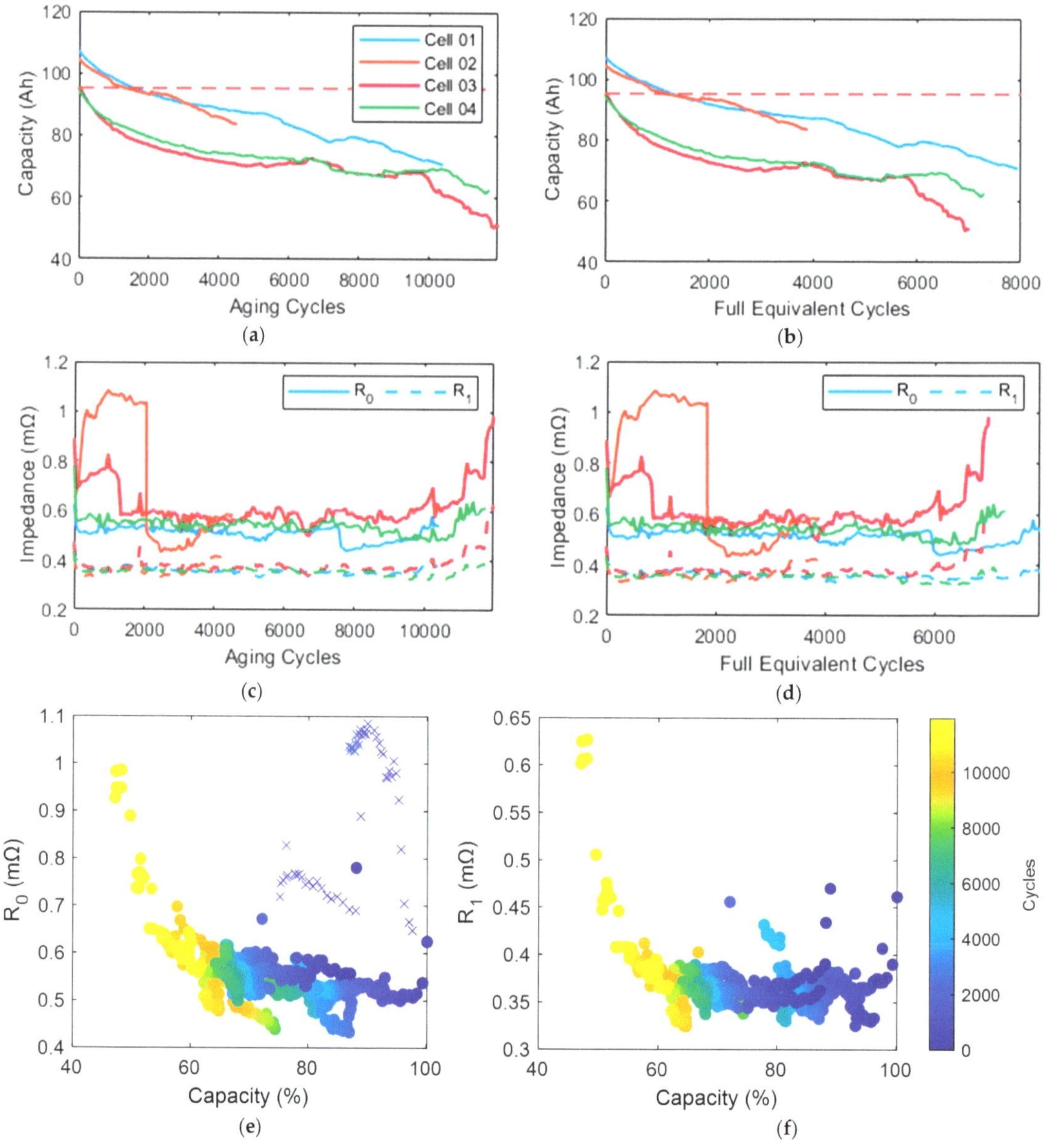

Figure 3. Battery aging test results: (**a**) capacity trajectories versus test cycles; (**b**) capacity trajectories versus full equivalent cycles; (**c**) internal resistance trajectories versus test cycles; (**d**) internal resistance trajectories versus full equivalent cycles. (**e**) Correlation between capacity and internal resistance R_0 (The cross represents the abnormalities of Cells 02 and 03 in the first 1000 to 2000 cycles.); (**f**) correlation between capacity and internal resistance R_1. The dotted lines in (**a**,**b**) indicate the initial capacity level of retired Cells 03 and 04. The solid line in (**c**,**d**) are the internal resistance R_0 while the dotted line in (**c**,**d**) represents R_1.

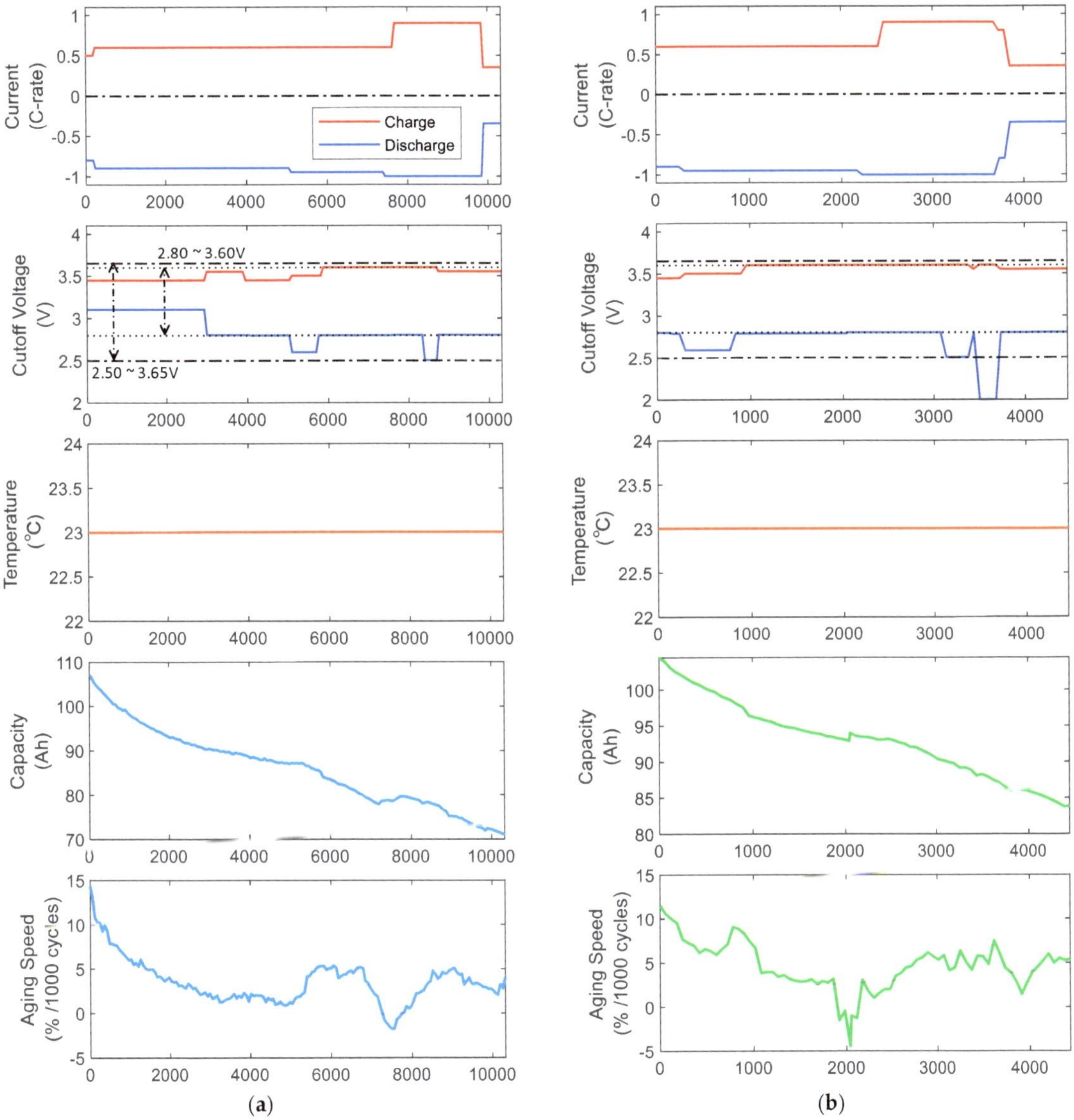

Figure 4. *Cont.*

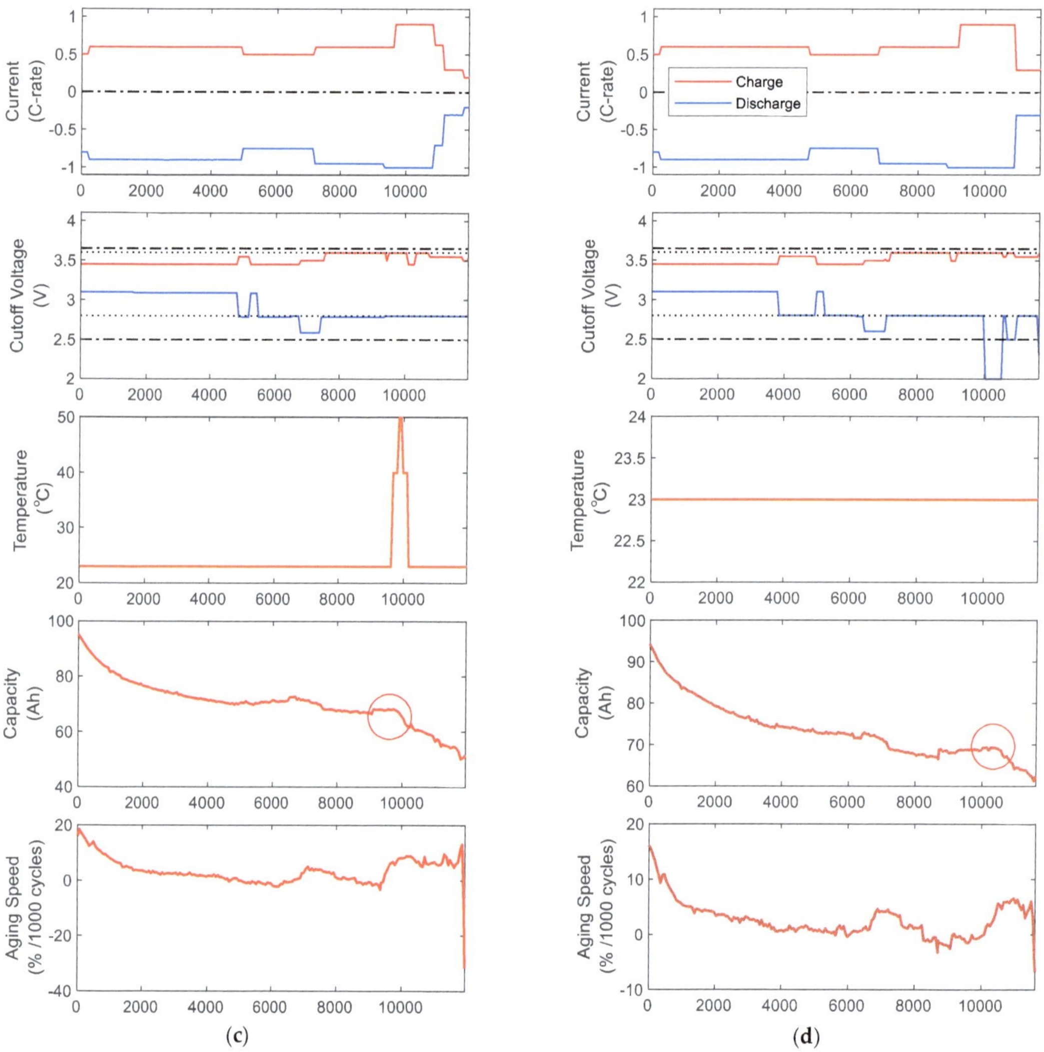

Figure 4. Battery cycling conditions, capacity trajectories, and aging speeds: (**a**) Cell 01, (**b**) Cell 02, (**c**) Cell 03, (**d**) Cell 04.

3.1.2. Internal Resistance Analysis

The internal resistance of the battery cells remains consistently low over 10,000 cycles. R_0 ranges between 0.5 mΩ and 0.6 mΩ, and R_1 is around 0.36 mΩ. The internal resistance starts to increase after 10,000 cycles. The internal resistances of Cells 02 and 03 show an abnormal increase in the first 1000 to 2000 cycles and return to normal levels afterward. Figure 3e,f show the correlation between the impedance and capacity. The impedances of the batteries do not show a noticeable increase until 60% SoH. After this, the battery impedance increases drastically with the capacity degradation. The cross marks in Figure 3e indicate the abnormalities of Cells 02 and 03 in the first 1000 to 2000 cycles.

Low internal resistance ensures good power performance. CALB LFP batteries have Coulomb efficiency of nearly 100%, and the round-trip energy efficiency is about 94% to 95% at 0.6 C charging and 1 C discharge currents. The energy efficiency can be increased to 98% at a 0.2 C current. The energy efficiency does not change across 10,000 aging cycles because of the lack of increase in the internal resistance.

3.1.3. Incremental Capacity Analysis (ICA) to Identify Aging Modes

Incremental capacity analysis (ICA) is a useful tool in identifying battery degradation modes [28,29]. Figure 5 illustrates the dQ/dV curves for Cells 01, 02, and 03, where three distinct peaks are visible. The changes in the intensity and position of these peaks suggest different degradation mechanisms, such as the loss of lithium inventory (LLI), loss of active material (LAM) at the negative electrode (NE), and LAM at the positive electrode (PE) [30,31].

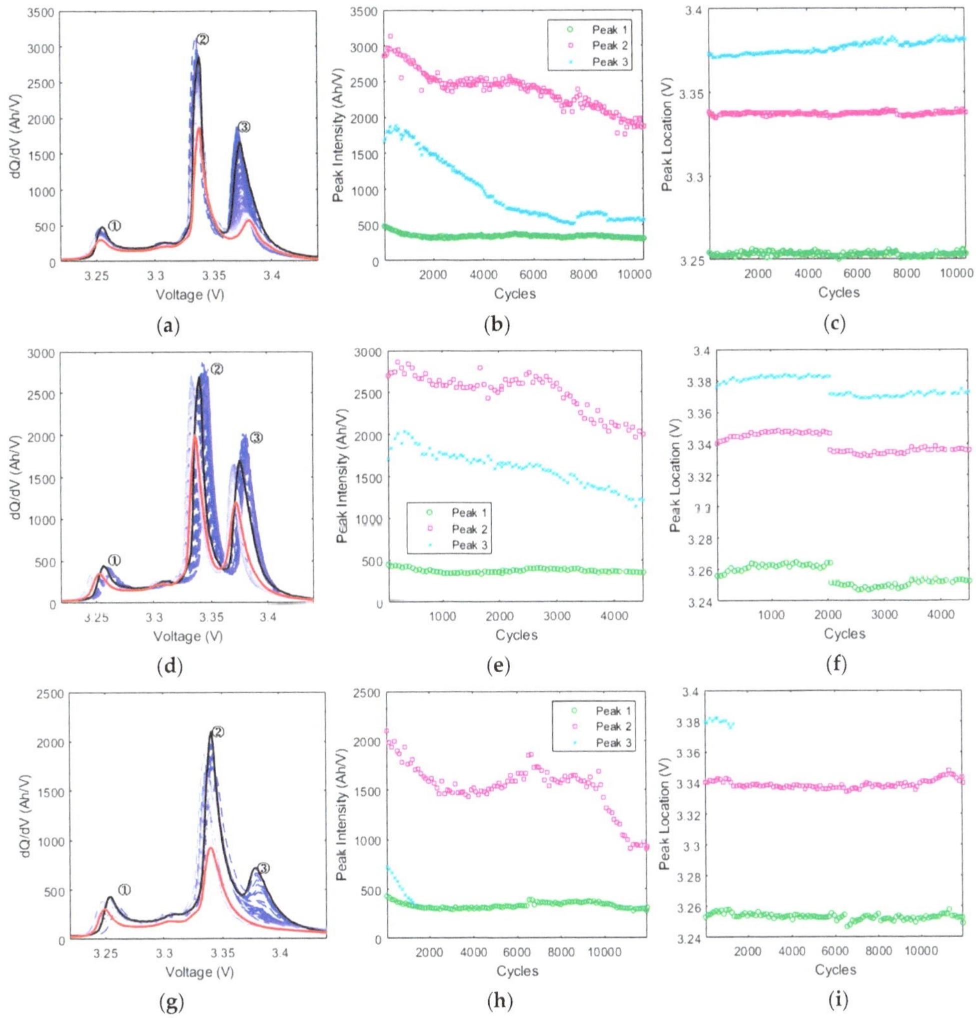

Figure 5. ICA analysis: (**a**,**d**,**g**) dQ/dV curves of battery Cells 01, 02, and 03 (black curves represent the start cycle, red curves represent the end cycle, and blue curves fade as the number of cycles increases). (**b**,**e**,**h**) Peak intensities versus cycle numbers for Cells 01, 02, and 03. (**c**,**f**,**i**) Peak positions versus cycle numbers for Cells 01, 02, and 03.

For new Cells 01 and 02, Peaks 1 and 2 decrease monotonously throughout cycling, implying that a combination of LAM at the NE and LLI causes the capacity to fade. The intensity of Peak 3 grows to a maximum in the first 400 to 500 cycles and then drops rapidly in a linear manner. This is because the LAM at the NE makes the excess relative capacity of the NE available for intercalation [32]. As the LAM at the NE increases, the NE eventually reaches a point where its relative capacity is entirely within the range of the PE. Consequently, Peak 3 stops growing and decreases rapidly. The positions of all three peaks have a similar trend to the internal resistance. Especially for Cell 02, the significant increase in internal resistance in the first 2000 cycles causes the peaks to shift toward higher voltages. No LAM at the PE is inferred from ICA.

For Cell 03, as a retired battery, its Peak 3 does not increase initially but decreases linearly and rapidly until it disappears. The disappearance of Peak 3 coincides with the onset of a slow change in the intensity of Peak 2, implying an adequate lithium inventory. However, the high temperature between 9680 and 10,160 cycles destabilizes the SEI film on graphite, and the film continuously breaks down and reforms, contributing to LLI and a rapid capacity decline after 10,000 cycles [33,34]. Moreover, the shift in Peak 1 indicates that the LLI is accompanied by the LAM of the PE.

3.2. BYD Battery Module Test Results

3.2.1. Module Capacity before and after Balancing

Figure 6 presents the capacities of the four modules before and after balancing. The initial capacities of the four battery modules are 68.2%, 67.3%, 53.9%, and 53.4%, respectively. We assume that the estimated cycle number of the batteries in their first-life use is approximately 1800 (5 years × 365 cycles/year), and the battery cell capacity degradation speed ranges between 8.5% and 15.8% per 1000 cycles.

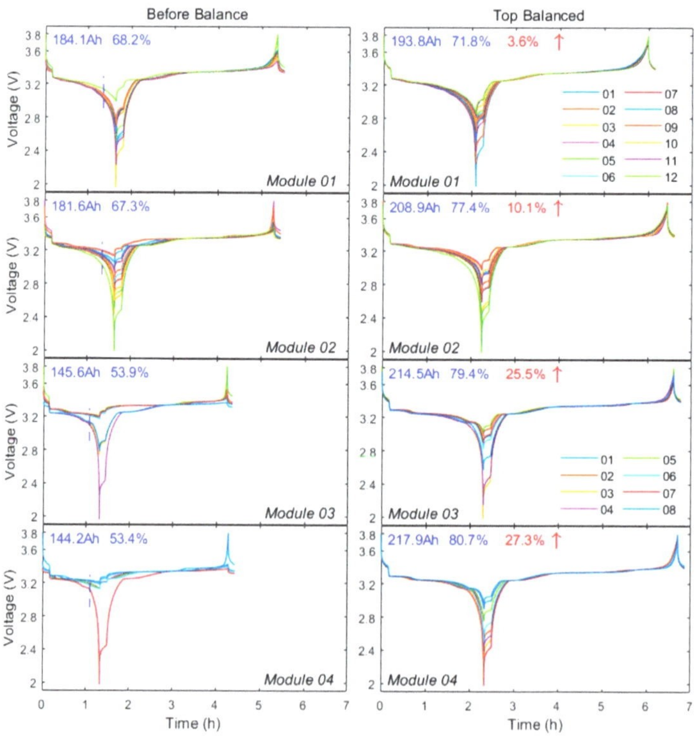

Figure 6. BYD battery module test comparison before and after balancing.

Through the balancing process, their capacities are recovered by 3.6%, 10.1%, 25.5%, and 27.3%, respectively. When breaking down the capacity loss, Modules 03 and 04 exhibit more severe balance issues upon retirement compared to Modules 01 and 02. Module 01 lost 31.8% of its capacity, with 28.2% due to battery cell degradation and 3.6% due to balance issues. Module 02 lost 32.7% of its capacity, with 22.6% due to battery cell degradation and 10.1% due to balance issues. Module 03 lost 46.1% of its capacity, with 20.6% due to battery cell degradation and 25.5% due to balance issues. Module 04 lost 46.6% of its capacity, with 19.3% due to battery cell degradation and 27.3% due to balance issues.

Although the BYD BMS is equipped with a passive balance system, Modules 03 and 04 still exhibited severe balance issues, suggesting that the passive balance system cannot effectively balance the LFP battery pack. In other words, with an effective balance system, the capacities of Modules 03 and 04 could be recovered to about 80%, allowing the battery pack to function for three more years before the capacity drops to 70%.

3.2.2. Analysis of Cell Health Conditions

Figure 7 shows the distribution of the cell capacity, voltage, ohmic resistance R_0, and diffusion resistance R_1 in each module. The values of R_0 and R_1 at about 35% SoC are used for comparison. The maximum and minimum cell capacities and R_0 values are listed in Table 5. Because the internal resistance of the new BYD battery is unknown, the lowest R_0 and R_1 of all of the battery cells, 0.24 mΩ, is considered the benchmark for comparison.

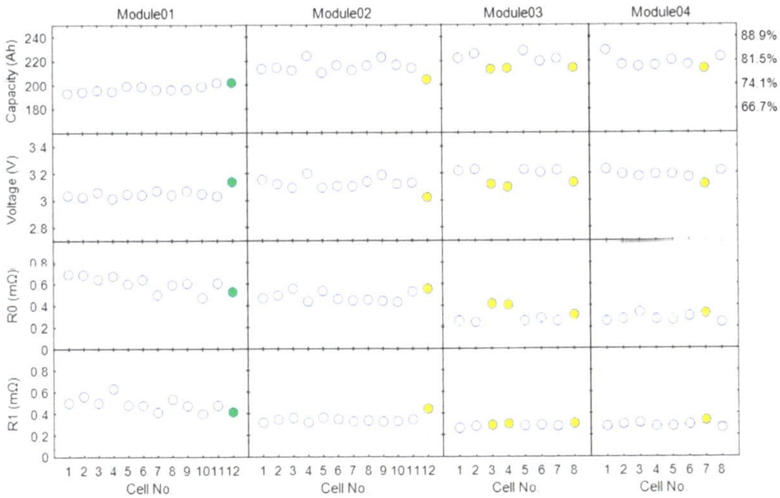

Figure 7. Distribution of cell capacity, voltage, and resistance in each module.

Table 5. BYD battery cell paramcter statistics.

Module Number	01	02	03	04
Maximum Cell Capacity	201.9 Ah (74.8%)	224.2 Ah (83.0%)	228.4 Ah (84.6%)	228.7 Ah (84.7%)
Minimum Cell Capacity	193.2 Ah (71.5%)	206.2 Ah (76.4%)	213.5 Ah (79.1%)	215.3 Ah (79.8%)
Capacity Difference	8.7 Ah (3.2%)	18.0 Ah (6.7%)	15.0 Ah (5.5%)	13.4 Ah (5.0%)
Maximum R_0 (mΩ)	0.67	0.55	0.40	0.33
Minimum R_0 (mΩ)	0.46	0.41	0.24	0.24

In terms of cell capacities, Module 03 and Module 04 have the highest values, followed by Module 02, and Module 01 has the lowest values. The cell capacity difference in a battery module ranges from 3.2% to 6.7%. Regarding the internal resistance, Module 01 has the highest, followed by Module 02 and Module 03, and Module 04 has the lowest. The difference in internal resistance ranges from 30% to 60%. Higher internal resistance leads to reduced round-trip energy efficiency. For example, the round-trip energy efficiency of Module 04-Cell 01 (the best battery cell) is about 95.9% at 0.25 C charging and 0.5 C discharge currents, which is normal. The internal resistance of Module 01-Cell 01 (the worst battery cell) increases to 2.81 times the lowest value, reducing the round-trip energy efficiency to 92.6%, which is considerably low.

Additionally, Figure 7 shows that the internal resistance of BYD batteries has a negative correlation with the capacity. Cells with a lower SoC have lower capacity and vice versa. For example, Cell 12 in Module 01 has the highest SoC, the highest capacity, and the lowest internal resistance. In Module 03, the SoC of Cells 03, 04, and 08 is obviously lower than that of other cells, with correspondingly lower capacity and higher internal resistance.

From the cell perspective, Module 01 is in the worst health condition since its cells have the lowest capacity and the highest internal resistance. The health condition of Module 02 is better than that of Module 01. The cells in Modules 03 and 04 have the highest capacity and lowest internal resistance. However, due to the balance issues, the available capacity of Modules 03 and 04 is lower. Therefore, Modules 03 and 04 are less used than Modules 01 and 02, resulting in the slower degradation of the batteries.

3.2.3. Development of Balance Issues

Multiple factors contribute to the development of LFP battery balance issues, including differences in their internal resistance, temperature, and Coulomb efficiency. During the charging process, higher internal resistance results in a higher voltage and increased heat generation. A higher voltage and temperature facilitate side reactions in the batteries, slightly reducing the Coulomb efficiency and consequently causing SoC variance and balance issues. In addition, LFP battery modules in EVs can experience significant state of charge (SoC) estimation uncertainties due to their voltage characteristics. These uncertainties lead to frequent balance issues [35].

After production, battery cells are tested for their capacity and internal resistance, and only cells with similar values are placed in the same battery pack. Thus, new battery packs usually have good parameter consistency, and balance issues develop slowly in the first few hundred cycles over 2 to 3 years. While an ideally designed balancing system should theoretically achieve near-perfect balance at the end of every full charge event, practical challenges such as cell aging, variations in internal resistance, and operational temperature differences can lead to imbalances that necessitate more frequent correction. This suggests that the current BYD balancing system, although effective under certain conditions, may require further optimization to address these challenges more robustly over the extended lifetime of the battery pack.

In aged LFP battery packs with existing internal resistance differences, balance issues re-emerge faster than in new packs, even after rebalancing. Balance issues become more pronounced as the battery pack ages and require periodic intervention, often more frequently than once a year, especially in second-life applications. This poses a challenge when using second-life LFP batteries in battery energy storage systems (BESSs), highlighting the need for a more effective balance system.

3.3. *Second-Life Evaluation of the LFP Batteries*

Although the BYD and CALB batteries are both LFP batteries, they exhibit different aging characteristics. BYD batteries have a higher energy density of 116 Wh/kg compared to CALB batteries' 100 Wh/kg, but this comes at the cost of inferior aging performance. When the BYD battery's capacity degrades to 70%, the internal resistance increases to at least 2.8 times the benchmark, and the energy efficiency drops to 92.6%. However, CALB batteries maintain consistently low internal resistance until the capacity degrades to 60%. For example, despite the capacity of Cell 03 degrading to 60%, it still shows decent performance, and the round-trip energy efficiency is 94.5% at a 0.5 C current. An abundant electrolyte is crucial in ensuring a long battery cycle life and low internal resistance. However, increasing the energy density of LFP batteries will result in using less electrolyte, thereby reducing the battery's aging performance.

For the second-life usage of CALB batteries, if the working voltage range remains between 2.80 V and 3.55 V (10%~90% SoC), and high working temperatures are avoided, the charge current is 0.5 C, and the discharge current is 1 C, CALB batteries can have long life cycles. The aging speed of LFP batteries in their second life is expected to be 2% to 4% per 1000 aging cycles. If the batteries are repurposed at 80% SoH, they can be used for 5000 to 10,000 cycles before their capacity degrades to 60%, with an estimated second life of 14 to 28 years, assuming one charge and discharge cycle per day. Even when the capacity degrades below 60%, the battery remains still usable with lower currents.

Although LFP battery cells have a very long cycle life, balance issues remain a significant challenge for LFP battery packs. These issues can induce severe pack capacity reductions. In this study, the useful capacity of the four battery modules could be recovered to 71~80% if properly balanced. Assuming that the battery aging speed for second life can be maintained below 4% per 1000 cycles, the estimated second life of the battery modules is 5000 to 7500 cycles before the capacity drops to 50%. If the battery is charged once daily, it can operate for 14 to 20 years.

4. Morphology Characterization

Two CALB cells, Cell 03 (with 50% capacity) and Cell 05 (with 88% SoH), were disassembled for material analysis. A QUANTA FEG 450 scanning electron microscope (SEM) was employed to examine the material morphology at the micrometer scale, utilizing an accelerating voltage of 20 kV. The SEM imaging was conducted under vacuum conditions.

Figure 8 presents the macroscope and SEM images of the positive and negative electrode materials of the two cells. A visual analysis of Cells 03 and 05 in Figure 8a reveals no surface deposits or mechanical changes to the positive electrode between the two cells. In contrast, a dark grey deposit on the separator at the graphite side is found in Cell 03 but not in Cell 06, indicating the exfoliation of the negative electrode materials.

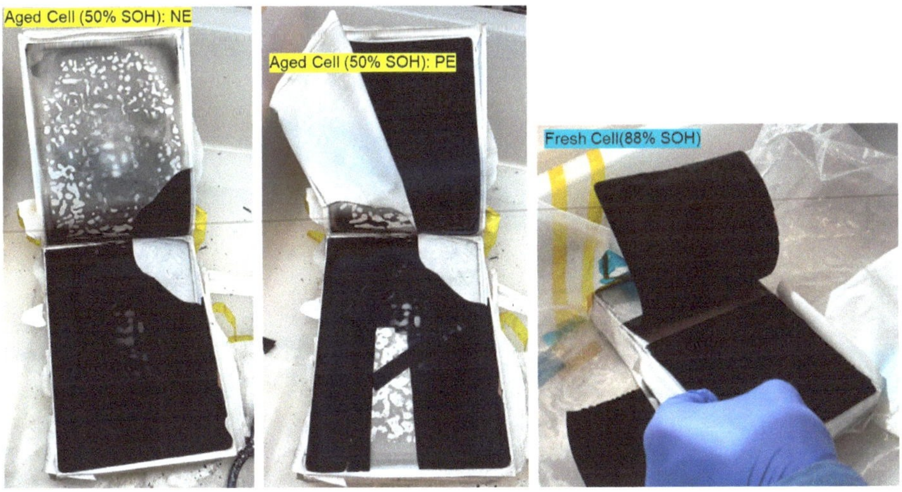

(a)

Figure 8. *Cont.*

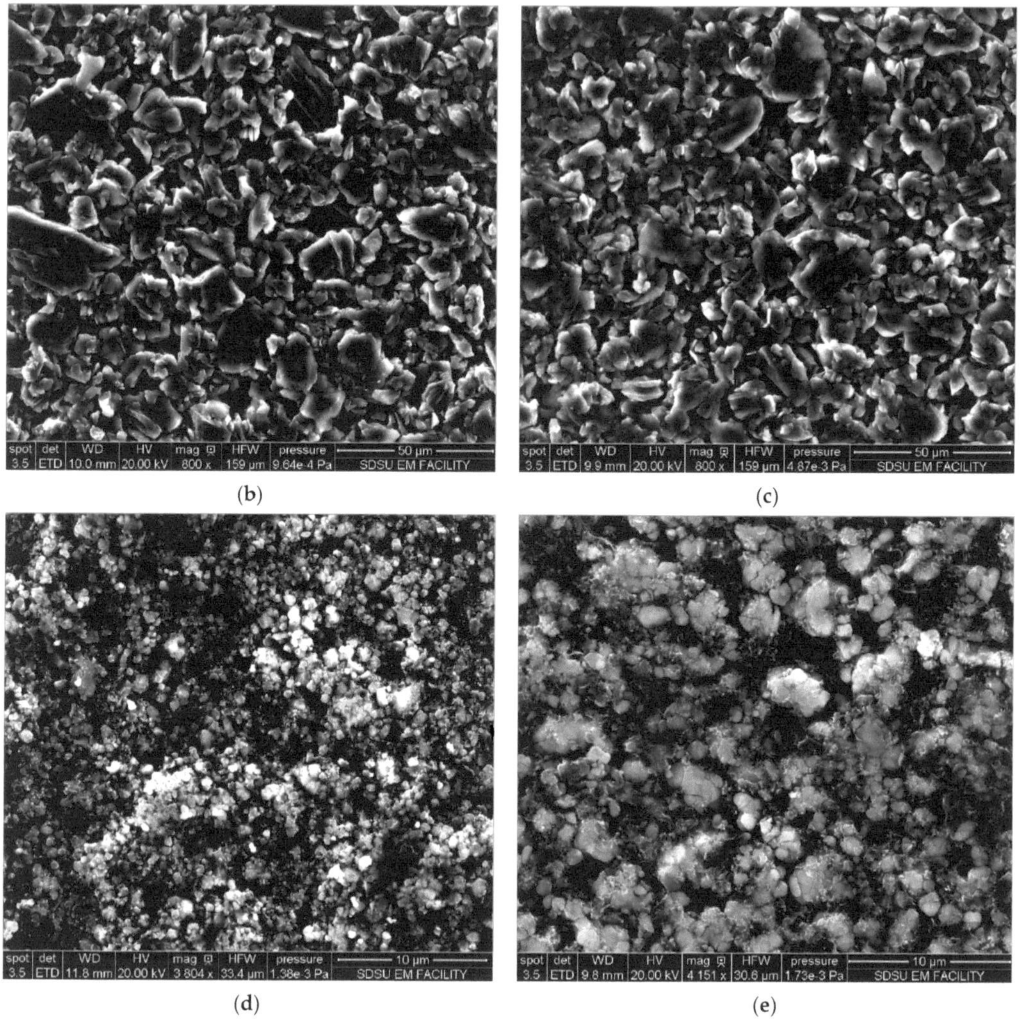

Figure 8. Post-mortem analysis of two CALB LFP cells. (**a**) The macroscopic analysis of the PE and NE materials; (**b**) the SEM image of the fresh cell's graphite electrode; (**c**) the SEM image of the aged cell's graphite electrode; (**d**) the SEM image of the fresh cell's positive electrode; (**e**) the SEM image of the aged cell's positive electrode.

In Figure 8b, the edges of the fresh graphite flakes are slightly more angular compared to those in Figure 8c, which show smoother and more rounded edges. The surface layer is inferred to be SEI reformed in high-temperature cycles at the expense of the lithium inventory. The LFP particles of the fresh electrode in Figure 8d are generally smaller than those of the aged electrode in Figure 8e, and some cracking is visible in the LFP particles of the aged electrode. The high temperatures cause the LFP particles to crack and subsequently aggregate, leading to a reduction in surface area and decreased electrochemical activity. These agglomerates are responsible for the LAM of the PE and the increase in the internal resistance. These findings agree with those in the literature [36,37].

The post-mortem analysis aligns with the aging modes revealed by ICA, indicating that LAM and LLI contribute to the capacity fade during cycling.

5. Conclusions

This work conducted two case studies on retired LFP batteries at the cell and module levels, aiming to specify the aging performance of the battery cells, assess the health conditions of the battery modules, and evaluate their second-life potential. The key findings of this study include the following.

1. LFP battery cells have a very long cycle life. For example, Cell 01 loses 33.9% of its capacity after 10,000 aging cycles, with an aging rate of only 3.26% per 1000 cycles.
2. The battery cells' impedance did not increase in the first 10,000 cycles, which is impressive.
3. For second-life use, if the working voltage range remains between 2.80 V and 3.55 V (10~90% SoC), and high working temperatures are avoided, the charge current is 0.5 C, and the discharge current is 1 C, CALB batteries can exhibit very stable aging performance and achieve sustainable and efficient second-life use.
4. High temperatures induce battery aging knee and cause an impedance increase.
5. LFP batteries generally exhibit excellent performance at the cell level, as demonstrated by the CALB 100 Ah cells. However, the BYD module encounters balancing issues at the pack level, making an effective balancing system essential for LFP batteries.

In summary, this study clarified the prospects for the second-life utilization of LFP batteries and proposed strategies to ensure their long second life and performance. It provides recommendations for both the first-life and second-life operation of EV batteries.

Author Contributions: Conceptualization, C.M. and W.G.; methodology, W.G. and C.M.; software, W.G. and Z.C.; validation, W.G., Y.F, Z.C., C.T. and J.G.; formal analysis, W.G. and Z.C.; investigation, W.G.; resources, C.M. and J.G.; data curation, W.G., Y.F. and Z.C.; writing—original draft preparation, Z.C. and W.G.; writing—review and editing, Z.C., C.M. and N.V.K.; visualization, Z.C., C.T. and W.G.; supervision, C.M.; project administration, C.M.; funding acquisition, C.M. All authors have read and agreed to the published version of the manuscript.

Funding: This research was funded by the California Energy Commission, grant number EPC-19-053.

Data Availability Statement: Data are available on request from the corresponding author.

Conflicts of Interest: The authors declare no conflicts of interest.

References

1. King, N. EVs Forecast to Account for Two Thirds of Global Light-Vehicle Sales in 2035. Available online: https://ev-volumes.com/news/ev/evs-forecast-to-account-for-two-thirds-of-global-light-vehicle-sales-in-2035/ (accessed on 29 July 2024).
2. Hu, X.; Deng, X.; Wang, F.; Deng, Z.; Lin, X.; Teodorescu, R.; Pecht, M.G. A Review of Second-Life Lithium-Ion Batteries for Stationary Energy Storage Applications. *Proc. IEEE* **2022**, *110*, 735–753. [CrossRef]
3. Gao, W.; Cao, Z.; Kurdkandi, N.V.; Fu, Y.; Mi, C. Evaluation of the Second-Life Potential of the First-Generation Nissan Leaf Battery Packs in Energy Storage Systems. *eTransportation* **2024**, *20*, 100313. [CrossRef]
4. Yang, X.-G.; Leng, Y.; Zhang, G.; Ge, S.; Wang, C.-Y. Modeling of Lithium Plating Induced Aging of Lithium-Ion Batteries: Transition from Linear to Nonlinear Aging. *J. Power Sources* **2017**, *360*, 28–40. [CrossRef]
5. Hossain, E.; Murtaugh, D.; Mody, J.; Faruque, H.M.R.; Sunny, H.M.S.; Mohammad, N. A Comprehensive Review on Second-Life Batteries: Current State, Manufacturing Considerations, Applications, Impacts, Barriers & Potential Solutions, Business Strategies, and Policies. *IEEE Access* **2019**, *7*, 73215–73252. [CrossRef]
6. Hua, Y.; Liu, X.; Zhou, S.; Huang, Y.; Ling, H.; Yang, S. Toward Sustainable Reuse of Retired Lithium-Ion Batteries from Electric Vehicles. *Resour. Conserv. Recycl.* **2021**, *168*, 105249. [CrossRef]
7. Nitta, N.; Wu, F.; Lee, J.T.; Yushin, G. Li-Ion Battery Materials: Present and Future. *Mater. Today* **2015**, *18*, 252–264. [CrossRef]
8. Ohneseit, S.; Finster, P.; Floras, C.; Lubenau, N.; Uhlmann, N.; Seifert, H.J.; Ziebert, C. Thermal and Mechanical Safety Assessment of Type 21700 Lithium-Ion Batteries with NMC, NCA and LFP Cathodes-Investigation of Cell Abuse by Means of Accelerating Rate Calorimetry (ARC). *Batteries* **2023**, *9*, 237. [CrossRef]
9. Sockeel, N.; Gafford, J.; Manjrekar, M.; Mazzola, M. Economic Analysis of High-Longevity Battery Capacity Fades for Electric Vehicles Supercharger Buffering Application. *IEEE Trans. Transp. Electrif.* **2020**, *6*, 995–1002. [CrossRef]
10. Gao, W.; Cao, Z.; Fu, Y.; Turchiano, C.; Kurdkandi, N.V.; Gu, J.; Mi, C. Comprehensive Study of the Aging Knee and Second-Life Potential of the Nissan Leaf E+ Batteries. *J. Power Sources* **2024**, *613*, 234884. [CrossRef]
11. Attia, P.M.; Bills, A.; Planella, F.B.; Dechent, P.; dos Reis, G.; Dubarry, M.; Gasper, P.; Gilchrist, R.; Greenbank, S.; Howey, D.; et al. Review—"Knees" in Lithium-Ion Battery Aging Trajectories. *J. Electrochem. Soc.* **2022**, *169*, 060517. [CrossRef]

12. Safari, M.; Delacourt, C. Aging of a Commercial Graphite/$LiFePO_4$ Cell. *J. Electrochem. Soc.* **2011**, *158*, A1123. [CrossRef]
13. Sun, S.; Guan, T.; Shen, B.; Leng, K.; Gao, Y.; Cheng, X.; Yin, G. Changes of Degradation Mechanisms of $LiFePO_4$/Graphite Batteries Cycled at Different Ambient Temperatures. *Electrochim. Acta* **2017**, *237*, 248–258. [CrossRef]
14. Naumann, M.; Spingler, F.B.; Jossen, A. Analysis and Modeling of Cycle Aging of a Commercial $LiFePO_4$/Graphite Cell. *J. Power Sources* **2020**, *451*, 227666. [CrossRef]
15. Dubarry, M.; Liaw, B.Y. Identify Capacity Fading Mechanism in a Commercial $LiFePO_4$ Cell. *J. Power Sources* **2009**, *194*, 541–549. [CrossRef]
16. Dubarry, M.; Truchot, C.; Liaw, B.Y. Cell Degradation in Commercial $LiFePO_4$ Cells with High-Power and High-Energy Designs. *J. Power Sources* **2014**, *258*, 408–419. [CrossRef]
17. Jiang, Y.; Jiang, J.; Zhang, C.; Zhang, W.; Gao, Y.; Li, N. State of Health Estimation of Second-Life $LiFePO_4$ Batteries for Energy Storage Applications. *J. Clean. Prod.* **2018**, *205*, 754–762. [CrossRef]
18. Simolka, M.; Heger, J.-F.; Traub, N.; Kaess, H.; Friedrich, K.A. Influence of Cycling Profile, Depth of Discharge and Temperature on Commercial LFP/C Cell Ageing: Cell Level Analysis with ICA, DVA and OCV Measurements. *J. Electrochem. Soc.* **2020**, *167*, 110502. [CrossRef]
19. Wheeler, W.; Bultel, Y.; Venet, P.; Sari, A.; Riviere, E. Postmortem Analysis of 18650 Graphite/LFP Cells in a Long-Term Aging Study for Second-Life Applications. *Batteries* **2024**, *10*, 119. [CrossRef]
20. Zhu, Y.; Zhu, J.; Jiang, B.; Wang, X.; Wei, X.; Dai, H. Insights on the Degradation Mechanism for Large Format Prismatic Graphite/$LiFePO_4$ Battery Cycled under Elevated Temperature. *J. Energy Storage* **2023**, *60*, 106624. [CrossRef]
21. Etxandi-Santolaya, M.; Canals Casals, L.; Corchero, C. Extending the Electric Vehicle Battery First Life: Performance beyond the Current End of Life Threshold. *Heliyon* **2024**, *10*, e26066. [CrossRef]
22. Ramirez-Meyers, K.; Rawn, B.; Whitacre, J.F. A Statistical Assessment of the State-of-Health of $LiFePO_4$ Cells Harvested from a Hybrid-Electric Vehicle Battery Pack. *J. Energy Storage* **2023**, *59*, 106472. [CrossRef]
23. Jiang, Y.; Jiang, J.; Zhang, C.; Zhang, W.; Gao, Y.; Guo, Q. Recognition of Battery Aging Variations for $LiFePO_4$ Batteries in 2nd Use Applications Combining Incremental Capacity Analysis and Statistical Approaches. *J. Power Sources* **2017**, *360*, 180–188. [CrossRef]
24. CALB CA100 3.2V 100Ah LiFePO4 Lithium Ion Battery—LiFePO4 Battery. Available online: https://www.lifepo4-battery.com/Products/CALB-Battery/CALB-CA100-100Ah.html (accessed on 26 July 2024).
25. Hu, X.; Li, S.; Peng, H. A Comparative Study of Equivalent Circuit Models for Li-Ion Batteries. *J. Power Sources* **2012**, *198*, 359–367. [CrossRef]
26. Wang, X.; Wei, X.; Zhu, J.; Dai, H.; Zheng, Y.; Xu, X.; Chen, Q. A Review of Modeling, Acquisition, and Application of Lithium-Ion Battery Impedance for Onboard Battery Management. *eTransportation* **2021**, *7*, 100093. [CrossRef]
27. Gao, W.; Zou, Y.; Sun, F.; Hu, X.; Yu, Y.; Feng, S. Data Pieces-Based Parameter Identification for Lithium-Ion Battery. *J. Power Sources* **2016**, *328*, 174–184. [CrossRef]
28. Dubarry, M.; Anseán, D. Best Practices for Incremental Capacity Analysis. *Front. Energy Res.* **2022**, *10*, 1023555. [CrossRef]
29. Dubarry, M.; Liaw, B.Y.; Chen, M.-S.; Chyan, S.-S.; Han, K.-C.; Sie, W.-T.; Wu, S.-H. Identifying Battery Aging Mechanisms in Large Format Li Ion Cells. *J. Power Sources* **2011**, *196*, 3420–3425. [CrossRef]
30. Dubarry, M.; Truchot, C.; Liaw, B.Y. Synthesize Battery Degradation Modes via a Diagnostic and Prognostic Model. *J. Power Sources* **2012**, *219*, 204–216. [CrossRef]
31. Dubarry, M.; Beck, D. Perspective on Mechanistic Modeling of Li-Ion Batteries. *Acc. Mater. Res.* **2022**, *3*, 843–853. [CrossRef]
32. Anseán, D.; Dubarry, M.; Devie, A.; Liaw, B.Y.; García, V.M.; Viera, J.C.; González, M. Operando Lithium Plating Quantification and Early Detection of a Commercial $LiFePO_4$ Cell Cycled under Dynamic Driving Schedule. *J. Power Sources* **2017**, *356*, 36–46. [CrossRef]
33. Andersson, A.M.; Edström, K. Chemical Composition and Morphology of the Elevated Temperature SEI on Graphite. *J. Electrochem. Soc.* **2001**, *148*, A1100. [CrossRef]
34. Guo, Z.; Chen, Z. High-Temperature Capacity Fading Mechanism for $LiFePO_4$/Graphite Soft-Packed Cell without Fe Dissolution. *J. Electroanal. Chem.* **2015**, *754*, 148–153. [CrossRef]
35. Halfi, I. NMC vs LFP: Safety and Performance in Operation. Available online: https://powerup-technology.com/nmc-vs-lfp-safety-and-performance-in-operation/ (accessed on 29 July 2024).
36. Liu, P.; Wang, J.; Hicks-Garner, J.; Sherman, E.; Soukiazian, S.; Verbrugge, M.; Tataria, H.; Musser, J.; Finamore, P. Aging Mechanisms of $LiFePO_4$ Batteries Deduced by Electrochemical and Structural Analyses. *J. Electrochem. Soc.* **2010**, *157*, A499. [CrossRef]
37. Simolka, M.; Heger, J.-F.; Kaess, H.; Biswas, I.; Friedrich, K.A. Influence of Cycling Profile, Depth of Discharge and Temperature on Commercial LFP/C Cell Ageing: Post-Mortem Material Analysis of Structure, Morphology and Chemical Composition. *J. Appl. Electrochem.* **2020**, *50*, 1101–1117. [CrossRef]

batteries

MDPI

Article

Modelling of a Cylindrical Battery Mechanical Behavior under Compression Load

Adrian Daniel Muresanu and Mircea Cristian Dudescu *

Department of Mechanical Engineering, Faculty of Automotive, Mechatronics and Mechanical Engineering, Technical University of Cluj-Napoca, 103–105 Muncii Boulevard, 400641 Cluj-Napoca, Romania; adrian.muresanu@campus.utcluj.ro
* Correspondence: mircea.dudescu@rezi.utcluj.ro; Tel.: +40-264-401663

Abstract: The extensive utilization of lithium-ion (Li-ion) batteries within the automotive industry necessitates rigorous measures to ensure their mechanical robustness, crucial for averting thermal runaway incidents and ensuring vehicle safety. This paper introduces an innovative methodology aimed at homogenizing the mechanical response of Li-ion batteries under compression load, using Finite Element Method (FEM) techniques to improve computational efficiency. A novel approach is proposed, involving the selective application of compression loads solely to the Jelly Roll and its casing, achieved by cutting the battery heads. Through this method, distinct mechanical behaviors are identified within the battery force displacement curve: an elastic region, a zone characterized by plastic deformation, and a segment exhibiting densification. By delineating these regions, our study facilitates a comprehensive understanding of the battery's mechanical response under compression. Two battery models were employed in this study: one representing the battery as a solid volume, and another featuring the jelly roll as a solid volume enclosed by a shell representing the casing. The material utilized was LS Dyna MAT24, chosen for its piecewise characteristics' definition, and its validation was primarily conducted through the curve fitting method applied to the force–displacement curve, taking in account the three regions of the compression force behavior. This approach not only optimizes computational resources but also offers insights crucial for enhancing the mechanical stability of Li-ion batteries in automotive applications.

Keywords: Li-ion battery; experimental compression test; material models; quasi-static simulation

1. Introduction

Li-ion batteries are used in many consumer electronic devices, such as laptops and smartphones, as well as in electric vehicles (EVs) and other applications where their high energy density, low self-discharge, and relatively low maintenance make them a suitable choice. However, like any type of battery, Li-ion batteries have the potential to fail, which can lead to hazards, such as thermal runaway, fire and explosion, well documented in [1,2]. One important aspect of Li-ion battery design is crashworthiness, or the ability of the battery to withstand impact, vibration, and other mechanical stresses. This is especially important in EVs, where the battery is a critical component of the vehicle and must be able to withstand the rigors of the road. Crashworthiness is also important in consumer electronics, where drops and other accidental impacts are common.

In the pursuit of enhancing the crashworthiness and safety of Li-ion batteries, a range of mechanical loading scenarios are investigated to evaluate their structural response and potential risks in various conditions. One crucial aspect is the investigation of nail penetration tests, where external objects puncture the battery casing, simulating potential scenarios, such as mishandling or manufacturing defects. A safe, time-efficient, and cost-effective method for studying the nail penetration problem is proposed in [3,4]. Additionally, dropping tests are conducted to simulate accidental falls or impacts during transportation and handling [5]. In the context of automotive safety, Li-ion batteries are subjected to compression loads that replicate the forces experienced during car crashes; the applied crush force emulates a vehicle accident, or any external load force that may damage the battery enclosure and cause its deformation. This subject is discussed extensively in the safety and regulation review papers of [6,7]. The comprehensive investigations found in the literature aid in understanding the battery's behavior under different mechanical loads, identifying potential failure modes, and designing safety measures to mitigate risks associated with thermal runaway, short circuits, and containment breaches [8–15]. In [13], short-circuit criteria are investigated and compared, based on stress, strain, and geometry in simulations with FEM, demonstrating that a combination of radial and axial geometric criteria is optimal; furthermore, these criteria are implemented into a post-processing tool for efficient short-circuit analysis across various loadings and potential mechanical integration in vehicle crash safety.

A Li-ion battery is composed of several essential structural components that collectively enable its energy storage function. The anode and cathode, or the current collectors, are fundamental electrodes that store and release lithium ions during charging and discharging cycles. They are typically made of copper or aluminum and exhibit mechanical properties such as anisotropy, strain hardening, ductile fracture, and rate-dependence. The separator layer, situated between the anode and cathode, serves as an insulating barrier to prevent direct contact between the electrodes. This separator often consists of polymer with orthotropy and elasto-viscoplasticity as mechanical properties. The case shell is made from steel or aluminum sheet, with such mechanical attributes as anisotropy, strain hardening, and ductile fracture, and plays an important part in the structural integrity of the battery [16].

The interaction between these layers can affect the battery's response to mechanical loading, influencing factors such as deformation, stress distribution, and failure modes. Simulating the mechanical properties of these components consumes a great deal of the computing resources in large-scale and complex FEM models. In a detailed model of the battery, there are more than 100,000 elements [16] and a Tesla Model S, for example, has about 7000 cells [17] of type 18650, that will count 700 million elements. In [17], the optimization of the FEM model is investigated by dividing it into beam elements representing different components with separate mechanical properties and reducing the computation time by 90%. This paper proposes a further step, with homogenization of the battery in a simple representative volume for simulations and analyses of the battery's behavior under compression loads, and can be used in larger scale vehicle model safety simulations.

In recent years, computational methods have played an increasingly vital role in investigating the mechanical properties of complex structures, including Li-ion batteries [18]. Finite Element Analysis (FEA) is a powerful numerical technique that can simulate the behavior of materials and structures under various loading conditions. LS Dyna version R11, a commercial software package, is widely employed for such simulations due to its versatility and accuracy in capturing complex material responses [19]. The focus of this study is the mechanical homogenization of a Li-ion cylindrical battery under compression load using LS Dyna Finite Element Analysis [8,20], specifically employing a plastic material model within the software. Plastic material models are commonly used to represent the behavior of materials that undergo large deformation and energy absorption under compression. Efficient computational methods are paramount in large-scale simulations, such as Li-ion battery analysis. The homogenization approach adopted in this study offers a unique advantage by reducing the complexity of the modeling of individual battery components, resulting in significant computational savings. This reduction in computational burden allows for quicker simulations, enabling researchers and engineers to explore a wider range of scenarios and design options. Although the advantages of the homogenized model are evident in terms of computational efficiency, there is a limitation, particularly in accurately predicting jelly roll layer failure. Homogenized models, by simplifying the battery's multi-layered structure, lose some of the granularity required to capture intricate failure mechanisms such as internal short circuits or localized deformation of layers under extreme loading. This limitation is mitigated by supplementing the homogenized models with insights from detailed models or experimental data. In these studies, parameters such as deformation limits and failure thresholds are derived from high-fidelity models or direct testing [20]. These values are then applied to set safety margins in the larger computational models, ensuring that the simulations remain both computationally efficient and reflective of critical safety conditions. This hybrid approach enables researchers to use homogenized models for large-scale safety analysis while still addressing the detailed failure modes that could compromise the integrity of the battery pack under crash conditions.

The specific focus on the mechanical behavior of the homogenization of cylindrical Li-ion batteries using a plastic material model and its computational benefits remains a gap in the current literature. The objective of this paper is to address this gap by systematically studying the mechanical behavior of a Li-ion cylindrical battery under compression load, using a homogenization approach with a plastic material model in LS Dyna. Through a comprehensive simulation study, we aim to provide insights into the macroscopic mechanical properties of the battery, enhance our understanding of its behavior under compression and improve computational efficiency through the homogenization technique, contributing to the broader knowledge of Li-ion battery mechanics.

2. Physical Experiment Description

2.1. Test Description

In the physical test, 18650 Li-ion batteries sourced from Tesla Model S vehicles were subjected to compression testing to ascertain their material properties, followed by validation using Finite Element Method (FEM) modeling. Compression testing offers insights into the batteries' response to external forces, generating a force–deformation curve that unveils pivotal mechanical characteristics. By subjecting the batteries to controlled compression, the material model can be validated against real-world behavior, facilitating accurate simulations for compression scenarios. The validation of material models for Finite Element simulation is very important in ensuring accurate prediction of mechanical behaviors. In this context, the validation of a very simplified model is achieved through a twofold approach, with the emphasis on analyzing the force–deformation curve and comparing the overall deformation patterns. The force deformation curve, obtained from compression testing, serves as a fundamental benchmark. By comparing the simulation results with the experimental force–deformation data, the simplified FEM model's accuracy in capturing the battery's response to external loading is assessed. The combined validation of force–deformation responses and deformation patterns establishes a robust framework for utilizing simplified FEM simulations as predictive tools for optimizing the design, safety, and performance of 18650 Li-ion batteries in more complex, pack or full vehicle safety simulation.

For the 18650 battery the external dimensions are 18 mm radius and 65 mm length. The main relevant structural composition is the shell case, and in the case of Tesla batteries two case shells; a robust enclosure enveloping the battery's core components acts as the initial line of defense against external impacts, fortifying the battery's inner workings. Inside the case shell, there is a jelly roll of the current collectors, anode, and cathode materials, each performing a distinct yet synergistic role. The current collectors are rolled more than 20 times with a cylindrical gap in the middle. Nestled between these electrodes lies the separator, which is rolled together with the current collector (Figure 1a). The battery's mechanically relevant components and their thickness are presented in Table 1.

(a) (b)

Figure 1. The 18650-type battery: (**a**) top view inside the battery, (**b**) battery without the top and bottom cap.

Table 1. Mechanically relevant battery components.

Component	Material	Thickness (mm)
Case	Steel	0.32
Anode	Copper	0.025
Cathode	Aluminum	0.025
Separator	Polyethylene	0.025

In the pursuit of a more refined understanding of the mechanical behavior of the 18650 Li-ion battery under compression, a deliberate and strategic modification was undertaken. Specifically, the upper and lower caps of the battery were cut precisely with the lathe to help ensure precision and a clean cut. The cuts were taken 5 mm from the top, where the shoulder is, and 2 mm from the bottom, and precision was paramount as mm in case length could influence the load results. The battery was discharged for the cutting, but some residual exothermal reaction took place without influencing the main structural components. This surgical alteration was motivated by the recognition that the composition of these caps with steel components could significantly skew the resultant compression forces exerted upon the battery during testing. By adeptly removing this external influence, the focus of investigation was meticulously narrowed to the intrinsic interaction between the battery's jelly roll—housing the anode, cathode, and separator materials—and the robust outer casing. This strategic excision, while seemingly reductive, was extremely purposeful, serving to isolate and elucidate the core mechanical response of the battery's internal components and casing when subjected to compression. The resulting insights, untainted by external cap interactions (Figure 1b), foster a more nuanced comprehension of the interplay between structural elements, furthering the discourse on the battery's mechanical dynamics within the ambit of enhanced safety and operational performance.

The uniaxial compression testing campaign was executed on three samples using an Instron 3366 (10 kN) machine. The compression was executed until a discernible point of densification was achieved, substantiated by the abrupt and pronounced escalation in the force profile. Throughout the entirety of the compression process, a vigilant monitoring of the battery's voltage was meticulously maintained, as a noticeable drop in the value could predict a short-circuit, in order to prevent a hazardous situation, but also to make sure that a short circuit was not taking place before the densification (Figure 2).

Figure 2. Battery uniaxial compression test performed on universal testing machine.

The uniaxial compression test conducted on the 18650 Li-ion battery resulted in a non-linear load–displacement curve of important significance, offering valuable insights into the battery's mechanical behavior under external pressure. This curve can be distinctively categorized into three pivotal zones, each revealing distinct mechanical phenomena, shown in Figure 3. The initial section of the curve, up to an applied load of approximately 2 kN, shows the curve's elastic domain. Here, the response of the battery is characterized by linear elasticity. As force is applied, the battery's deformation remains reversible upon unloading, showcasing its ability to withstand external stress while readily returning to its original shape. This zone illustrates the structural resilience of the battery, encapsulating its capacity to temporarily store mechanical energy before yielding to plastic deformation. The deformation in this zone is the smallest, with displacements typically of around 0.5 mm. The elastic module measured is 565, 431 and 551 MPa, respectively, for the three samples with an average of 515 MPa. As an observation, the knee around 0.2 mm could be the yield factor in the plastic deformation of the case, the jelly roll remaining in the elastic domain. Beyond the elastic limit, Zone 2 commences and extends to the onset of the plateau, signifying a critical juncture in the battery's response. Here, the applied load surpasses the battery's elastic limit, leading to plastic deformation of the structural components. This plastic deformation is notable, as it occurs without a significant increase in the load, resulting in a discernible load plateau. This phase suggests that the battery's core structure undergoes irreversible changes while the load remains relatively constant. It is in this zone that the battery's mechanical integrity begins to be compromised, and deformations are the most significant, reaching values of around 2.5 mm. Zone 3 represents the culmination of the load–displacement curve as the battery undergoes a pronounced densification process. Here, the load increases substantially, reflecting the compression-induced compaction of the battery layers. The battery's internal components experience significant rearrangement and compaction, resulting in a steep rise in the applied load. This phase signifies a critical point in the test, as it indicates the threshold beyond which the battery's structural components reach maximum density. This densification zone unveils the limits of the battery's mechanical tolerance, offering insights into its structural behavior under extreme compression conditions. The test in this zone was stopped at around 7 kN, because densification at that point became a certainty. In summary, the load–displacement curve derived from the uniaxial compression test on the 18650 Li-ion battery illuminates the battery's complex mechanical response. The distinct zones delineate the battery's elasticity, plastic deformation, and ultimate densification, all of which contribute to a comprehensive understanding of its mechanical characteristics and inform crucial aspects of design, safety, and performance optimization in various applications.

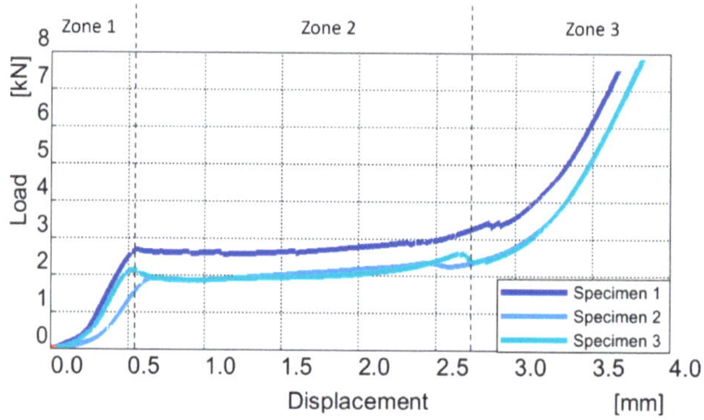

Figure 3. Uniaxial compression of the battery Load vs. Displacement curve.

In the same manner the Jelly Roll of the battery was tested (Figure 4), with similar results as seen in Figure 5. The elastic module measured is 308 and 176 MPa, respectively, for the two samples, with an average of 242 MPa.

Figure 4. 18640 Battery Jelly Roll uniaxial compression test.

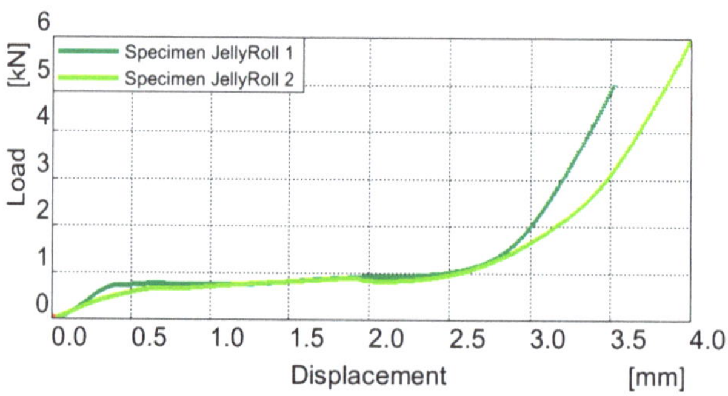

Figure 5. Uniaxial compression of the jelly roll Load vs. Displacement curve.

2.2. Densification and Failure

The process of densification observed during the uniaxial compression test on the 18650 Li-ion battery layers is a phenomenon documented in the existing literature, as corroborated by similar studies [8,20]. This pivotal stage, marked by a pronounced increase in load, serves as a pivotal juncture where the structural elements within the battery experience a remarkable transformation. To delve deeper into this phenomenon, high-resolution light microscopy was employed, capturing detailed images that elucidate the dynamic interplay between the battery's internal layers, while also affirming the presence of densification (Figure 6).

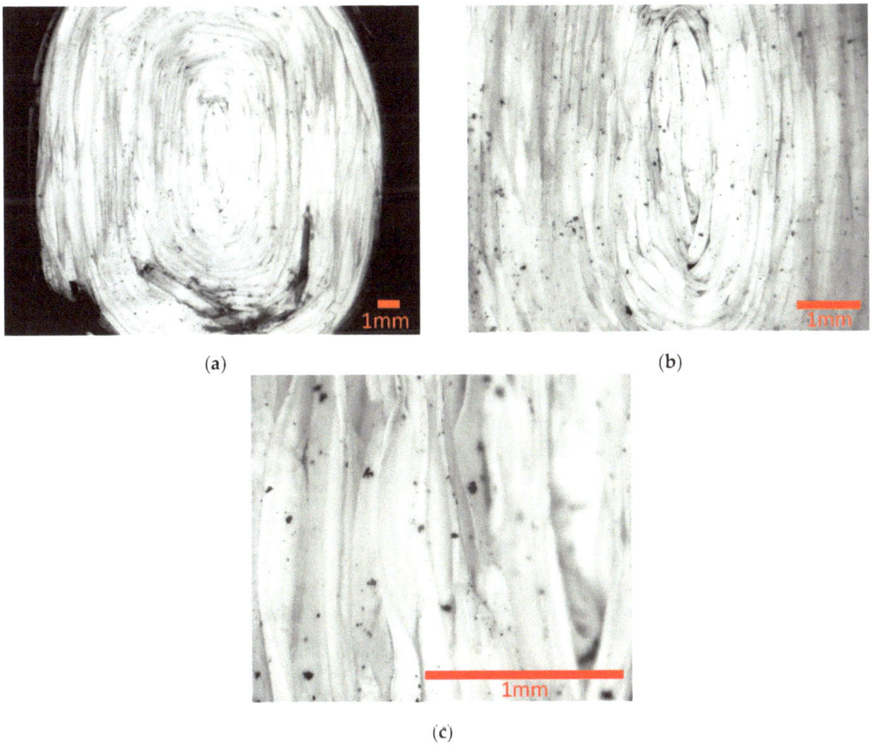

Figure 6. Microscopic pictures of the battery densification: (**a**) overall section; (**b**) middle of the battery section; (**c**) layers up close.

A visual journey into the densification process of the 18650 Li-ion battery layers was facilitated through light microscopy after unloading, capturing three key snapshots (a, b, c), from Figure 6. In picture (a), we gain an overarching perspective, providing an overall view of the battery section. Transitioning to picture (b), we venture further into the core of the battery, a region subjected to relentless compression forces. Notably, the middle hole is conspicuously absent, underscoring the compaction and deformation incurred during the test. Finally, picture (c) takes us to the heart of the matter—a close-up view that offers a high-resolution look at the battery's internal layers. What is particularly interesting in this image is the absence of visible space between these layers. This tangible evidence poignantly captures the culmination of the densification process, as the once-distinct layers now tightly interlock, illuminating the profound structural changes unfolding within the battery during compression. These micrographic snapshots, when examined collectively, serve as an invaluable visual record, unraveling the intricate metamorphosis of the battery's layers and underlining the significance of the densification phenomenon in the realm of Li-ion battery mechanics.

In this stage of densification, failures within the battery jelly roll layers were also observed. These failures (Figure 7) shed light on the critical points of mechanical vulnerability, which, in turn, are intrinsic to understanding the battery's structural limitations.

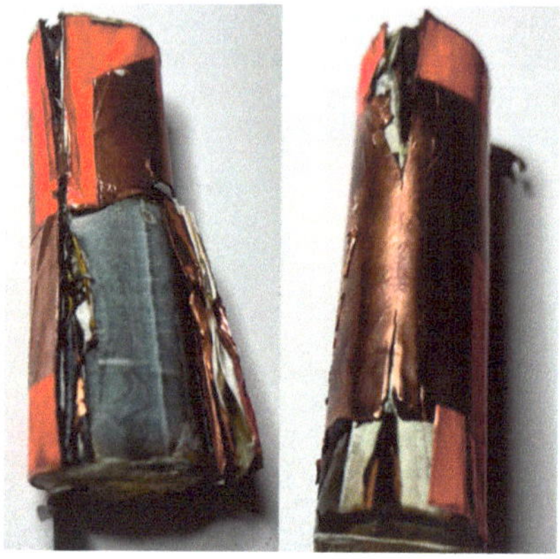

Figure 7. Failure in the Jelly Roll layers.

3. Model and Material Definition

In this simulation conducted with the LS-DYNA solver, the focus was on a volume representation of an 18650 battery, a widely used cylindrical lithium-ion battery cell, which was then employed to simulate the behavior of the battery under compression mechanical loading. This comprehensive simulation aimed to provide a comparable finite element model with the component test, contributing valuable data for further computing optimization and research into a new method of defining an 18650 battery for large-scale and resource-consuming simulation.

3.1. Model Definition

The computational model intricately replicates the experimental test scenario, featuring a rigid component exerting pressure on the battery, with a rigid wall serving as fixed boundary to emulate the lower support plate of the press. The battery geometry is represented using solid elements within a singular part, characterized by material properties derived from empirical data obtained during the experimental phase. This singular part design aims to encapsulate the nuanced and intricate behavior of the battery in a manner that is both straightforward and comparable. Notably, key parameters, such as the stress–strain curve and compressive force were crucial for accurately capturing the dynamic response of the battery. Given the inherent complexity of the actual system, parameters including density and Young's modulus underwent optimization through the simulation process to align the model with the intricate characteristics observed in the real-world scenario.

The simulation model comprises a rigid barrier, defined as a shell part consisting of eight quadrilateral elements, each assigned an LS Dyna rigid material property. This rigid barrier is in contact with a battery, which is situated on an LS Dyna rigid wall surface specifically defined as a flat infinite plane (Figure 8a).

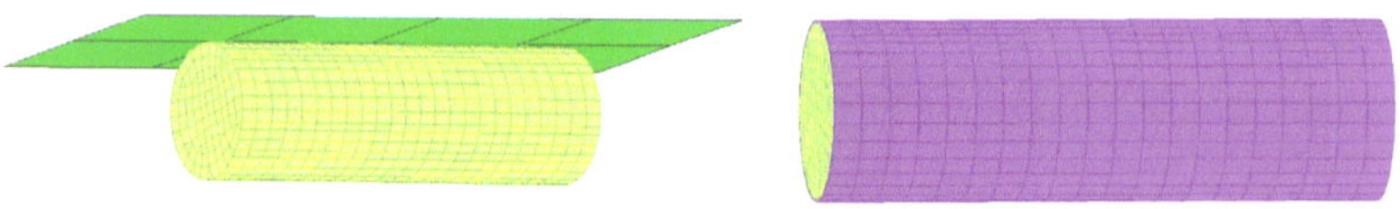

(a) (b)

Figure 8. FE model of the battery cell: (**a**) volume model in contact with a barrier; (**b**) battery model composed of the jelly roll volume and a case.

The force reading function of the rigid wall is activated to monitor and record the forces exerted during the simulation. The battery, a solid part, is represented by 2400 hexahedral elements, each with dimensions of approximately 2.5 × 1.7 × 1.7 mm. These elements collectively form a detailed and comprehensive representation of the battery geometry. This is a homogenization model of the hole battery without the caps. The boundary conditions are established by the rigid wall surface, and the force reading function allows for the real-time tracking of the forces applied to the battery during the simulation. This setup, seen in Figure 8a, enables a thorough examination of the structural behavior and response of the battery under the influence of the rigid barrier, providing important insights into the mechanical aspects of the system.

The same simulation set up is done for the Jelly Roll, but with a diameter offset to match the real dimensions. To add more credibility to the model the case from [21] is added in a battery model composed of the jelly roll volume and a case (Figure 8b) tied together with an LS Dyna contact TIED_SURFACE_SURFACE_OFFSET. The thickness of the case is doubled compared to [21], in order to mimic the usage of the cases in Tesla batterie to 0.64 mm. LS Dyna MAT24 MAT_PIECEWISE_LINEAR_PLASTICITY was used to obtain robust results.

3.2. Material Defining and Parameters' Calibration

Utilizing LS-DYNA's MAT24 material model facilitates a comprehensive analysis of structural integrity and deformation behavior in the 18650 lithium-ion battery under compression. MAT24 in LS-DYNA is a versatile material model that combines linear elasticity with viscoelasticity. It is suitable for simulating the time-dependent response of materials, making it valuable for analyzing the dynamic behavior of structures. The material properties in MAT24 are defined in a piecewise manner, meaning that different regions of the stress–strain curve and Prony series can be specified to accurately represent different phases of material behavior. The Prony series allows MAT24 to capture the time-dependent response of the material, essential for simulating dynamic loading conditions. MAT24 uses the Young modulus for the elastic area of deformation but also incorporates a stress–strain curve, allowing for a detailed representation of the material's plastic deformation characteristics. This curve is defined in a piecewise manner, providing flexibility to capture different phases of material behavior. The material also includes a damping coefficient that accounts for energy dissipation within the material and density.

MAT24 lacks the capability to define anisotropy. However, this limitation has not been a central concern in the present study, as cylindrical lithium-ion batteries primarily exhibit deformation in only two directions. The focus of this paper is specifically on the most critical direction of the battery, during high-impact vehicle crashes, where lateral forces dominate. Axial deformation, while important, occurs less frequently in these high-impact events and is thus outside the scope of this research.

The parameters associated with this material were defined based on the raw testing data and the numerical fitting of the load curve and deformation of the battery. LS SYNA MAT24, a plasticity material model, serves as a foundational element in our simulation framework. This model is adept at simulating the plastic deformation of metals and provides an isotropic, kinematic hardening plasticity model. A notable feature employed in this study is MAT_PIECEWISE_LINEAR_PLASTICITY, which allows for the definition of a piecewise linear stress–strain curve, enhancing the model's precision in mimicking nonlinear material responses. It is crucial to highlight that, for this homogenized battery model, a stress–strain curve derived from hardware tests was utilized as base guideline. Notably, the piecewise definition was scaled with the method of curve fitting as for the rest of the parameters, the resulting values shown in Table 2. Additional material parameters include a Young's modulus of 400 MPa scaled down from the average 515 MPa measured by curve fitting in the elastic area, a Poisson ratio of 0.4, and a density of 2×10^{-6} (kg/mm^3). For the Jelly Roll, the same method is used but with a difference in E Modulus: the value uses is 200 MPa compared with the average measured 242 MPa. The case material, as also found in [22], uses an E Modulus of 200 GPa. The plastic domain is defined by a stress–strain curve starting from the yield point of 360 MPa.

Table 2. Piecewise values for Battery and Jelly Roll volume.

Piecewise Value	Effective Plastic Strain Battery Volume (mm/mm)	Corresponding Yield Stress Values Battery Volume (MPa)	Effective Plastic Strain Jelly Roll Volume (mm/mm)	Corresponding Yield Stress Values Jelly Roll Volume (MPa)
1	0.1466388	3.34656	0.077	1.54624
2	0.257	3.34656	0.1904	1.54624
3	0.26	1.34656	0.19247	19
4	0.263	1.34656	-	-
5	0.266	18.2952	-	-

It is crucial to acknowledge the inherent challenges associated with the homogenization of a complex system like a Li-ion battery. While the stress–strain curve parameters derived from hardware tests, with modifications in stress and strain scaling, were incorporated into our simulation, it is essential to recognize that these parameters may not possess direct and easily interpretable physical meanings in the context of the homogenized model. The intricacies of a Li-ion battery's internal structure and composition make it inherently challenging to directly correlate individual material properties from hardware tests with the behavior of the homogenized model. Therefore, while these parameters provide a basis for simulation, their direct interpretation may be limited in the broader context of the complex and multifaceted mechanical interactions within the homogenized Li-ion battery model. The study aims to navigate these challenges and extract valuable insights into the overall compression behavior of Li-ion batteries through careful consideration of the homogenized model and its associated material parameters.

3.3. Discussion

In comparing the experimentally derived load–displacement curve from the uniaxial compression test on the 18650 Li-ion battery with a simulated counterpart, a comprehensive evaluation can be conducted across the three distinctive zones that characterize the mechanical response of the battery. The validation regards the three zones of the force curve described in chapter 2 (Figures 3 and 5). In the elastic domain Zone 1, where the battery exhibits linear elasticity, the comparison is focused on the level of force applied and the corresponding displacement. The simulation ideally replicates the reversible deformation observed in this zone, showcasing the battery's ability to withstand external stress without undergoing irreversible changes. Moving to Zone 2, which marks the transition from elastic to plastic deformation, the comparison is centered around the critical juncture where plastic deformation occurs. The simulation captures the onset of plastic deformation without a significant increase in load, leading to a discernible load plateau. Attention is given to ensuring that the simulated structural compromise aligns with the experimental observations in terms of force level at the displacement of critical points, and the absence of a proportional increase in load. In the final densification zone, Zone 3, the comparison is critical in evaluating the simulated battery's response to compression-induced compaction. The focus here is on the substantial increase in load as the battery undergoes pronounced densification. The simulation mirrors the experimental findings in terms of force magnitude, displacement characteristics, and the rapid rise in applied load.

Additionally, the rate at which the force increases the stiffness of the battery in each zone is carefully compared between the simulated and experimental curves. Given that we are homogenizing a complex model, a pragmatic approach to analysis was taken. We consider the curve profile to be piecewise, as we are primarily interested in key stiffness points within the compression phenomenon. These points include the beginning of the plastic zone, the onset of densification, and the intensity of densification. Discrepancies in the stiffness application could indicate an error deviation in the simulated material properties or structural behavior. This systematic evaluation based on the criteria of force levels, displacement characteristics, and stiffness across the three zones is pivotal for validating the simulation model. It ensures reliability in informing design, safety, and performance optimization considerations across various applications, providing a robust understanding of the battery's mechanical behavior.

In the examination of the battery volume deformation, specific focus is directed towards the compression behavior seen in the alignment of the force curves seen in Figure 9 and the analytical values from Table 3.

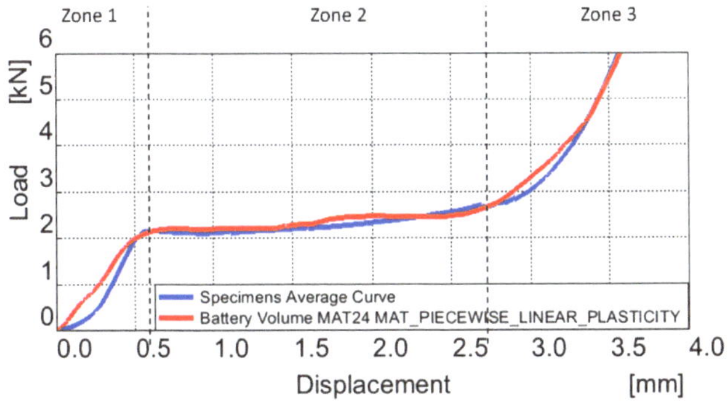

Figure 9. Comparison between Battery Volume model MAT24 and real test Load vs. Displacement curve.

Table 3. Battery volume simulation results.

	Zone 1 (Elastic)		Zone 2 (Plastic Deformation)		Zone 3 (Densification)	
Force–Disp. Curve	Real	Mat24	Real	Mat24	Real	Mat24
Force peak (kN)	2.14	2.14	2.67	2.64	5.5	5.5
Start point (mm)	0	0	0.6	0.6	2.7	2.7
Finish point (mm)	0.6	0.6	2.7	2.7	3.5	3.5
Stiffness (kN/mm)	3.56	3.56	0.98	0.97	1.57	1.57

In the actual response curve in the elastic region, designated as Zone 1, the termination of elastic deformation occurs at 0.6 mm press displacement after the initial contact with the battery, registering a force of 2.1 kN and a stiffness of 3.56 kN/mm. Observing Figure 9, it is discerned that the MAT24 battery volume force curve demonstrates a commendable alignment with the actual curve, but with a more linear behavior. However, comparative analysis, as delineated in Table 3, also reveals an analytical view of the elastic region with no error noted in the values.

Within the plastic deformation zone, denoted as Zone 2, specific metrics characterize the deformation behavior. The displacement in this region extends to 2.7 mm, resulting in a maximum end force of 2.67 kN and a stiffness of 0.98 kN/mm. Figure 9 visually indicates a favorable alignment between the force curve and the MAT24 battery volume model. In a quantitative assessment, the MAT24 battery volume model exhibits a peak end force of 2.64 kN with a 1.1% error, and a stiffness of 0.97 kN/mm, reflecting a small 1% discrepancy.

In the densification zone, denoted as Zone 3, a controlled displacement limit of 3.5 mm was imposed to facilitate the calculation of densification rates. Key parameters considered in this context include the initiation point and the subsequent rate of densification. In the actual model, densification commences at 2.7 mm, exhibiting a stiffness of 1.57 kN/mm.

The MAT24 battery volume model indicates a stiffness of 1.57 kN/mm, reflecting no analytical deviation. The alignment between the curves during the densification phase is evident in Figure 9.

In the examination of battery Jelly Roll deformation, as in the battery volume model, specific focus is directed towards the compression behavior observed in the alignment of force curves (Figure 10) and analytical values (Table 4).

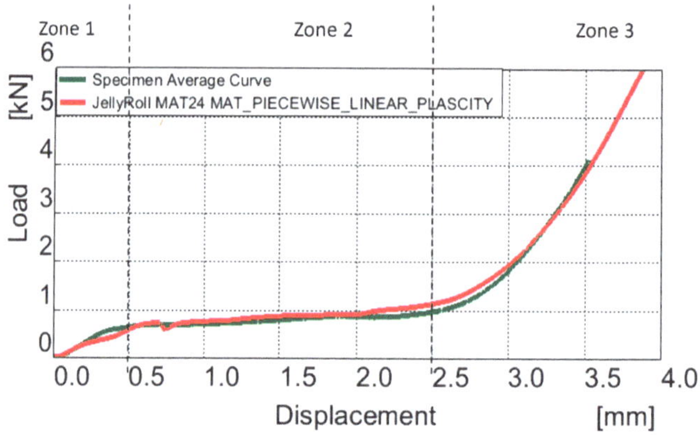

Figure 10. Comparison between MAT24 jelly roll volume and real test, Load vs. Displacement curve.

Table 4. Jelly roll volume simulation results.

	Zone 1 (Elastic)		Zone 2 (Plastic Deformation)		Zone 3 (Densification)	
Force–Disp. Curve	Real	Mat24	Real	Mat24	Real	Mat24
Force peak (kN)	0.6	0.6	0.99	1.16	4	3.98
Start point (mm)	0	0	0.5	0.5	2.5	2.5
Finish point (mm)	0.5	0.5	2.5	2.5	3.5	3.5
Stiffness (kN/mm)	1.2	1.2	0.39	0.46	1.14	1.13

In the elastic region (Zone 1), termination of elastic deformation occurs at 0.5 mm displacement, with a force of 0.6 kN and a stiffness of 1.2 kN/mm in the actual model. Notably, the MAT24 Jelly Roll volume model demonstrates commendable alignment with the actual curve. However, comparative analysis reveals that the model registers a force of 0.6 kN and a stiffness of 1.2 kN/mm, resulting in a no difference in force and stiffness.

Transitioning to the plastic deformation zone (Zone 2), displacement extends to 2.5 mm, resulting in a maximum end force of 0.99 kN and a stiffness of 0.39 kN/mm. Figure 10 visually indicates favorable alignment between the force curve and the MAT24 Jelly Roll volume model. However, quantitatively, the model displays values of a peak end force of 1.16 kN, a 17% error, and a stiffness of 0.46 kN/mm, a 17.9% discrepancy. A potential enhancement for future versions of the model would be to introduce more piecewise points within the densification zone. This would allow for a finer resolution of the material's response during this critical phase, leading to better representation of the compaction process and increased accuracy.

In the densification zone (Zone 3), with a displacement of 3.5 mm, densification commences at 2.5 mm in the actual model, at a peak force of 4, with a stiffness of 1.14 kN/mm. The MAT24 Jelly Roll volume model has a peak force of 3.98, with a 0.5% error, accompanied by a stiffness of 1.13 kN/mm, reflecting 0.8% deviation. This good alignment of the curves in Zone 2 and 3 can also be seen in Figure 10.

When analyzing the Jelly Roll volume alongside the battery case model, particularly in Zone 1, which denotes the elastic region, distinct areas of elastic deformation become evident. The battery case exhibits higher stiffness initially, followed by a combination of Jelly Roll elasticity and case plasticity, as seen in Figure 11. This conclusion is not seen in the analytical results presented in Table 5 because, in cases of high deformation, the point of the force and displacement at the beginning of the plastic area is in focus.

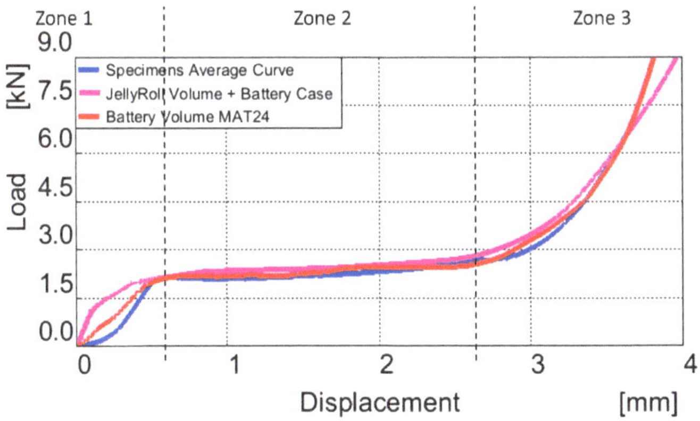

Figure 11. Comparison between battery models and real test, Load vs. Displacement curve.

Table 5. Jelly roll volume with battery case simulation results.

	Zone 1 (Elastic)		Zone 2 (Plastic Deformation)		Zone 3 (Densification)	
Force–Disp. Curve	Real	Mat24	Real	Mat24	Real	Mat24
Force peak (kN)	2.14	2.14	2.67	2.9	5.5	5.5
Start point (mm)	0	0	0.6	0.6	2.7	2.7
Finish point (mm)	0.6	0.6	2.7	2.7	3.5	3.5
Stiffness (kN/mm)	3.56	3.56	0.98	1.07	1.57	1.57

The comparison between various modeling approaches for the battery can be elucidated through the examination of Figures 11 and 12.

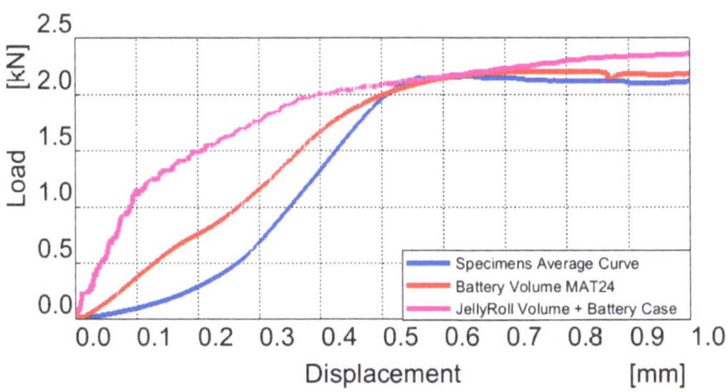

Figure 12. Comparison between battery models elastic zone and real test, Load vs. Displacement curve.

Figure 12 suggests that the elasticity of the battery volume model aligns more closely with the actual curve in the elastic zone. This can be attributed to the simplified definition of material properties within a homogeneous volume model compared to the complexities involved in modeling materials for heterogeneous models. The deviations observed in the plastic region of the model are also influenced by the interaction with the battery case, though the case tends to reduce these discrepancies to 8.6% and 9.1%, respectively.

3.4. Model Sensibility Test

To ensure the model's validity and applicability across different configurations, a sensitivity analysis was performed on a battery model composed of a jelly roll volume and its protective case. The analysis involved scaling the plastic strain and case thickness to observe how these modifications affect the densification process and the load curve.

When the plastic strain was scaled down by 10% and 20%, the densification occurred at lower deformations, as expected. Specifically, the initial point of densification shifted from 2.7 mm (original) to 2.5 mm and 2.3 mm, respectively. This indicates that reducing plastic strain sensitivity decreases the deformation required for densification to begin, altering the material response during high compression loads (Figure 13).

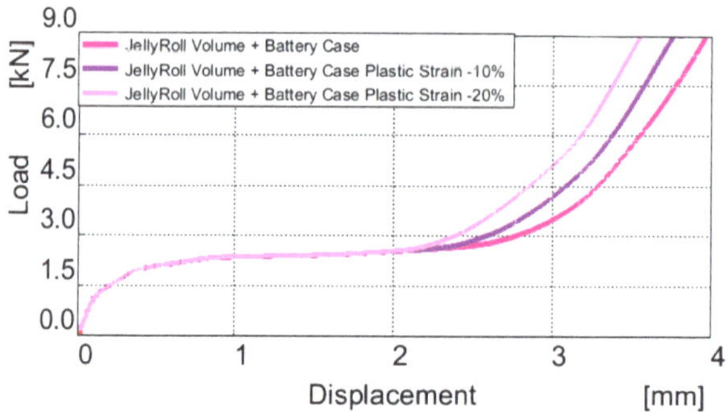

Figure 13. Sensibility test of the Jelly Roll plastic strain piecewise parameters.

For the case thickness, scaling it down to half, which is representative of a thinner, commercially available Li-ion cell casing, resulted in a nearly halved load curve, reflecting the lower force resistance of the thinner case. Conversely, doubling the case thickness led to an exponential increase in force resistance, with the load curve scaling nearly threefold. This notable effect can be seen in the elastic region, where the increased thickness led to the elasticity of the case having a higher weight in the elasticity of the battery, as shown in Figure 14. These sensitivity tests highlight how both plastic strain and casing thickness directly influence the battery's overall mechanical behavior under compression.

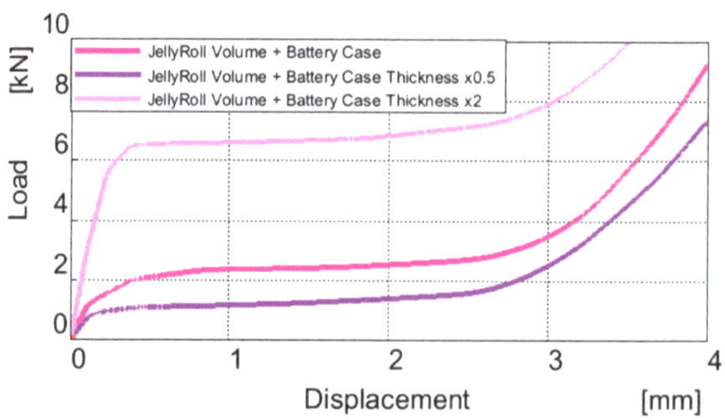

Figure 14. Sensibility test of the case thickness.

4. Conclusions

The Finite Element Method (FEM) model is employed in this study of LS Dyna MAT24 (MAT_PIECEWISE_LINEAR_PLASTICITY). Calibration of these materials is achieved through a meticulous comparison with the force curve obtained from the real-life compression tests. The real-life compression experiments are conducted on three battery cell specimens, with the cap removed, and two jelly rolls. The force curve obtained from these experiments delineates three discernible states: an initial elastic response, a plastic deformation plateau, and a subsequent phase marked by rapid densification. Notably, the model is represented as a single entity, homogenizing the entire battery structure into solids. The LS DYNA material MAT24 (MAT_PIECEWISE_LINEAR_PLASTICITY) is calibrated meticulously to replicate the experimentally obtained force curve. This calibration process ensures that the FEM model accurately reflects the mechanical behavior observed in the real-life compression tests.

MAT24 proves invaluable in defining suitable materials for both the entire battery volume and the Jelly Roll component. The effectiveness of the Jelly Roll model is further validated by its robust performance when integrated with the battery case, demonstrating its practical utility. The validity of the model is further strengthened by the predictable sensibility study of the plastic parameters and the variation of the case thickness.

Suggestions for future work should focus on incorporating more experimental data, particularly under dynamic load conditions, where the behavior of the elastic, plastic, and failure zones can vary significantly. Dynamic testing would provide a more comprehensive understanding of how the material behaves across different strain rates, especially in high-impact scenarios. Using these experimental insights, the MAT24 material model could be refined by introducing strain rate-dependent parameters, which would enhance the accuracy of the battery deformation predictions. Without such adjustments, the model risks defaulting to worst-case deformation scenarios in case of high-impact simulations. Another valuable direction would be to reintroduce the battery caps in the model and investigate load cases from the axial direction to assess anisotropy across the entire battery. Additionally, integrating the model into a full vehicle crash simulation would allow for evaluating its performance in a holistic crash environment, providing insights into how the battery pack responds when subjected to real-world, complex loading, vehicle crash dynamics. In conclusion, the comprehensive comparison between the experimentally derived load-displacement curve of a uni-axial compression test on the 18650 Li-ion battery and its simulated counterparts provides robust insights across three distinct mechanical response zones. The piecewise approach in analyzing the curves might overlook some details in the elastic zone; however, the focus of this model is primarily on predicting behavior under high deformation conditions, plastic zone and densification, which are more critical for assessing crashworthiness and safety in scenarios involving large impacts or compressive forces. The overall overlaying of curves, as depicted in Figures 9–11, reveals a comparable profile between the simulated model and the actual test data.

Author Contributions: Conceptualization: A.D.M. and M.C.D.; methodology: M.C.D.; software: A.D.M.; validation: A.D.M. and M.C.D.; formal analysis: M.C.D.; investigation: A.D.M. and M.C.D.; resources: A.D.M. and M.C.D.; data curation: M.C.D.; writing—original draft preparation: A.D.M.; writing—review and editing: M.C.D.; visualization: A.D.M.; supervision: M.C.D. All authors have read and agreed to the published version of the manuscript.

Funding: This research received no external funding.

Data Availability Statement: The data presented in this study are available in this manuscript.

Conflicts of Interest: The authors declare no conflicts of interest.

References

1. Liu, B.; Jia, Y.; Yuan, C.; Wang, L.; Gao, X.; Yin, S.; Xu, J. Safety issues and mechanisms of lithium-ion battery cell upon mechanical abusive loading: A review. *Energy Storage Mater.* **2020**, *24*, 85–112. [CrossRef]

2. Gong, C.; Liu, J.; Han, Y.; Hu, Y.; Yu, H.; Zeng, R. Safety of electric vehicles in crash conditions: A review of hazards to occupants, regulatory activities, and technical support. *IEEE Trans. Transp. Electrif.* **2021**, *8*, 3870–3883. [CrossRef]
3. Liu, B.H.; Yin, S.; Xu, J. Integrated computation model of lithium-ion battery subject to nail penetration. *Appl. Energy* **2016**, *183*, 278–289. [CrossRef]
4. Vyroubal, P.; Kazda, T. Finite element model of nail penetration into lithium ion battery. *J. Energy Storage* **2018**, *20*, 451–458. [CrossRef]
5. Xia, Y.; Wierzbicki, T.; Sahraei, E.; Zhang, X. Damage of cells and battery packs due to ground impact. *J. Power Sources.* **2014**, *267*, 78–97. [CrossRef]
6. Ruiz, V.; Pfrang, A.; Kriston, A.; Omar, N.; van den Bossche, P.; Boon-Brett, L. A review of international abuse testing standards and regulations for lithium ion batteries in electric and hybrid electric vehicles. *Renew. Sustain. Energy Rev.* **2018**, *81*, 1427–1452. [CrossRef]
7. Kotak, B.; Kotak, Y.; Brade, K.; Kubjatko, T.; Schweiger, H.-G. Battery crush test procedures in standards and regulation: Need for augmentation and harmonization. *Batteries* **2021**, *7*, 63. [CrossRef]
8. Lai, W.-J.; Ali, M.Y.; Pan, J. Mechanical behavior of representative volume elements of lithium-ion battery cells under compressive loading conditions. *J. Power Sources* **2014**, *245*, 609–623. [CrossRef]
9. Ali, M.Y.; Lai, W.-J.; Pan, J. Computational models for simulations of lithium-ion battery cells under constrained compression tests. *J. Power Sources* **2013**, *242*, 325–340. [CrossRef]
10. Xu, J.; Liu, B.; Wang, L.; Shang, S. Dynamic mechanical integrity of cylindrical lithium-ion battery cell upon crushing. *Eng. Fail. Anal.* **2015**, *53*, 97–110. [CrossRef]
11. Sheikh, M.; Elmarakbi, A.; Rehman, S. A combined experimental and simulation approach for short circuit prediction of 18650 lithium-ion battery under mechanical abuse conditions. *J. Energy Storage* **2020**, *32*, 101833. [CrossRef]
12. Avdeev, I.; Gilaki, M. Structural analysis and experimental characterization of cylindrical lithium-ion battery cells subject to lateral impact. *J. Power Sources* **2014**, *271*, 382–391. [CrossRef]
13. Jantscher, K.; Breitfuß, C.; Miklau, M.; Ismail, K.; Dobusch, P. Virtual Detection of Mechanically Induced Short Circuits in a Cylindrical Lithium-Ion Battery Cell Based on Finite Element Simulation. *Batteries* **2021**, *7*, 79. [CrossRef]
14. Madani, S.S.; Schaltz, E.; Kær, S.K. Characterization of the Compressive Load on a Lithium-Ion Battery for Electric Vehicle Application. *Machines* **2021**, *9*, 71. [CrossRef]
15. Hahn, A.; Doose, S.; Saathoff, D.; Kwade, A. Effect of External Compression on the Thermal Runaway of Lithium-Ion Battery Cells during Crush Tests: Insights for Improved Safety Assessment. *Batteries* **2023**, *9*, 404. [CrossRef]
16. Zhu, J.; Wierzbicki, T.; Li, W. A review of safety-focused mechanical modeling of commercial lithium-ion batteries. *J. Power Sources* **2018**, *378*, 153–168. [CrossRef]
17. Raffler, M.; Sevarin, A.; Ellersdorfer, C.; Heindl, S.F.; Breitfuss, C.; Sinz, W. Finite element model approach of a cylindrical lithium ion battery cell with a focus on minimization of the computational effort and short circuit prediction. *J. Power Sources* **2017**, *360*, 605–617. [CrossRef]
18. Abada, S.; Marlair, G.; Lecocq, A.; Petit, M.; Sauvant-Moynot, V.; Huet, F. Safety focused modeling of lithium-ion batteries: A review. *J. Power Sources* **2016**, *306*, 178–192. [CrossRef]
19. Abedini, M.; Zhang, C. Performance assessment of concrete and steel material models in ls-dyna for enhanced numerical simulation, a state of the art review. *Arch. Comput. Methods Eng.* **2021**, *28*, 2921–2942. [CrossRef]
20. Sarkar, A.; Shrotriya, P.; Chandra, A. Modeling of separator failure in lithium-ion pouch cells under compression. *J. Power Sources* **2019**, *435*, 226756. [CrossRef]
21. Muresanu, A.D.; Dudescu, M.C. Numerical and experimental evaluation of a battery cell under impact load. *Batteries* **2022**, *8*, 48. [CrossRef]
22. Szabo, I.; Scurtu, L.I.; Raboca, H.; Mariasiu, F. Topographical Optimization of a Battery Module Case That Equips an Electric Vehicle. *Batteries* **2023**, *9*, 77. [CrossRef]

batteries

MDPI

Review

Recent Advancements in Artificial Intelligence in Battery Recycling

Subin Antony Jose, Connor Andrew Dennis Cook, Joseph Palacios, Hyundeok Seo, Christian Eduardo Torres Ramirez, Jinhong Wu and Pradeep L. Menezes *

Department of Mechanical Engineering, University of Nevada-Reno, Reno, NV 89557, USA; subinj@unr.edu (S.A.J.); connorcook500@gmail.com (C.A.D.C.); josep.pala9852@gmail.com (J.P.); hyundeoks@unr.edu (H.S.); christiantorres@unr.edu (C.E.T.R.); jinhongw@unr.edu (J.W.)

* Correspondence: pmenezes@unr.edu

Abstract: Battery recycling has become increasingly crucial in mitigating environmental pollution and conserving valuable resources. As demand for battery-powered devices rises across industries like automotive, electronics, and renewable energy, efficient recycling is essential. Traditional recycling methods, often reliant on manual labor, suffer from inefficiencies and environmental harm. However, recent artificial intelligence (AI) advancements offer promising solutions to these challenges. This paper reviews the latest developments in AI applications for battery recycling, focusing on methodologies, challenges, and future directions. AI technologies, particularly machine learning and deep learning models, are revolutionizing battery sorting, classification, and disassembly processes. AI-powered systems enhance efficiency by automating tasks such as battery identification, material characterization, and robotic disassembly, reducing human error and occupational hazards. Additionally, integrating AI with advanced sensing technologies like computer vision, spectroscopy, and X-ray imaging allows for precise material characterization and real-time monitoring, optimizing recycling strategies and material recovery rates. Despite these advancements, data quality, scalability, and regulatory compliance must be addressed to realize AI's full potential in battery recycling. Collaborative efforts across interdisciplinary domains are essential to develop robust, scalable AI-driven recycling solutions, paving the way for a sustainable, circular economy in battery materials.

Keywords: battery recycling; artificial intelligence; computer vision; lithium ion battery

1. Introduction

1.1. Overview of Battery Recycling

Since their introduction in the early 1990s, lithium ion batteries (LIBs) have seen a significant surge in usage. With projections indicating further growth in the coming decade, LIBs have become the preferred energy storage technology due to their high energy density and relatively low cost. Increased LIB production has enhanced manufacturing efficiencies and reduced costs, further driving demand. LIBs are now widely used in applications ranging from electric vehicles (EVs) to portable electronics [1,2]. However, the finite supply of raw materials, such as lithium, essential for LIB production presents a significant challenge. This scarcity underscores the need for effective recycling strategies to optimize the use of these critical materials [3,4]. Recycling not only conserves resources but also minimizes the environmental impact associated with LIB production and disposal [5].

Current LIB recycling methods include pyrometallurgy, electrolytic recycling, and bioleaching, each offering varying degrees of success. Bioleaching shows considerable promise, although it remains experimental [6,7]. The global demand for LIB recycling is increasing, driven by the environmental and health risks posed by battery waste. As LIBs continue to play a crucial role in reducing reliance on pollutive energy sources, advancements in recycling technology are essential to mitigate economic and environmental costs [1].

The global market for LIB recycling was valued at USD 8.10 billion in 2023 and is projected to increase to USD 10.26 billion in 2024. As shown in Figure 1, it is anticipated to grow to approximately USD 85.69 billion by 2033, reflecting a compound annual growth rate (CAGR) of 26.6% from 2024 to 2033 [8].

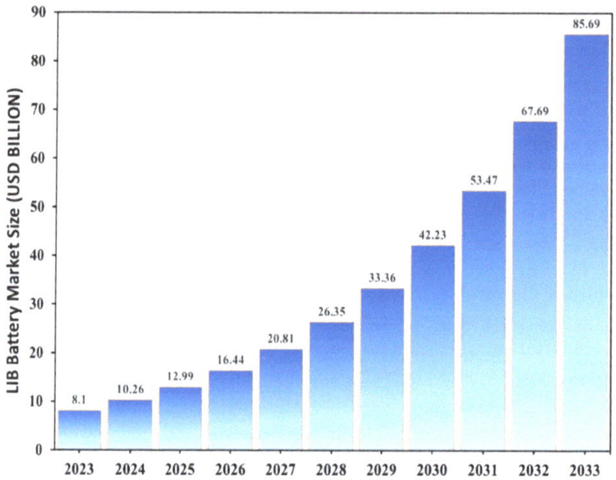

Figure 1. LIB recycling market size 2023 to 2033 (USD billion). Adapted from [8].

The economic benefits of LIB recycling are substantial. These include resource conservation and the creation of new economic opportunities [9]. Global demand for LIBs is projected to surge over the next decade, with the required capacity expected to grow from approximately 700 GWh in 2022 to about 4.7 TWh by 2030 [10]. Europe's current recycling infrastructure boasts a recovery rate of over 85% for materials like aluminum, copper, cobalt, manganese, and nickel and between 35% and 42% for iron and lithium. Hydrometallurgy has been identified as Europe's most economically viable recycling strategy, generating the highest revenue while maintaining the lowest operational costs. However, the existing infrastructure may struggle to meet the anticipated increase in demand by 2030 [11].

Recycling LIBs also reduces waste generation from used batteries and decreases the demand for large-scale mining operations. For instance, the Indonesian Ministry of Energy and Mineral Resources projects a significant rise in EV sales over the next 15 years. The waste generated by EVs is tied to the lifespan of their batteries, which can last up to eight years [12]. Recycling these batteries would significantly reduce waste, contributing to environmental sustainability by cutting greenhouse gas emissions and preventing harmful materials from contaminating soil and water.

1.2. Role of Artificial Intelligence in Recycling

Artificial intelligence (AI) is rapidly transforming various industries, and its impact on environmental sustainability is particularly noteworthy. AI's capabilities extend from simple tasks like baking cookies to complex programming, offering immense potential to benefit society. One of AI's most promising applications lies in recycling, which enhances the efficiency and accuracy of sorting processes, a critical challenge in waste management.

AI detection systems utilize visual inspection to assess quality, identify defects, and ensure the correct placement of materials [13]. This technology, already widely used across multiple industries, is now being applied to address the biggest challenge in recycling: effectively sorting recyclables from waste [14]. For instance, companies like Waste Vision have developed AI-driven systems that can be mounted on garbage trucks to detect overflows and contamination in recycling bins and monitor waste quantities [15]. This early detection helps prevent improper disposal practices right at the source.

Later in the recycling process, AI continues to refine the sorting accuracy. Technologies like Neuron's deep learning algorithms enable continuous improvement in identifying and categorizing materials such as paper, plastics, and metals based on color, size, shape, and brand characteristics [16]. AMP Robotics is already deploying AI systems in recycling centers to sort recyclable materials more efficiently and accurately, ensuring that different materials are correctly separated for reuse [14].

The primary advantage of AI in recycling lies in its detection accuracy, which ranges from 72.8% to 99.95% [17]. This accuracy variability depends on the sensors and camera capabilities, efficiency of garbage classifying robots, and AI algorithms [18]. This high level of precision is crucial, given the vast amount of waste generated daily. The average U.S. resident produces 4–5 lbs. of waste per day, underscoring the need for a recycling process that is both fast and accurate [19]. AI-driven systems meet these demands and improve continuously as they process more data, making them an essential tool in advancing recycling efforts and promoting environmental sustainability.

2. Artificial Intelligence Techniques in Battery Sorting and Identification

2.1. Computer Vision for Battery Recognition

The global shift from a wasteful economy to a more sustainable, circular economy is increasingly evident, particularly in the rise of EVs. However, the production of EVs significantly increases the demand for critical raw materials like lithium, cobalt, and rare earth metals essential for LIB. Projections indicate that the European Union will need 18 times more lithium by 2030 and nearly 60 times more by 2050 [20]. Similarly, cobalt demand is expected to grow fivefold by 2030 and 15-fold by 2050 [21]. Effective battery waste (B-waste) recycling is crucial to meet this demand sustainably.

Computer vision (CV) is a branch of AI that enables computers to interpret and analyze visual data, allowing them to recognize, classify, and understand images or videos of batteries. Widely used in areas like object detection and industrial automation, computer vision enhances battery recycling by automating tasks such as battery identification, sorting, and monitoring in B-waste management [22]. CV is anticipated to transform B-waste management by significantly enhancing the identification, classification, collection, sorting, segregation, and monitoring of B-waste [23]. Figure 2 shows the diverse application of computer vision technology across various sectors.

In 2019, the world generated 53.6 million tons of B-waste, which contains valuable materials like lithium and cobalt [24]. Addressing this, the RoboCRM System was proposed in 2022. This system uses AI, robotics, machine learning (ML), and optical imaging to identify devices containing batteries, which are then separated for proper recycling. Utilizing deep learning, the system can distinguish devices based on a vast database of optical images, enabling precise sorting at processing or B-waste collection facilities [25].

Effective waste sorting is vital for achieving a circular economy. Currently, sorting methods include manual and automated processes, using support vector machines (SVMs) and convolutional neural networks to classify materials [26]. A novel approach involves Radio Frequency Identification (RFID) technology, allowing contactless identification and waste sorting [27]. In Finland, ZenRobotics' ZRR2 robot has demonstrated the potential of AI in waste management, efficiently categorizing construction waste using deep learning and computer vision [28].

Integrating AI into B-waste sorting could revolutionize recycling, making a closed circular economy a tangible reality. By enhancing the efficiency and accuracy of material recovery, these technologies hold the key to a more sustainable future.

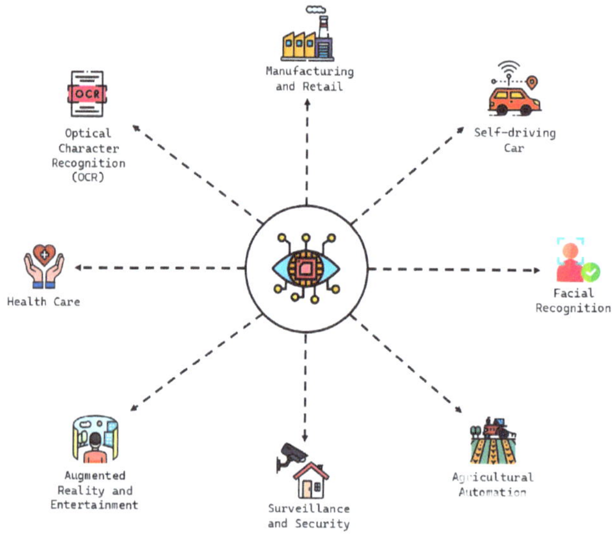

Figure 2. Broad applications of computer vision across multiple sectors. Reproduced with permission from [23].

2.2. Machine Learning for Material Composition Analysis

As technology advances, portable electronic gadgets have become integral to daily life. The invention of the battery revolutionized how we use these devices, enabling portability without needing a constant connection to a power source. However, the growing demand for batteries has led to a significant increase in battery waste. Batteries contain hazardous chemicals and valuable metals, necessitating proper recycling to manage their end-of-life cycle and recover useful materials safely.

ML has transformed the battery recycling industry by enhancing the accuracy of material analysis within batteries. AI and ML algorithms can predict the recycling potential of batteries and optimize resource recovery [29]. By processing large datasets from experiments and simulations, ML can identify desirable battery properties, such as high energy density, stability, and conductivity, accelerating improvements in energy storage capacity and battery life predictions [30].

LIBs, the most commonly used type, involve several key recycling steps. The process begins with the separation of the metallic shell, copper (Cu), and aluminum (Al) foils, followed by the collection of the cathode and anode materials [31]. The recovered materials are then purified to extract valuable elements, regenerate new batteries, or synthesize other functional materials. ML plays a crucial role in predicting metal recovery from these batteries. Open-source databases, like the Inorganic Crystal Structure Database and the Materials Project, provide vital resources on energy bands, crystal structures, and other physical properties, which researchers can input into ML algorithms for data analysis [32].

ML-guided research has led to significant advancements in the LIB recycling market. Logistic regression models, for example, offer higher accuracy and faster identification of materials with high ionic conductivity. ML-assisted robotic systems have been developed to address the challenge of diverse battery packaging during recycling disassembly. These systems use ML algorithms to navigate various battery casings efficiently, increasing the disassembly speed and enhancing recycling efficiency [29]. Additionally, ML can analyze data from charge–discharge cycles to predict a battery's remaining life and model engineering challenges encountered during recycling.

Feature engineering further enhances ML by selecting, transforming, extracting, and manipulating raw data to create relevant datasets for modeling. In lithium battery recycling, ML algorithms enable rapid screening and prediction of material properties, such as electrode voltage and electrolyte conductivity, by establishing relationships between atomic structures and electrochemical performance. ML can efficiently manage large datasets from experimental and simulation results, facilitating the discovery of novel materials for electrodes, electrolytes, and promoters. Moreover, ML improves mesoscale characterizations through image segmentation and labeling, automating the analysis of battery components' spatial distribution and morphology. This approach revolutionizes traditional characterization methods, providing deeper insights into material behavior and enabling more informed battery design and optimization [32].

The application of ML in battery recycling has significantly improved the accuracy and speed of material identification. By processing vast amounts of data faster than any human could, ML accelerates the development of safer, higher energy density, and longer-lasting batteries, pushing the boundaries of what is possible in battery technology.

3. Intelligent Robotics in Battery Dismantling

3.1. Robotic Systems for Automated Dismantling

As the demand for EVs continues to rise, so does the need for LIBs, which power these vehicles. The production of these batteries relies on critical materials such as lithium, cobalt, and rare earth metals—resources in limited supply. Recycling these valuable materials is essential for sustainable and efficient battery production. Dismantling LIBs in the automotive industry is largely manual, with robotics playing only a limited role in assisting human workers or performing simple tasks. These manual processes are slow and require highly skilled personnel, making them costly and potentially unprofitable, with the added risk of environmental pollution [33].

In contrast, automated systems offer several advantages over manual disassembly, including greater efficiency, lower costs, reduced workplace injuries, and the ability to scale up for higher volumes [34]. However, the primary challenge facing automated dismantling systems is the significant variation in EV battery designs, which differ not only between manufacturers but also between car models [35].

One experiment involving a six-degree-of-freedom industrial robotic system demonstrated the potential for robotics to improve the efficiency of battery disassembly. The robotic system was evaluated on benchmark tasks such as cutting and gripping to assess its effectiveness in disassembly. The robot arm cut battery tabs in EV battery modules in one test. The robot completed the operation in 112 s, nearly twice as fast as trained technicians, who took 220 s to perform the same task. This demonstrates the robot's ability to enhance the speed and accuracy of battery disassembly significantly [36].

Another study explored the feasibility of developing a hybrid human–robot workstation for dismantling an Audi Q5 Hybrid battery system. The robot was assigned the task of unscrewing screws located anywhere on the battery. Two approaches were tested: physical demonstration and camera-based detection. While the physical demonstration method allowed the robot to detect the fasteners' locations, it required considerable setup time. The camera-based detection method showed promise in speeding up the process, though further development was needed to implement a more efficient algorithm [37]. These studies indicate that while intelligent robotics have made significant strides in battery disassembly, achieving the optimal balance between human and robot collaboration still requires further refinement.

AI-assisted decision-making is crucial in dismantling processes across various industries, from manufacturing to environmental remediation. AI algorithms can analyze vast amounts of data to identify potential risks associated with dismantling processes, considering structural integrity, environmental hazards, and safety protocols. This allows AI systems to help prioritize tasks and allocate resources more effectively. AI can also create optimized dismantling plans by factoring in various constraints and objectives, such as scheduling tasks, coordinating equipment and personnel, and minimizing downtime.

ML algorithms can continuously refine these plans based on real-time data and feedback, improving efficiency.

AI enhances resource utilization by analyzing usage patterns and suggesting optimal allocation strategies, ensuring that materials, equipment, and personnel are used efficiently throughout dismantling. AI-powered predictive maintenance systems can monitor equipment health in real time, detecting potential failures before they occur, thus minimizing downtime and improving overall efficiency. Additionally, AI can assess the environmental impact of dismantling processes by analyzing emissions, waste generation, and ecological footprints, enabling stakeholders to make informed decisions that minimize environmental harm and ensure regulatory compliance [38].

AI also enhances dismantling safety by monitoring real-time conditions and alerting workers to potential hazards. For example, computer vision systems can detect unsafe behaviors or conditions, while wearable devices can provide personalized safety recommendations based on individual risk factors. Furthermore, AI-driven robotics and automation technologies enable the remote operation of dismantling equipment in hazardous environments, reducing human exposure to risks such as radiation, toxic chemicals, or unstable structures [39].

Finally, AI algorithms can analyze data collected during dismantling to identify trends, optimize workflows, and generate comprehensive reports. This information empowers stakeholders to evaluate performance, identify areas for improvement, and make data-driven decisions for future projects. Integrating AI and robotics in battery disassembly and other dismantling processes holds great promise for improving efficiency, safety, and sustainability in these critical operations.

3.2. AI-Assisted Decision-Making in Dismantling Processes

Robotic systems have become integral to automated disassembly lines, significantly enhancing both the efficiency of material recovery and the speed of disassembly. Integrating AI-assisted decision-making technology has further revolutionized this process, enabling the automation of complex operations, including the disassembly of batteries. As AI technology advances, driven by its ability to increase efficiency and reduce costs, its application in automation has grown substantially. The availability of diverse data sources has also spurred the development of more sophisticated ML algorithms, making AI-assisted programs an essential component in optimizing the disassembly of LIBs [40]. These systems boost production rates, reduce errors, and enhance resource recovery.

Deep learning is one of the key AI techniques that can be leveraged for decision-making in automated disassembly. Deep learning techniques have proven effective in improving the efficiency and accuracy of decision-making processes across various fields, including automation. By incorporating large datasets into deep learning algorithms, AI systems can identify meaningful patterns, leading to more intelligent decision-making support [41]. This technology has already streamlined operations in the automation of vehicles, and its application to battery disassembly promises similar benefits. Deep learning techniques can be crucial in advancing sustainable practices by optimizing the disassembly process and recovering valuable materials.

Automated disassembly processes face challenges, particularly due to the variability of recycling materials, which can complicate the design of efficient automated systems. This variability introduces uncertainty in the disassembly time for different products. The disassembly process can be modeled to address these challenges using a multi-objective, multi-product robotic disassembly line-balancing problem (MMRDP), which helps maximize profit and minimize energy consumption [42]. Research by Xu et al. [42] demonstrated the effectiveness of an improved algorithm based on the Pareto rule, Pareto-improved multi-objective brainstorming optimization (PIMBO), which incorporates a stochastic simulation approach to solve the MMRDP, resulting in enhanced disassembly efficiency. Similarly, research by Gulivindala et al. [43] highlights the use of genetic algorithms (GA), a common AI technique, to create optimized disassembly sequence plans. These algorithms, as shown in

Tech Science Press (TSP) studies, can improve real-time product disassembly by designing optimized solutions, further enhancing the efficiency of the disassembly process [43].

AI technology also plays a crucial role in error reduction during disassembly. Research by Li et al. [44]. demonstrated that AI can be used for fault detection in automated systems, allowing for real-time monitoring of disassembly sequences through distributed information systems. This capability not only reduces the occurrence of errors but also provides timely notifications to users when an error is detected or anticipated, thereby enhancing system security and reducing maintenance and operational costs. However, challenges remain, such as ensuring the quality of data input into AI systems and AI algorithms' precision [45]. Despite these challenges, AI continues to improve the modeling and execution of disassembly processes, optimizing material recovery and further increasing the efficiency of disassembly operations.

Table 1 summarizes AI applications in the battery recycling process, comparing the challenges of traditional methods.

Table 1. AI Applications in Battery Recycling Processes.

Process	Traditional Method Challenges	AI-Driven Solutions	Benefits
Sorting	Labor-intensive, prone to errors	Machine learning algorithms for automated sorting	Higher accuracy, reduced manual labor
Classification	Manual classification, inconsistent quality	Computer vision and deep learning models	Improved precision, faster processing
Disassembly	Risky for workers, time-consuming	Robotic disassembly using AI	Enhanced safety, efficiency
Material characterization	Limited accuracy with manual testing	AI-driven spectroscopy and X-ray imaging	Precise material identification
Real-time monitoring	Lack of adaptability to variable battery types	AI-based adaptive monitoring systems	Dynamic adjustment, increased recovery rate

4. Predictive Maintenance and Process Optimization

4.1. AI-Based Predictive Maintenance for Recycling Equipment

AI-based predictive maintenance is transforming the management of recycling equipment, offering unprecedented insights into equipment health, performance trends, and maintenance requirements. Recycling facilities face the ongoing challenge of maintaining diverse equipment that experiences continuous wear and tear due to processing various materials. Traditional reactive maintenance approaches are costly and disruptive, often leading to unplanned downtime and operational inefficiencies [46]. However, by utilizing AI algorithms to analyze sensor data, historical records, and real-time performance metrics, predictive maintenance can anticipate potential failures, enabling proactive intervention and significantly reducing downtime.

The implementation of AI-based predictive maintenance offers several compelling benefits to recycling facilities. Primarily, it drastically reduces downtime by predicting equipment failures before they occur, ensuring uninterrupted operations and maximizing productivity. This proactive approach minimizes the need for emergency repairs, leading to substantial cost savings that improve recycling companies' financial sustainability. These savings can be reinvested in technology upgrades, enhancing equipment performance and longevity [47].

One of the primary advantages of AI-based predictive maintenance is its ability to optimize maintenance schedules based on real-time insights into equipment health. AI algorithms can identify patterns and anomalies that indicate potential equipment failures by analyzing historical data and current performance metrics. This allows maintenance teams to prioritize tasks, schedule interventions during planned downtime, and allocate resources effectively, minimizing operational disruptions. Additionally, AI-driven predictive analytics provide valuable insights into equipment components' remaining useful life (RUL), enabling timely replacements and upgrades to prevent catastrophic failures [48].

AI-based predictive maintenance also contributes to enhanced workplace safety in recycling facilities. Sudden equipment failures or malfunctions can pose significant risks to worker safety, leading to accidents or injuries. By proactively identifying potential failure points and scheduling maintenance tasks accordingly, AI algorithms help mitigate these risks, creating a safer working environment [49]. Furthermore, reducing the need for reactive maintenance, which often requires technicians to work under time pressure and in hazardous conditions, improves overall occupational safety standards.

In addition to reducing downtime and improving safety, AI-driven predictive maintenance enhances resource utilization in recycling facilities. Organizations can allocate workforce, spare parts, and other resources by optimizing maintenance schedules and identifying critical needs. This reduces inventory costs, minimizes resource wastage, and improves operational efficiency. Moreover, predictive maintenance enables recycling companies to extend equipment lifespan, maximize performance, and reduce energy consumption, thereby contributing to environmental sustainability efforts [50].

The integration of AI-based predictive maintenance is poised to revolutionize the recycling industry. Ongoing advancements in the Internet of Things (IoT) integration, edge computing, and sensor technologies are transforming the landscape of data collection and analysis [51]. These developments will improve prediction accuracy and provide actionable insights that drive operational efficiencies and cost savings. Incorporating digital twins—a digital replica of a physical entity that maintains a strong connection to the original—will play a pivotal role in this evolution, allowing for the testing of maintenance strategies in a risk-free virtual environment and enabling real-time performance optimization [52]. Figure 3 provides a schematic diagram of the workflow of a digital twin system for a real battery EV with a framework to enable comprehensive battery lifecycle management. Battery data is digitized in real time and uploaded to a cloud-based database via the IoT, allowing for online monitoring of their health and operational status using AI. Robots can assess, screen, and sort incoming batteries by analyzing data from decommissioned batteries, following a structured management approach to facilitate efficient recycling. The insights generated by the smart battery design system enable tracking the flow of LIBs, facilitating more focused recycling initiatives.

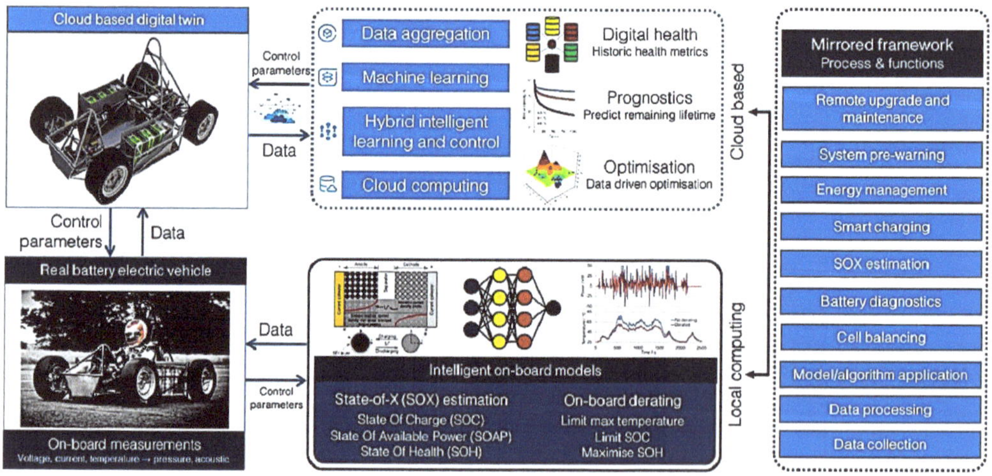

Figure 3. Workflow of a digital twin system for comprehensive battery lifecycle management. Reproduced with permission from [53].

As AI algorithms evolve, predictive maintenance capabilities will expand even further. Techniques such as deep learning, reinforcement learning, and predictive modeling are increasingly being integrated into AI systems, enabling them to analyze complex data sets, detect patterns, and accurately predict equipment failures. These advancements will reduce downtime and maintenance costs and extend the lifespan of recycling equipment, contributing to a more sustainable and efficient recycling ecosystem.

However, several challenges must be addressed to ensure the widespread adoption and successful implementation of AI-based predictive maintenance in recycling facilities. Data privacy and security concerns necessitate robust protocols and encryption mechanisms to protect sensitive information collected from equipment sensors and maintenance records. Ensuring algorithm transparency and interpretability is crucial for building stakeholder trust and enabling informed decision-making based on AI-driven insights. Additionally, scalability issues related to managing large volumes of sensor data, integrating diverse systems, and deploying AI solutions across multiple sites must be addressed to fully realize the potential of predictive maintenance in the recycling industry.

In conclusion, AI-based predictive maintenance is set to revolutionize the recycling industry by driving efficiency, sustainability, and operational excellence. By leveraging AI algorithms to anticipate equipment failures, optimize maintenance schedules, and enhance resource utilization, recycling facilities can minimize downtime, improve safety standards, and contribute to environmental conservation efforts. With ongoing technological advancements and strategic investments in AI-driven solutions, the future of predictive maintenance in recycling equipment is bright, offering unparalleled opportunities for innovation and growth in the waste management sector.

4.2. Process Optimization Through AI Algorithms

Recent advancements in AI algorithms are revolutionizing the recycling process, offering new ways to optimize efficiency and sustainability. As the global population grows, so does the volume of waste communities generate. To maintain a clean and sustainable environment, recycling processes must be modernized better to manage the ratio between waste and recycled materials. According to the United States Environmental Protection Agency, 75% of solid waste that could be recycled is currently lost or wasted, causing significant environmental harm that ultimately affects human quality of life [54]. By leveraging AI algorithms, the accuracy of classifying recyclable materials can be significantly enhanced, leading to higher recycling rates.

The initial step in recycling is sorting waste items by category, a process that can greatly benefit from AI-driven automation. ML algorithms can accurately determine and sort the waste category when paired with sensors using robotic systems [55]. One such technique is the convolutional neural network (CNN), which utilizes image processing to classify objects. CNNs have proven successful in training multi-layer network structures and can be integrated with robotic systems equipped with suction, grippers, and RGB sensors to sort materials based on their composition. For instance, a CNN-equipped model has demonstrated an overall classification accuracy of 96%. When applied in recycling facilities, CNNs improve the rate at which materials are categorized and reduce the costs associated with the recycling process [54,55].

While effective, manual sorting processes can be costly and pose health risks to workers in hazardous environments. The physical and mental exhaustion associated with manual labor can also lead to errors and the loss of recyclable materials. AI-driven sorting systems, on the other hand, operate without fatigue, ensuring consistent performance. A case study on the application of AI in a recycling production line found that a CNN-equipped model trained on video feeds of waste could detect and classify materials in real-time with an accuracy of 92.43% [54]. This demonstrates the potential of AI to enhance both the efficiency and cost-effectiveness of recycling operations.

AI-driven recycling systems also have the potential to significantly reduce the amount of waste that is discarded without being reused. For example, a city in China implemented an AI network to manage waste across 20 selected sites. The results were impressive: the sites saved 357,000 yuan (approximately 50,000 US dollars) per recycling cycle and could reuse 98.25% of recyclable waste [56]. This case study highlights AI's economic and environmental benefits in waste management. Similarly, AI and ML technologies have been adopted in Latin America and the Caribbean to enhance municipal solid waste generation prediction, optimize collection routes, and improve resource management [57]. The ML models identified trends and patterns in waste generation, enabling more informed and efficient decision-making. This demonstrates how AI can accelerate the transition toward a circular economy by optimizing resource utilization and minimizing waste.

However, despite the promise of AI in waste management, some challenges need to be addressed. These include the quality of data, the availability of computational resources, the integration of AI with existing systems, and the need for expertise and training [57]. ML requires large, high-quality datasets to train systems effectively, but such data can be difficult to obtain in waste management. Additionally, many small waste management facilities may lack the financial resources to adopt AI technologies. Integrating AI with existing systems can also be time-consuming and costly, and many facilities may not have staff with the necessary expertise in ML.

Despite these challenges, AI's recycling and waste management benefits are significant. ML algorithms can predict waste generation, forecast equipment failures, and optimize operational efficiency, all of which reduce costs. AI can also classify waste into recyclable, biodegradable, or hazardous categories, improving the recycling rate. Additionally, AI can enhance waste-to-energy production by analyzing data on waste generation patterns. AI systems offer tremendous potential to improve the efficiency and sustainability of recycling processes and waste management.

5. Data Analytics for Lifecycle Assessment

5.1. Environmental Impact Assessment Through Data Analytics

Recycling is a crucial process that significantly reduces the negative environmental impacts on our planet. While recycling has been practiced for many years, recent advancements in AI detection have opened new avenues for optimizing this process. AI's ability to accurately identify recyclable materials ensures that fewer new materials need to be produced, as recycled materials can be effectively reused. For instance, AI systems can recognize plastic water bottles, vegetable waste, cardboard, and black bags with an accuracy of around 95% [58]. This high accuracy rate means more recyclable materials can be correctly processed, reducing the need for additional manufacturing and lowering the overall environmental footprint.

The environmental benefits of recycling extend beyond the reuse of materials; they encompass the entire lifecycle of the recycling process. Although recycling still involves remanufacturing and transportation, its environmental impact is significantly lower than extracting and processing raw materials. For example, the mining industry generates between 1.9 and 5.1 gigatons of CO_2 equivalent (CO_2e) of greenhouse gas emissions annually, with coal mining alone responsible for a significant portion of this pollution [59,60]. In contrast, the emissions associated with recycling are minimal. We can achieve substantial environmental gains by excluding the carbon-intensive raw material extraction processes and focusing on the more efficient recycling lifecycle.

Recycling primarily involves three main steps: collection, processing, and remanufacturing into new products [59]. While these steps produce some emissions, particularly during transportation and remanufacturing, the overall environmental impact is far less than that of the initial production of materials from raw resources. With AI detection systems achieving 95% accuracy, the likelihood of non-recyclable materials entering the recycling stream is minimized, ensuring that most transported items are recyclable. This streamlines the process and reduces the environmental burden associated with waste management [61].

The benefits of recycling are further highlighted when considering the alternative—landfills and combustion centers. If materials were not recycled and instead had only one lifecycle, landfills would quickly overflow, combustion centers would have to operate continuously, and the demand for newly mined materials would increase. Estimates indicate that greenhouse gas emissions from municipal solid waste (MSW) combustion facilities in the U.S. range from 10 to 20 million metric tons annually, a small fraction compared to the nearly six billion tons emitted by fossil fuel combustion [62]. By diverting recyclables from these facilities, we can prevent significant amounts of pollution from entering the atmosphere.

However, there are areas within the recycling lifecycle that can be improved. The most significant opportunities for improvement lie in the transportation and remanufacturing stages. In the United States, trucks travel a staggering 93.5 billion miles annually, contributing nearly 6 million tons of CO_2 per year to the atmosphere [63]. Transitioning to emissions-free vehicles for transportation would have a substantial positive impact on the environment. Additionally, by enhancing telematics and AI technologies to optimize routes and reduce the time spent carrying empty loads, we can further decrease the carbon footprint of recycling logistics [64].

Another key area for improvement is the remanufacturing process. Ideally, recycling centers should integrate remanufacturing capabilities on-site, allowing materials to be processed and transformed into new products at the same location. AI detection systems can facilitate this by efficiently sorting materials and directing them to the appropriate remanufacturing facilities within the same site. Many recycling centers in the U.S. are already moving toward this model, as it reduces economic costs and improves emission control, making the entire recycling process more sustainable.

In summary, while recycling is already a beneficial practice, integrating AI technology and targeted improvements in transportation and remanufacturing can further enhance its environmental benefits. By focusing on these areas, we can continue to reduce the negative impacts of waste on our planet and move closer to a truly sustainable future.

5.2. Economic Analysis and Decision Support

Recent advancements in AI have significantly transformed the economic analysis of the battery recycling industry, particularly by enhancing data analytics and decision support systems. These technologies revolutionize battery recycling processes by enabling more efficient resource allocation, process optimization, and real-time decision-making.

AI-driven data analytics play a crucial role in assessing the lifecycle of batteries, from production to disposal. With AI tools, stakeholders can analyze vast amounts of data collected throughout a battery's lifecycle, including manufacturing processes and recycling methods. Many companies leverage AI-powered Application Programming Interfaces (APIs) to streamline production and simplify operations [65]. Continuous data monitoring through AI analytics provides valuable insights that can lead to more efficient resource use and improved process outcomes. For instance, Guoan Wei [66] demonstrated that the application of Building Information Modeling (BIM) technology, combined with AI-driven Genetic Algorithms, resulted in an optimized scheme that maximized the use of raw materials while delivering the highest economic benefits to the company. This approach in construction management reduced the number of processors needed and minimized working hours, showcasing the economic viability of AI-enhanced processes.

AI algorithms also identify product usage patterns and degradation trends, including batteries. AI can optimize recycling processes and reduce consumer costs by monitoring these trends. The ability to perform comprehensive cost-benefit analyses using AI allows for real-time resource analysis, helping companies optimize resource allocation and maximize profitability. For example, BIM has been used in the automated specification of steel reinforcement in construction, optimizing the efficiency of reinforced concrete (RC) flat slabs [67]. Eleftheriadis's research shows that this modeling has improved material efficiency and reduced costs, highlighting the financial benefits of AI in assessing new recycling technologies and expanding existing facilities [67].

A significant advantage of AI-driven economic modeling is its ability to adapt to rapidly changing markets and regulatory requirements. Unlike traditional economic models, AI models can continuously analyze market trends and operational data, providing real-time solutions for complex challenges. Roodsari [68], for example, has demonstrated the use of system dynamics in financial modeling to simulate and optimize decision-making under uncertainty, allowing companies to make informed decisions in volatile environments. This adaptability enhances the efficiency and sustainability of battery recycling, as AI can quickly respond to market shifts and regulatory changes.

Decision support systems powered by AI offer invaluable tools for recycling facility managers to optimize operations. AI systems can develop automated solutions for planning and scheduling key stages of large-scale industrial projects, such as shop fabrication and on-site construction studies, which have shown that these systems can reduce cycle times in automated processes by 4.8% to 12% [69]. Applying such AI-driven decision-making support to battery recycling can further enhance process efficiency. These systems integrate data from sensor networks, supply chain logistics, and regulatory databases to provide real-time recommendations and solutions. AI algorithms can optimize various recycling facility processes, including production planning, predictive maintenance, and quality control, ensuring reduced costs, high precision, and increased efficiency [70]. Additionally, AI can quickly analyze incoming materials, accurately identifying recyclable components, thus improving automation, reducing labor costs, and minimizing waste.

In conclusion, integrating AI into the battery recycling industry revolutionizes economic analysis and process optimization. By leveraging AI-driven data analytics and decision support systems, companies can enhance resource efficiency, reduce costs, and adapt to changing market and regulatory landscapes, ultimately leading to a more sustainable and economically viable recycling industry.

5.3. AI/ML Algorithms and Techniques in Battery Recycling

Rapid growth in battery use, especially in electric vehicles and renewable energy systems, has created an urgent need for efficient recycling processes that can recover valuable materials such as lithium, cobalt, and nickel. AI and ML play a transformative role in optimizing battery recycling by automating workflows, improving material recovery rates, and minimizing waste. These technologies enable intelligent sorting systems, predictive maintenance, process optimization, and innovative strategies for resource recovery. This section explores the various AI/ML algorithms and techniques, their applications in battery recycling, and their comparative capabilities.

The most conventional approach to machine learning, Supervised Learning, relies on training models with labeled datasets to predict or classify events or outcomes. Algorithms such as Random Forests, Neural Networks, and Gradient Boosting Machines have been applied to battery recycling to predict material recovery rates depending on process parameters or to classify battery types to optimize recycling strategies [71,72]. While these approaches provide very high accuracy, they also require well-qualified, labeled datasets to be available, which may be difficult to obtain in this domain [73].

In battery recycling, Random Forest is a valuable ML tool for predicting material recovery yields by analyzing parameters like temperature, pressure, and chemical reagents. It aids in optimizing processes to recover critical materials like lithium, cobalt, and nickel, supporting the circular economy by enhancing resource reuse and reducing reliance on virgin materials [74]. Combining multiple decision trees, Random Forest offers robust and reliable predictions across diverse datasets, helping to identify conditions that maximize recovery efficiency while minimizing environmental impact. Despite interpretability challenges, tools like feature importance analysis make it possible to extract actionable insights, fostering sustainable growth in the battery industry [75].

Unsupervised learning methods, such as clustering algorithms like K-means, DB-SCAN, and Principal Component Analysis, are useful in data exploration, especially when unstructured data is not labeled. These methods can find patterns in battery waste streams, classify unknown compositions, and detect anomalies in recycling process data. These techniques are particularly good at data exploration but often require domain expertise to interpret their results effectively [76,77].

Clustering algorithms like K-means are great for situations where data labels are not available. These unsupervised methods can automatically group battery waste streams based on their chemical or physical properties [78]. For example, batteries with different chemistries—like lithium-ion, lead-acid, or nickel-cadmium—can be clustered together, making the recycling process more efficient. The simplicity and scalability of K-means make it especially useful for handling large datasets, allowing for quicker and more effective organization of recycling tasks [79].

Reinforcement learning (RL) is another promising AI technique that excels in dynamic and adaptive environments. RL algorithms, such as Deep Q-Networks and Policy Gradient Methods, have been used to optimize disassembly sequences for maximum material recovery and to adaptively control recycling processes to minimize energy consumption [80]. Although RL offers significant potential, it is computationally intensive and requires careful parameter tuning [81].

6. Challenges and Ethical Considerations

6.1. Technical Challenges in AI-Driven Battery Recycling

The rapid rise in EVs, renewable energy systems, and portable electronic devices has driven a corresponding surge in demand for battery technologies [82]. However, this exponential growth in battery production also presents significant environmental challenges, particularly in recycling. AI has emerged as a promising solution to enhance battery recycling processes, but it faces complex technical hurdles that must be skillfully addressed. This section explores the intricacies of AI-driven battery recycling, examining the technological challenges it encounters and potential strategies for overcoming them.

Battery chemistries are highly intricate, involving nuanced chemical interactions and degradation mechanisms that extend beyond superficial differences. For example, LIBs feature complex electrode materials and electrolyte compositions, requiring advanced AI algorithms to identify optimal recycling pathways while minimizing resource waste [83]. Similarly, nickel–cadmium and lead–acid batteries introduce unique challenges, such as managing toxic materials and optimizing recovery efficiencies. AI's adaptability is crucial in navigating these complexities, necessitating continuous advancements in data analytics, ML models, and predictive simulations to effectively address the diverse chemistries encountered in battery recycling.

The effectiveness of AI models in battery recycling depends heavily on access to high-quality data for training and inference. However, obtaining comprehensive datasets encompassing various battery types, states of charge, degradation levels, and recycling outcomes is a formidable challenge. Furthermore, the lack of data standardization across different recycling facilities and geographical regions hampers the development of universally applicable AI-driven solutions, making it difficult to create models that can be effectively deployed globally [84].

AI-driven battery recycling systems must operate in real-time to optimize efficiency and minimize environmental impact. Achieving this requires seamlessly integrating AI algorithms with sensor networks, robotics, and automation technologies [85]. Ensuring synchronization and reducing latency between data acquisition, processing, and decision-making presents a significant technical challenge, demanding sophisticated AI architectures capable of rapid analysis and response.

A critical technical dilemma is balancing operational efficiency with safety and environmental responsibility. While enhancing efficiency is essential for maximizing productivity and cost-effectiveness, it must not come at the expense of safety standards or environmental sustainability. Therefore, AI-driven battery recycling efforts must prioritize the development of algorithms and systems that streamline operations while upholding stringent safety protocols and eco-friendly practices. This delicate balance underscores the complexity of AI implementation in the recycling industry. It highlights the need for ongoing innovation and refinement of AI-driven solutions to meet operational and sustainability goals [86].

Handling hazardous materials and chemicals during battery recycling further necessitates the implementation of comprehensive safety protocols and environmental sustainability measures. AI-driven systems play a pivotal role in this context by integrating predictive analytics capabilities. These algorithms enable anticipating and identifying potential safety hazards, allowing for proactive measures that optimize resource utilization and minimize waste, thereby reducing the ecological footprint of battery recycling processes.

As the demand for battery recycling grows, AI-driven solutions must demonstrate scalability and adaptability to accommodate increasing volumes of batteries with diverse chemistries and conditions. Scalable AI architectures leveraging cloud computing, edge computing, and distributed processing frameworks can enhance computational efficiency and meet dynamic operational requirements [87]. Additionally, adaptive ML algorithms capable of continuous learning and optimization are essential for addressing the evolving challenges of battery recycling.

The success of AI-driven battery recycling initiatives depends on interdisciplinary collaboration and the integration of knowledge across domains such as materials science, chemistry, engineering, data analytics, and environmental science. Bridging the gap between domain-specific expertise and AI proficiency requires cohesive teamwork, effective communication channels, and shared repositories of domain knowledge and best practices [88]. Facilitating collaboration among researchers, industry practitioners, policymakers, and environmental advocates is essential for overcoming technical challenges and driving innovation in AI-driven battery recycling.

AI-driven battery recycling holds immense potential for addressing the global demand for sustainable energy storage solutions. However, it is not without its technical challenges. From navigating the complexities of battery chemistry to ensuring regulatory compliance and ethical integrity, AI-driven recycling systems must overcome multifaceted hurdles. The path toward efficient and eco-friendly AI-driven battery recycling can be forged through concerted efforts in data quality enhancement, real-time monitoring integration, safety optimization, scalability, and interdisciplinary collaboration. Embracing these challenges as opportunities for innovation and sustainability can propel the advancement of AI technologies in battery recycling, contributing significantly to a greener and more sustainable future.

6.2. Ethical Considerations in AI Implementation

Implementing AI technology has revolutionized many industries, offering enhanced efficiency and innovation. However, while AI can significantly improve recycling processes, it presents various ethical challenges that companies must navigate carefully. Data privacy and security stand out as critical concerns in our increasingly interconnected world.

AI systems rely heavily on vast amounts of data to learn and execute tasks effectively. This dependence on data raises significant privacy and security risks, as AI often gathers information by tracking and monitoring personal data through IoT devices [89]. For instance, IoT-enabled products, such as smart shoes, can track users' locations for remanufacturing purposes, exposing sensitive geospatial data. In some Circular Economy models, data sharing among a company's stakeholders is necessary, but this can lead to security breaches and compromise personal information.

Another ethical concern is the potential for AI to perpetuate or even exacerbate bias. AI algorithms learn from existing data, which means that if the input data is biased, the AI system may replicate and reinforce those biases. A notable example is Amazon's AI system used for candidate selection in hiring, which unintentionally discriminated against women due to biases present in the training data [90,91]. This situation highlights the broader issue of algorithmic bias, which can manifest in various ways, such as predictive policing or university ranking systems, where AI may unintentionally favor certain responses based on the data it has been trained on.

Moreover, the transparency of AI in decision-making is a growing concern. There have been instances where AI systems have been accused of manipulating consumer behavior or influencing voter intentions, raising questions about the ethical use of such technology. While AI offers tremendous potential, it is still in the developmental stage, and there are numerous possibilities for ethical violations concerning personal data and privacy [92,93].

As the recycling industry expands alongside population growth, making the process more efficient is crucial. AI systems could be transformative in this regard, but their implementation must be performed responsibly. AI's ability to process vast amounts of data and make decisions is both its strength and its risk. The accuracy of AI-driven models depends heavily on the quality of the data on which they are trained. Poor-quality data—whether due to errors, biases, or misinformation—can lead to flawed predictions and biased outcomes [94]. Therefore, ensuring data reliability is a critical responsibility in the development of AI technology.

Good experimental practices like well-designed experiments and procedures are essential for maintaining dataset reliability. In the context of recycling, AI systems require accurate information about the types and conditions of waste materials. Unsupervised ML, in particular, can pose challenges in evaluating model quality and interpreting results, making human oversight and assessment vital to ensure the quality of the data and the learning process.

Ultimately, while AI has the potential to revolutionize the recycling industry, companies must approach its application with caution and responsibility. Ensuring the ethical use of AI, maintaining data accuracy, and addressing the challenges of algorithmic bias are essential steps in harnessing AI's full potential while safeguarding privacy and security.

7. Research Gaps, Opportunities, and Future Directions

The advancements in AI within the battery recycling industry have been remarkable, yet there is still considerable room for further improvement. As environmental degradation and climate change concerns intensify, AI presents immense potential to drive sustainable practices. However, significant gaps remain in our understanding of how AI can be effectively leveraged to address environmental challenges. Future research must explore the application of AI in optimizing resource management, enhancing energy efficiency, and mitigating environmental risks. For instance, the electric double-layer effect, discussed in a paper by Sikiru [95], offers a promising avenue for improving LIB storage and recycling methods. This effect involves ions rearranging themselves around charged surfaces in an electrolyte to form an electric double layer, which differs from the bulk composition of the electrolyte and varies with electrode voltage. Leveraging this effect, techniques are being developed to use solvents that break down battery components while preserving the electric double layer, allowing for potential reuse. In another study, Bhar et al. [96] examined improved recycling methods that could enhance the quality of recycled batteries. His research focuses on the pretreatment process of end-of-life batteries, including diagnosis, sorting, storage, various cell discharge methods, black mass recovery, and mechanical dismantling. The article emphasized that disassembling battery modules at the cellular level, combined with AI-based automated segregation, could be highly beneficial. However, the authors noted that the physical dismantling process is still in its infancy and largely confined to laboratory-scale experiments. Further research is needed to scale these processes for industrial applications, which could catalyze unprecedented growth in the portable electronics and automotive sectors. Given the rapid pace of technological advancement, the future of AI in battery recycling is difficult to predict. In one study, Lipu et al. [97] explored how AI can be integrated into advanced battery management systems (BMS) for EVs. The article discussed enhancing BMS algorithms to optimize battery performance, ensuring efficient energy use and prolonged battery life. The authors concluded that exploring these areas could revolutionize EV technology, significantly improving reliability, safety, and longevity.

Transitioning toward a circular economy—where resources are reused, recycled, and regenerated—is essential for sustainable development. AI can be crucial in optimizing resource flows, designing eco-friendly products, and implementing closed-loop systems. However, interdisciplinary research is needed to integrate AI with circular economy principles, driving systemic changes across industries.

The convergence of AI with high-throughput experimentation techniques is set to revolutionize materials science. Future research will likely see AI and materials discovery coming together, accelerating the development of next-generation battery materials with superior performance and sustainability. ML algorithms will be instrumental in predicting material properties and guiding researchers toward novel compositions and structures.

Automating recycling processes through robotics and AI holds great promise for improving the efficiency and accuracy of material recovery from end-of-life batteries. Future trends will likely involve the development of autonomous recycling systems equipped with advanced sensors, robotics, and AI algorithms for sorting, disassembling, and extracting valuable materials. These systems will streamline recycling operations and minimize human intervention, enhancing economic viability and environmental sustainability.

Blockchain technology also offers a decentralized and transparent framework for tracking batteries' lifecycles, from manufacturing to recycling. Future research will explore the integration of blockchain with AI and IoT devices to establish robust traceability systems for batteries. By recording key information such as origin, usage history, and recycling pathways, blockchain-enabled traceability can enhance accountability, incentivize responsible disposal practices, and facilitate the circular economy.

Leveraging AI to orchestrate cognitive recycling networks represents a transformative trend in battery recycling research. These networks, powered by ML algorithms, will dynamically optimize battery waste collection, transportation, and processing based on real-time data and predictive analytics. By allocating resources and coordinating stakeholders intelligently, cognitive recycling networks will enhance battery recycling operations' efficiency, scalability, and sustainability.

8. The Benefits and Future Potential of AI-Driven Recycling

Extensive research has confirmed that AI-driven recycling offers significant societal, environmental, and occupational benefits. As we continue to develop and refine AI, its applications in the recycling industry are becoming increasingly evident. For example, the total generation of MSW in the United States in 2018 was 292.4 million tons, or 4.9 pounds per person per day [98]. Approximately 69 million tons of this waste were recycled, and 25 million tons were composted—a study conducted before AI made significant inroads into the recycling sector [98]. With the advent of AI, these numbers have the potential to improve significantly as AI enhances the efficiency and accuracy of waste sorting and processing. One particularly impactful area where AI makes a difference is battery sorting. Batteries contain toxic elements, including cadmium, lead, mercury, nickel, and lithium, which can pose serious environmental hazards if not properly disposed of. As batteries corrode, they leach toxic chemicals into the soil and water systems, with lithium potentially igniting and emitting hazardous chemicals that can burn underground for years [99]. AI detection systems can be crucial in identifying and sorting batteries before they contaminate the environment, providing a significant environmental safeguard. Another key benefit of AI in recycling is its contribution to environmental life cycles. By accurately detecting and sorting recyclable materials, AI helps reduce the need for additional manufacturing by enabling the reuse of materials. For instance, the Life Cycle Assessment of North American Aluminum Cans found that greenhouse gas emissions from aluminum beverage can production have dropped by more than 40% since 1991 and 7% since 2012, largely due to increased recycling efforts [100]. With AI detection, recycling centers can further boost the number of aluminum cans processed, continuing this positive trend.

The accuracy of AI in identifying and sorting waste has grown significantly since its initial implementation, with current systems achieving accuracy rates ranging from 72.8% to 99.95% [47]. This growing accuracy allows AI to be implemented across various detection areas. In battery recycling, for instance, AI can use image recognition techniques to identify, localize, and determine the size of waste, with studies showing that container filling levels can be determined with 99.8% accuracy using advanced classifiers [47]. As AI continues to evolve, it will improve its ability to detect batteries based on updated shapes, colors, and other characteristics, further enhancing its positive impact on public health, the environment, and overall welfare.

Standardizing AI processes in recycling will open new avenues for even more efficient detection methods. As with many machines, proactive maintenance and preventive problem detection are essential for optimal performance. Improved data collection will enable researchers to program AI systems to sort through a broader range of examples, increasing their accuracy and efficiency. Government recycling centers, in particular, stand to benefit from these advancements as they strive to reduce pollution and promote sustainability [101].

Overall, integrating AI into recycling processes offers tremendous benefits across multiple sectors. As AI technology evolves and more data become available, its potential to drive further improvements in recycling and environmental protection will only grow.

Table 2 summarizes the challenges and solutions in AI for battery recycling.

Table 2. Challenges and solutions in AI for battery recycling.

Challenge	Description	Potential AI Solutions	Future Directions
Data quality	Inconsistent data from varied battery sources	Data preprocessing, synthetic data generation	Standardized data collection
Scalability	Difficulty in adapting solutions to large scales	Modular AI systems, scalable architectures	Cloud-based AI for distributed applications
Regulatory Compliance	Variability in global regulations	AI-driven compliance monitoring	Collaboration with regulatory bodies
Integration with Existing Systems	Complexity in merging AI with traditional methods	Hybrid AI systems for gradual integration	Development of AI-compatible recycling systems
Cost	High upfront costs for AI integration	Cost reduction through automation	Government incentives and subsidies

9. Conclusions

Integrating AI into battery recycling processes marks a significant leap forward in addressing environmental challenges and advancing sustainable practices. Recent advancements in AI have demonstrated immense potential in optimizing battery sorting, enhancing resource recovery, and minimizing the environmental impact of battery waste. By improving the accuracy and efficiency of waste detection, AI has proven to be a critical tool in mitigating the risks associated with improper battery disposal, including the contamination of soil and water by toxic elements. AI-driven systems have increased the precision of sorting recyclable materials and contributed to reducing greenhouse gas emissions by enabling the reuse of materials like aluminum. The potential of AI to further revolutionize battery management, particularly in electric vehicles, underscores its growing importance in the transition toward a circular economy. AI's ability to enhance battery management systems, optimize energy utilization, and extend battery life could significantly boost future technologies' reliability, safety, and sustainability. Moreover, as AI technology advances, its role in automating recycling processes through robotics and predictive analytics will become increasingly pivotal. The development of autonomous recycling systems equipped with AI algorithms for sorting, disassembly, and material recovery promises to streamline operations, reduce human intervention, and improve the economic viability of recycling efforts. Integrating AI with emerging technologies such as blockchain and the IoT can create robust traceability systems, ensuring greater accountability and promoting responsible disposal practices. By facilitating cognitive recycling networks, AI can dynamically optimize the entire recycling process, from collection to material recovery, enhancing the scalability and sustainability of battery recycling operations.

In summary, the recent advancements in AI in battery recycling represent a transformative shift in managing and mitigating the environmental impacts of battery waste. As AI continues to evolve, its potential to drive recycling and resource management innovations will be crucial in achieving a sustainable future. The ongoing research and development in this field will be essential to unlocking new opportunities and addressing the challenges that lie ahead.

Author Contributions: Conceptualization, P.L.M.; writing—original draft preparation, S.A.J., C.A.D.C., J.P., H.S., C.E.T.R. and J.W.; writing—review and editing, S.A.J. and P.L.M.; visualization, P.L.M.; supervision, P.L.M. All authors have read and agreed to the published version of the manuscript.

Funding: This research received no external funding.

Conflicts of Interest: The authors declare no conflict of interest.

References

1. Zanoletti, A.; Carena, E.; Ferrara, C.; Bontempi, E. A Review of Lithium-Ion Battery Recycling: Technologies, Sustainability, and Open Issues. *Batteries* **2024**, *10*, 38. [CrossRef]
2. Ralls, A.M.; Leong, K.; Clayton, J.; Fuelling, P.; Mercer, C.; Navarro, V.; Menezes, P.L. The Role of Lithium-Ion Batteries in the Growing Trend of Electric Vehicles. *Materials* **2023**, *16*, 6063. [CrossRef] [PubMed]
3. Kachate, N.; Sharma, D.; Baidya, K. Material Recovery in Li-Ion Battery Recycling: Case Study. *Mater. Today Proc.* **2023**, *72*, 1498–1502. [CrossRef]
4. Jose, S.A.; Stoll, J.L.; Smith, T.; Jackson, C.; Dieleman, T.; Leath, E.; Eastwood, N.; Menezes, P.L. Critical Review of Lithium Recovery Methods: Advancements, Challenges, and Future Directions. *Processes* **2024**, *12*, 2203. [CrossRef]
5. Bird, R.; Baum, Z.J.; Yu, X.; Ma, J. The Regulatory Environment for Lithium-Ion Battery Recycling. *ACS Energy Lett.* **2022**, *7*, 736–740. [CrossRef]
6. Antony Jose, S.; Dworkin, L.; Montano, S.; Noack, W.C.; Rusche, N.; Williams, D.; Menezes, P.L. Pathways to Circular Economy for Electric Vehicle Batteries. *Recycling* **2024**, *9*, 76. [CrossRef]
7. Liliu, Q.; Liu, C. Current Status of Battery Recycling and Technology. *J. Phys. Conf. Ser.* **2023**, *2608*, 012011. [CrossRef]
8. Lithium-Ion Battery Recycling Market Size, Report 2024 to 2033. Available online: https://www.precedenceresearch.com/lithium-ion-battery-recycling-market (accessed on 27 October 2024).
9. Deshwal, D.; Sangwan, P.; Dahiya, N. Economic Analysis of Lithium Ion Battery Recycling in India. *Wirel. Pers. Commun.* **2022**, *124*, 3263–3286. [CrossRef]
10. Lithium-Ion Battery Demand Forecast for 2030 | McKinsey. Available online: https://www.mckinsey.com/industries/automotive-and-assembly/our-insights/battery-2030-resilient-sustainable-and-circular (accessed on 26 October 2024).
11. Bruno, M.; Fiore, S. Material Flow Analysis of Lithium-Ion Battery Recycling in Europe: Environmental and Economic Implications. *Batteries* **2023**, *9*, 231. [CrossRef]
12. Wasesa, M.; Hidayat, T.; Andariesta, D.T.; Natha, M.G.; Attazahri, A.K.; Afrianto, M.A.; Mubarok, M.Z.; Zulhan, Z.; Putro, U.S. Economic and Environmental Assessments of an Integrated Lithium-Ion Battery Waste Recycling Supply Chain: A Hybrid Simulation Approach. *J. Clean. Prod.* **2022**, *379*, 134625. [CrossRef]
13. What Is AI Visual Inspection for Defect Detection? A Deep Dive. Available online: https://marutitech.com/ai-visual-inspection-for-defect-detection/ (accessed on 1 October 2024).
14. Artificial Intelligence for Recycling: AMP Robotics. Available online: https://www.ellenmacarthurfoundation.org/circular-examples/artificial-intelligence-for-recycling-amp-robotics (accessed on 1 October 2024).
15. AI Contamination Detection | WasteVision AI. Available online: https://wastevision.ai/ (accessed on 1 October 2024).
16. AMP Robotics' AI Platform and Capabilities | IRCAI. Available online: https://ircai.org/top100/entry/amp-robotics-ai-platform-and-capabilities/ (accessed on 3 December 2024).
17. The Importance of Artificial Intelligence in the Plastic Recycling Industry—RepetCo Innovations. Available online: https://www.repetco.com/the-importance-of-artificial-intelligence-in-the-plastic-recycling-industry/ (accessed on 1 October 2024).
18. Fang, B.; Yu, J.; Chen, Z.; Osman, A.I.; Farghali, M.; Ihara, I.; Hamza, E.H.; Rooney, D.W.; Yap, P.-S. Artificial Intelligence for Waste Management in Smart Cities: A Review. *Environ. Chem. Lett.* **2023**, *21*, 1959–1989. [CrossRef] [PubMed]
19. How Much Trash Does The United States Produce Each Day? Available online: https://www.trashcanswarehouse.com/how-much-trash-does-the-united-states-produce-each-day (accessed on 29 October 2024).

20. Wolters, L.; Brusselaers, J. The Energy Transition Paradox: How Lithium Extraction Puts Pressure on Environment, Society, and Politics. *Extr. Ind. Soc.* **2024**, *19*, 101498. [CrossRef]
21. Parliamentary Question | Risks Arising from the Increase in Demand for Cobalt | E-000997/2023 | European Parliament. Available online: https://www.europarl.europa.eu/doceo/document/E-9-2023-000997_EN.html#def1 (accessed on 1 October 2024).
22. Sharma, H.; Kumar, H.; Mangla, S.K. Enablers to Computer Vision Technology for Sustainable E-Waste Management. *J. Clean. Prod.* **2023**, *412*, 137396. [CrossRef]
23. Akbar, J.U.M.; Kamarulzaman, S.F.; Muzahid, A.J.M.; Rahman, M.A.; Uddin, M. A Comprehensive Review on Deep Learning Assisted Computer Vision Techniques for Smart Greenhouse Agriculture. *IEEE Access* **2024**, *12*, 4485–4522. [CrossRef]
24. Ankit; Saha, L.; Kumar, V.; Tiwari, J.; Sweta; Rawat, S.; Singh, J.; Bauddh, K. Electronic Waste and Their Leachates Impact on Human Health and Environment: Global Ecological Threat and Management. *Environ. Technol. Innov.* **2021**, *24*, 102049. [CrossRef]
25. Johnson, M.; Khatoon, A.; Fitzpatrick, C. Application of AI and Machine Vision to Improve Battery Detection and Recovery in E-Waste Management. In Proceedings of the 2022 International Conference on Electrical, Computer, Communications and Mechatronics Engineering (ICECCME), Maldives, Maldives, 16–18 November 2022; pp. 1–6.
26. Bhattacharya, S.; Kumar, A.; Krishav, K.; Panda, S.; Vidhyapathi, C.M.; Sundar, S.; Karthikeyan, B. Self-Adaptive Waste Management System: Utilizing Convolutional Neural Networks for Real-Time Classification. *Eng. Proc.* **2024**, *62*, 5. [CrossRef]
27. RFID Technology Examples on Sustainability and Circular Economy. Available online: https://checkpointsystems.com/blog/rfids-role-in-the-circular-economy-and-sustainability-journey/ (accessed on 1 October 2024).
28. ZenRobotics Robots Sort Waste Efficiently: "Artificial Intelligence Will Revolutionise Recycling"—Sitra. Available online: https://www.sitra.fi/en/cases/zenrobotics-robots-sort-waste-efficiently-artificial-intelligence-will-revolutionise-recycling/ (accessed on 1 October 2024).
29. Valizadeh, A.; Amirhosseini, M.H.; Ghorbani, Y. Predictive Precision in Battery Recycling: Unveiling Lithium Battery Recycling Potential through Machine Learning. *Comput. Chem. Eng.* **2024**, *183*, 108623. [CrossRef]
30. Lv, C.; Zhou, X.; Zhong, L.; Yan, C.; Srinivasan, M.; Seh, Z.W.; Liu, C.; Pan, H.; Li, S.; Wen, Y.; et al. Machine Learning: An Advanced Platform for Materials Development and State Prediction in Lithium-Ion Batteries. *Adv. Mater.* **2022**, *34*, 2101474. [CrossRef]
31. Niu, B.; Wang, X.; Xu, Z. Application of Machine Learning to Guide Efficient Metal Leaching from Spent Lithium-Ion Batteries and Comprehensively Reveal the Process Parameter Influences. *J. Clean. Prod.* **2023**, *410*, 137188. [CrossRef]
32. Mao, J.; Miao, J.; Lu, Y.; Tong, Z. Machine Learning of Materials Design and State Prediction for Lithium Ion Batteries. *Chin. J. Chem. Eng.* **2021**, *37*, 1–11. [CrossRef]
33. Wu, S.; Kaden, N.; Dröder, K. A Systematic Review on Lithium-Ion Battery Disassembly Processes for Efficient Recycling. *Batteries* **2023**, *9*, 297. [CrossRef]
34. Blankemeyer, S.; Wiens, D.; Wiese, T.; Raatz, A.; Kara, S. Investigation of the Potential for an Automated Disassembly Process of BEV Batteries. *Procedia CIRP* **2021**, *98*, 559–564. [CrossRef]
35. Rettenmeier, M.; Möller, M.; Sauer, A. Disassembly Technologies of End-of-Life Automotive Battery Packs as the Cornerstone for a Circular Battery Value Chain: A Process-Oriented Analysis. *Resour. Conserv. Recycl.* **2024**, *209*, 107786. [CrossRef]
36. Kay, I.; Farhad, S.; Mahajan, A.; Esmaeeli, R.; Hashemi, S.R. Robotic Disassembly of Electric Vehicles' Battery Modules for Recycling. *Energies* **2022**, *15*, 4856. [CrossRef]
37. Wegener, K.; Andrew, S.; Raatz, A.; Dröder, K.; Herrmann, C. Disassembly of Electric Vehicle Batteries Using the Example of the Audi Q5 Hybrid System. *Procedia CIRP* **2014**, *23*, 155–160. [CrossRef]
38. Alsabt, R.; Alkhaldi, W.; Adenle, Y.A.; Alshuwaikhat, H.M. Optimizing Waste Management Strategies through Artificial Intelligence and Machine Learning—An Economic and Environmental Impact Study. *Clean. Waste Syst.* **2024**, *8*, 100158. [CrossRef]
39. Yousif, I.; Samaha, J.; Ryu, J.; Harik, R. Safety 4.0: Harnessing Computer Vision for Advanced Industrial Protection. *Manuf. Lett.* **2024**, *41*, 1342–1356. [CrossRef]
40. Paul, S.; Yuan, L.; Jain, H.K.; Robert, L.P., Jr.; Sporher, J.; Lifshitz-Assaf, H. Intelligence Augmentation: Human Factors in AI and Future of Work. *AIS Trans. Hum. Comput. Interact.* **2022**, *14*, 426–445. [CrossRef]
41. Taherdoost, H. Deep Learning and Neural Networks: Decision-Making Implications. *Symmetry* **2023**, *15*, 1723. [CrossRef]
42. Xu, G.; Zhang, Z.; Li, Z.; Guo, X.; Qi, L.; Liu, X. Multi-Objective Discrete Brainstorming Optimizer to Solve the Stochastic Multiple-Product Robotic Disassembly Line Balancing Problem Subject to Disassembly Failures. *Mathematics* **2023**, *11*, 1557. [CrossRef]
43. Gulivindala, A.K.; Bahubalendruni, M.V.A.R.; Chandrasekar, R.; Ahmed, E.; Abidi, M.H.; Al-Ahmari, A. Automated Disassembly Sequence Prediction for Industry 4.0 Using Enhanced Genetic Algorithm. *Comput. Mater. Contin.* **2021**, *69*, 2531–2548. [CrossRef]
44. Li, H.; Xu, Q.; Wang, Q.; Tang, B. A Review of Intelligent Verification System for Distribution Automation Terminal Based on Artificial Intelligence Algorithms. *J. Cloud Comput.* **2023**, *12*, 146. [CrossRef]
45. Sharma, K.M.; Sreedhar, P.N. Intelligent Approach for Efficient Operation of Electrical Distribution Automation Systems. In Proceedings of the TENCON 2003: Conference on Convergent Technologies for Asia-Pacific Region, Bangalore, India, 15–17 October 2003; Volume 2, pp. 761–765.
46. The Relevance of Artificial Intelligence in the Recycling Industry. Available online: https://recyclinginside.com/the-relevance-of-artificial-intelligence-in-the-recycling-industry/ (accessed on 25 October 2024).

47. Pregowska, A.; Osial, M.; Urbańska, W. The Application of Artificial Intelligence in the Effective Battery Life Cycle in the Closed Circular Economy Model—A Perspective. *Recycling* **2022**, *7*, 81. [CrossRef]
48. Taşcı, B.; Omar, A.; Ayvaz, S. Remaining Useful Lifetime Prediction for Predictive Maintenance in Manufacturing. *Comput. Ind. Eng.* **2023**, *184*, 109566. [CrossRef]
49. Javaid, M.; Haleem, A.; Singh, R.P.; Suman, R. Artificial Intelligence Applications for Industry 4.0: A Literature-Based Study. *J. Ind. Integr. Manag.* **2022**, *07*, 83–111. [CrossRef]
50. Kuźnar, M.; Lorenc, A. A Hybrid Method for Technical Condition Prediction Based on AI as an Element for Reducing Supply Chain Disruptions. *Appl. Sci.* **2023**, *13*, 12439. [CrossRef]
51. Utilization AI for Predictive Maintenance in IoT-Enabled Industrial Systems | Journal of Artificial Intelligence and Development. Available online: https://edujavare.com/index.php/JAI/article/view/317 (accessed on 25 October 2024).
52. Huang, Z.; Shen, Y.; Li, J.; Fey, M.; Brecher, C. A Survey on AI-Driven Digital Twins in Industry 4.0: Smart Manufacturing and Advanced Robotics. *Sensors* **2021**, *21*, 6340. [CrossRef]
53. Wu, B.; Widanage, W.D.; Yang, S.; Liu, X. Battery Digital Twins: Perspectives on the Fusion of Models, Data and Artificial Intelligence for Smart Battery Management Systems. *Energy AI* **2020**, *1*, 100016. [CrossRef]
54. Ziouzios, D.; Baras, N.; Balafas, V.; Dasygenis, M.; Stimoniaris, A. Intelligent and Real-Time Detection and Classification Algorithm for Recycled Materials Using Convolutional Neural Networks. *Recycling* **2022**, *7*, 9. [CrossRef]
55. Erkinay Ozdemir, M.; Ali, Z.; Subeshan, B.; Asmatulu, E. Applying Machine Learning Approach in Recycling. *J. Mater. Cycles Waste Manag.* **2021**, *23*, 855–871. [CrossRef]
56. Yu, K.H.; Zhang, Y.; Li, D.; Montenegro-Marin, C.E.; Kumar, P.M. Environmental Planning Based on Reduce, Reuse, Recycle and Recover Using Artificial Intelligence. *Environ. Impact Assess. Rev.* **2021**, *86*, 106492. [CrossRef]
57. Bijos, J.C.B.F.; Zanta, V.M.; Morató, J.; Queiroz, L.M.; Oliveira-Esquerre, K.P.S.R. Improving Circularity in Municipal Solid Waste Management through Machine Learning in Latin America and the Caribbean. *Sustain. Chem. Pharm.* **2022**, *28*, 100740. [CrossRef]
58. Munir, M.T.; Li, B.; Naqvi, M. Revolutionizing Municipal Solid Waste Management (MSWM) with Machine Learning as a Clean Resource: Opportunities, Challenges and Solutions. *Fuel* **2023**, *348*, 128548. [CrossRef]
59. Lixo and Ficha Offer AI Tools to Make Life Easier for Waste Managers—Hello Future Orange. Available online: https://hellofuture.orange.com/en/data-driven-recycling-ai-sorts-through-the-waste/ (accessed on 25 October 2024).
60. The U.S. Recycling System | US EPA. Available online: https://www.epa.gov/circulareconomy/us-recycling-system (accessed on 25 October 2024).
61. Turning Trash Into Treasure: How AI Is Revolutionizing Waste Sorting. Available online: https://www.forbes.com/sites/ganeskesari/2024/05/31/turning-trash-into-treasure-how-ai-is-revolutionizing-waste-sorting/ (accessed on 1 December 2024).
62. Climate Risk and Decarbonization: What Every Mining CEO Needs to Know | McKinsey. Available online: https://www.mckinsey.com/capabilities/sustainability/our-insights/climate-risk-and-decarbonization-what-every-mining-ceo-needs-to-know (accessed on 25 October 2024).
63. Air Emissions from MSW Combustion Facilities | Energy Recovery from Waste | Municipal Solid Waste | US EPA. Available online: https://archive.epa.gov/epawaste/nonhaz/municipal/web/html/airem.html (accessed on 25 October 2024).
64. The Environmental Impact of Trucking—Teletrac Navman US. Available online: https://www.teletracnavman.com/resources/blog/the-environmental-impact-of-trucking (accessed on 25 October 2024).
65. [2312.14972] A Trade-off Analysis of Replacing Proprietary LLMs with Open Source SLMs in Production. Available online: https://ar5iv.labs.arxiv.org/html/2312.14972 (accessed on 25 October 2024).
66. Wei, G. Application of BIM Technology Combined with Artificial Intelligence in Construction Management: Artificial Intelligence in Construction Management. *Staveb. Obz. Civ. Eng. J.* **2021**, *30*, 535–536. [CrossRef]
67. Eleftheriadis, S.; Duffour, P.; Stephenson, B.; Mumovic, D. Automated Specification of Steel Reinforcement to Support the Optimisation of RC Floors. *Autom. Constr.* **2018**, *96*, 366–377. [CrossRef]
68. Yousefi Roodsari, S. Using System Dynamics for Simulation and Optimization of an Investment Decision System Under Uncertainty. *Process Integr. Optim. Sustain.* **2022**, *6*, 367–381. [CrossRef]
69. Hu, D. *Automated Planning and Scheduling for Industrial Construction Processes*; ERA: Edmonton, AB, USA, 2013. [CrossRef]
70. Elahi, M.; Afolaranmi, S.O.; Martinez Lastra, J.L.; Perez Garcia, J.A. A Comprehensive Literature Review of the Applications of AI Techniques through the Lifecycle of Industrial Equipment. *Discov. Artif. Intell.* **2023**, *3*, 43. [CrossRef]
71. Priyadarshini, J.; Elangovan, M.; Mahdal, M.; Jayasudha, M. Machine-Learning-Assisted Prediction of Maximum Metal Recovery from Spent Zinc–Manganese Batteries. *Processes* **2022**, *10*, 1034. [CrossRef]
72. Zhou, F.; Shi, D.; Mu, W.; Wang, S.; Wang, Z.; Wei, C.; Li, R.; Mu, T. Machine Learning Models Accelerate Deep Eutectic Solvent Discovery for the Recycling of Lithium-Ion Battery Cathodes. *Green Chem.* **2024**, *26*, 7857–7868. [CrossRef]
73. Daware, S.; Chandel, S.; Rai, B. A Machine Learning Framework for Urban Mining: A Case Study on Recovery of Copper from Printed Circuit Boards. *Miner. Eng.* **2022**, *180*, 107479. [CrossRef]

74. Acharya, S.; Viswesh, P.; Sridhar, M.K.; Pathak, A.D.; Sharma, H.; Nazir, A.; Kasbe, A.; Sahu, K.K. Chapter 21—Artificial Intelligence and Machine Learning in Battery Materials and Their Applications. In *Nanostructured Materials Engineering and Characterization for Battery Applications*; Gueye, A.B., Maria, H.J., Kalarikkal, N., Fall, M., Stephan, A.M., Thomas, S., Eds.; Elsevier: Amsterdam, The Netherlands, 2024; pp. 639–676. ISBN 978-0-323-91304-1.
75. Zhao, J.; Qu, X.; Wu, Y.; Fowler, M.; Burke, A.F. Artificial Intelligence-Driven Real-World Battery Diagnostics. *Energy AI* **2024**, *18*, 100419. [CrossRef]
76. Zhou, X.; Du, H.; Xue, S.; Ma, Z. Recent Advances in Data Mining and Machine Learning for Enhanced Building Energy Management. *Energy* **2024**, *307*, 132636. [CrossRef]
77. Esteban, A.; Zafra, A.; Ventura, S. Data Mining in Predictive Maintenance Systems: A Taxonomy and Systematic Review. *WIREs Data Min. Knowl. Discov.* **2022**, *12*, e1471. [CrossRef]
78. Liu, T.; Chen, X.; Peng, Q.; Peng, J.; Meng, J. An Enhanced Sorting Method for Retired Battery with Feature Selection and Multiple Clustering. *J. Energy Storage* **2024**, *87*, 111422. [CrossRef]
79. Xu, Z.; Wang, J.; Lund, P.D.; Fan, Q.; Dong, T.; Liang, Y.; Hong, J. A Novel Clustering Algorithm for Grouping and Cascade Utilization of Retired Li-Ion Batteries. *J. Energy Storage* **2020**, *29*, 101303. [CrossRef]
80. Chand, M.; Ravi, C. An Optimal Disassembly Sequence Planning for Complex Products Using Enhanced Deep Reinforcement Learning Framework. *SN Comput. Sci.* **2024**, *5*, 581. [CrossRef]
81. Nian, R.; Liu, J.; Huang, B. A Review On Reinforcement Learning: Introduction and Applications in Industrial Process Control. *Comput. Chem. Eng.* **2020**, *139*, 106886. [CrossRef]
82. Renewables—Energy System—IEA. Available online: https://www.iea.org/energy-system/renewables (accessed on 25 October 2024).
83. The Rising Complexity of the Battery Industry | GreenFleet. Available online: https://greenfleet.net/features/07082018/rising-complexity-battery-industry (accessed on 25 October 2024).
84. Data Quality in AI: Challenges, Importance & Best Practices. Available online: https://research.aimultiple.com/data-quality-ai/ (accessed on 25 October 2024).
85. Villegas-Ch, W.; García-Ortiz, J.; Sánchez-Viteri, S. Toward Intelligent Monitoring in IoT: AI Applications for Real-Time Analysis and Prediction. *IEEE Access* **2024**, *12*, 40368–40386. [CrossRef]
86. Zong, Z.; Guan, Y. AI-Driven Intelligent Data Analytics and Predictive Analysis in Industry 4.0: Transforming Knowledge, Innovation, and Efficiency. *J. Knowl. Econ.* **2024**. [CrossRef]
87. Scalability and Distributed Computing in NET for Large-Scale AI Workloads | Eduzone: International Peer Reviewed/Refereed Multidisciplinary Journal. Available online: https://www.eduzonejournal.com/index.php/eiprmj/article/view/513 (accessed on 25 October 2024).
88. Dellermann, D.; Calma, A.; Lipusch, N.; Weber, T.; Weigel, S.; Ebel, P. The Future of Human-AI Collaboration: A Taxonomy of Design Knowledge for Hybrid Intelligence Systems 2021. *arXiv* **2021**, arXiv:2105.03354.
89. Roberts, H.; Zhang, J.; Bariach, B.; Cowls, J.; Gilburt, B.; Juneja, P.; Tsamados, A.; Ziosi, M.; Taddeo, M.; Floridi, L. Artificial Intelligence in Support of the Circular Economy: Ethical Considerations and a Path Forward. *AI Soc.* **2024**, *39*, 1451–1464. [CrossRef]
90. Benjamins, R. A Choices Framework for the Responsible Use of AI. *AI Ethics* **2021**, *1*, 49–53. [CrossRef]
91. Stahl, B.C.; Wright, D. Ethics and Privacy in AI and Big Data: Implementing Responsible Research and Innovation. *IEEE Secur. Priv.* **2018**, *16*, 26–33. [CrossRef]
92. The Ethical Dilemma of AI in Marketing: A Slippery Slope. Available online: https://www.forbes.com/sites/elijahclark/2024/03/14/the-ethical-dilemma-of-ai-in-marketing-a-slippery-slope/ (accessed on 1 December 2024).
93. Park, P.S.; Goldstein, S.; O'Gara, A.; Chen, M.; Hendrycks, D. AI Deception: A Survey of Examples, Risks, and Potential Solutions. *Patterns* **2024**, *5*, 100988. [CrossRef]
94. Lombardo, T.; Duquesnoy, M.; El-Bouysidy, H.; Årén, F.; Gallo-Bueno, A.; Jørgensen, P.B.; Bhowmik, A.; Demortière, A.; Ayerbe, E.; Alcaide, F.; et al. Artificial Intelligence Applied to Battery Research: Hype or Reality? *Chem. Rev.* **2022**, *122*, 10899–10969. [CrossRef]
95. Sikiru, S.; Dele-Afolabi, T.T.; Ghotbi, M.Y.; Rehman, Z.U. Recent Advancements in Technology Projection on Electric Double Layer Effect in Battery Recycling for Energy Storage. *J. Power Sources* **2024**, *596*, 234056. [CrossRef]
96. Bhar, M.; Ghosh, S.; Krishnamurthy, S.; Kaliprasad, Y.; Martha, S.K. A Review on Spent Lithium-Ion Battery Recycling: From Collection to Black Mass Recovery. *RSC Sustain.* **2023**, *1*, 1150–1167. [CrossRef]
97. Lipu, M.S.H.; Miah, M.S.; Jamal, T.; Rahman, T.; Ansari, S.; Rahman, M.S.; Ashique, R.H.; Shihavuddin, A.S.M.; Shakib, M.N. Artificial Intelligence Approaches for Advanced Battery Management System in Electric Vehicle Applications: A Statistical Analysis towards Future Research Opportunities. *Vehicles* **2024**, *6*, 22–70. [CrossRef]
98. National Overview: Facts and Figures on Materials, Wastes and Recycling | US EPA. Available online: https://www.epa.gov/facts-and-figures-about-materials-waste-and-recycling/national-overview-facts-and-figures-materials (accessed on 25 October 2024).
99. Dobó, Z.; Dinh, T.; Kulcsár, T. A Review on Recycling of Spent Lithium-Ion Batteries. *Energy Rep.* **2023**, *9*, 6362–6395. [CrossRef]

100. Carbon Footprint of U.S. Aluminum Can Production Down 40% + Since 1990s | The Aluminum Association. Available online: https://www.aluminum.org/carbon-footprint-us-aluminum-can-production-down-40-1990s (accessed on 25 October 2024).
101. Artificial Intelligence and Machine Learning: How Can They Improve Observability? Available online: https://www.knowmadmood.com/en/blog/artificial-intelligence-and-machine-learning-how-can-they-improve-observability/ (accessed on 25 October 2024).

Disclaimer/Publisher's Note: The statements, opinions and data contained in all publications are solely those of the individual author(s) and contributor(s) and not of MDPI and/or the editor(s). MDPI and/or the editor(s) disclaim responsibility for any injury to people or property resulting from any ideas, methods, instructions or products referred to in the content.

batteries

MDPI

Article

Enhancing Mass Transport in Organic Redox Flow Batteries Through Electrode Obstacle Design

Joseba Martínez-López [1], Unai Fernández-Gamiz [1,*], Eduardo Sánchez-Díez [2], Aitor Beloki-Arrondo [3] and Íñigo Ortega-Fernández [3]

[1] Nuclear Engineering and Fluid Mechanics Department, University of the Basque Country UPV/EHU, Nieves Cano 12, 01006 Vitoria-Gasteiz, Spain; joseba.martinezl@ehu.eus

[2] Centre for Cooperative Research on Alternative Energies (CIC EnergiGUNE), Basque Research and Technology Alliance (BRTA), Alava Technology Park, Albert Einstein 48, 01510 Vitoria-Gasteiz, Spain; esanchez@cicenergigune.com

[3] TECNALIA Research & Innovation, Basque Research and Technology Alliance (BRTA), Mikeletegi Pasealekua 2, 20009 Donostia-San Sebastian, Spain; aitor.beloki@tecnalia.com (A.B.-A.); inigo.ortega@tecnalia.com (Í.O.-F.)

* Correspondence: unai.fernandez@ehu.eus

Abstract: This study examines the impact of incorporating obstacles in the electrode structure of an organic redox flow battery with a flow-through configuration. Two configurations were compared: a control case without obstacles (Case 1) and a modified design with obstacles to enhance mass transport and uniformity (Case 2). While Case 1 exhibited marginally higher discharge voltages (average difference of 0.18%) due to reduced hydraulic resistance and lower Ohmic losses, Case 2 demonstrated significant improvements in concentration uniformity, particularly at low state-of-charge (SOC) levels. The obstacle design mitigated local depletion of active species, thereby enhancing limiting current density and improving minimum concentration values across the studied SOC range. However, the introduction of obstacles increased flow resistance and pressure drops, indicating a trade-off between electrochemical performance and pumping energy requirements. Notably, Case 2 performed better at lower flow rates, showcasing its potential to optimize efficiency under varying operating conditions. At higher flow rates, the advantages of Case 2 diminished but remained evident, with better concentration uniformity, higher minimum concentration values, and a 1% average increase in limiting current density. Future research should focus on optimizing obstacle geometry and positioning to further enhance performance.

Keywords: organic redox flow battery; numerical model; concentration uniformity; limiting current density; pressure drop

1. Introduction

The energy scene is transforming into one moving toward a sustainable future. Renewable resources such as solar and wind are considered cleaner substitutes for fossil fuel, yet their intermittency and unpredictability introduce significant challenges toward grid stability [1,2]. While energy systems are evolving and incorporating increasing shares of variable renewable energy resources, energy storage solutions will be important to maintain the balance between supply and demand and ensure reliability to accomplish a successful energy transition [3,4]. Energy storage will play one of the leading roles in this transition, allowing energy to be captured and provided later at needed times. Among the different storage systems developed, redox flow batteries (RFBs) have emerged as one of the most promising technologies for wide application because of their modular design, scalability, and flexibility [5,6]. Unlike traditional batteries, RFBs allow independent scaling of storage capacity and power output due to the separation of the energy storage and power conversion components [7,8]. This feature makes them extremely suitable for grid-scale applications and integration with renewable energy sources.

Among the variants of RFBs, vanadium-based RFBs (VRFBs) are the most commercially developed technology [9]. These VRFBs have demonstrated promising performance in terms of long cycle life, chemical stability, and grid-scale deployments [10,11]. However, VRFBs also have their own challenges, such as the energy density being limited by the solubility constraints of vanadium species, and their reliance on vanadium, a critical metal with price and supply uncertainties [12,13]. Limitations in the existing systems has led to researchers exploring other chemistries such as organic redox flow batteries (ORFBs). This includes using organic active species derived from abundant elements in the Earth's crust to reduce some drawbacks in VRFBs. These molecules provide a sustainable solution for the substitution of inorganic compounds with environmentally more benign materials [14,15]. These systems provide multiple advantages, like higher solubility, tunability, and the potential for cost reduction, and hence, may lead to better sustainability and efficiency in energy storage [16].

In the area of organic redox chemistry, significant progress has been achieved in alkaline systems. Lin et al. [17] presented the synergy between 2,6-dihydroxyanthraquinone (2,6 DHAQ) as a negative electrolyte and ferricyanide ($[Fe(CN_6)]^{4-}/[Fe(CN_6)]^{3-}$) as a positive counterpart. This configuration enabled the achievement of power densities of 0.45 W cm^{-2} at room temperature and 0.7 W cm^{-2} at 45 °C. Other researchers have confirmed the effectiveness of anthraquinones and ferricyanides as excellent alternatives [18–20]. Other chemistries have furthered the state of the art in alkaline RFBs. Zou et al. [21] reported an S/Fe RFB with a remarkable volumetric capacity of 40.74 Ah L^{-1} that can achieve 99% coulombic efficiency and maintain an extremely low capacity fade of 0.0166% per cycle. Recent work has shown that Fe/Mn-based alkaline batteries have promise for sustainable energy storage. Shen et al. [22] reported a theoretical cell voltage of 1.43 V with close to ideal coulombic efficiency over 400 cycles; however, there was a gradual drop in voltage efficiency from 75.3% to 61.4%.

While experimental approaches provide critical insights, their high resource demands have shifted attention toward computational methods. Within these, computational fluid dynamics (CFD) has proved capable of modeling electrochemical devices successfully [23]. In this framework, numerous studies have been conducted to enhance mass transport using numerical simulations [24,25]. Pan et al. [26] applied a gradient decrease from the inlet to the outlet in channel width. The uniformity of active species was improved, which translated to higher power density and reduced polarization losses. Martinez-Lopez et al. [27] investigated the impact of electrode compression in convection using a 2D numerical model. Their findings revealed that applying 50% compression enhanced velocity profiles by 12.7%. Xu et al. [28] utilized a different distributor within a detailed 3D model. They evaluated the batteries both with and without a flow field and with serpentine and parallel patterns. They studied the performance metrics, overpotentials, pressure drops, and uniformity along the distributor, including a sensitivity analysis on the flow rate. Xu et al. also gave an overall efficiency metric for each geometry and then concluded that the serpentine flow pattern had a better performance and efficiency. Chu et al. [29] introduced a 3D numerical model utilizing 2,6-dihydroanthraquinone and ferrocyanide electrolytes, focusing on the effects of various electrode geometries, such as rectangular, trapezoidal, and sector shapes, on performance metrics like voltage, overpotentials, and efficiency. The sector-shaped electrode configuration showed the best results in mass transfer and power-based efficiency. Aparicio-Mauricio et al. [30] explored the influence of manifold design on electrolyte distribution, showing that uneven flow from dividing manifolds led to increased shear stress, higher pressure drops, and performance losses, including elevated voltage and current density deviations. Akuzum et al. [31] investigated the effects of channel obstructions and ramps on mass transport. They found that a ramped channel design enhances electrochemical performance by increasing electrolyte penetration at the electrode–membrane interface. However, adding obstructions worsened performance due to electrolyte bypassing. Notably, both tapered and obstructed channels result in lower pressure drops compared to an unmodified flow field. Messaggi et al. [32] studied the effects of obstruction placement on the flow channels. Results showed that electrolyte penetration was enhanced when the obstructions were placed on the channel side, whereas those placed at the electrode interface direct electrolyte better toward the channel outlet. Biomimetic flow fields inspired by leaf veins were proposed for VRFBs to improve electrolyte distribution by Liu et al. [33]. Among the designs, circular obstacles in the main channel achieved the best performance, with enhanced voltage efficiency and concentration uniformity, offering a promising approach for practical applications. The use of static mixers along variable compression has also been studied to improve performance on flow-through configurations [34,35]. The reactant distribution is shown to be enhanced, translating to cell voltage improvement, reduction of concentration overpotential and extended capacity.

Previous research has predominantly concentrated on flow-by configuration VRFBs, and the limited research on organic redox flow batteries has focused on clarifying the influence of electrode construction and manifold configuration. This works aims to expand the knowledge on the effects of introducing obstacles on flow-through configuration batteries. To address this gap, a two-dimensional stationary isothermal model is developed to analyze obstacle placement compared to a baseline design under different working conditions. The study evaluates the impact on discharge voltage, limiting current density, concentration uniformity, and pressure drop across different states of charge (SOC) and flow rates.

2. Model Description

A schematic representation of the ORFB is shown in Figure 1. The electrolytes are kept in two separate reservoirs. The ($[Fe(CN_6)]^{4-}/[Fe(CN_6)]^{3-}$) redox pair is stored in the posolyte tank, while 2,6 dihydroxyanthraquinone (2,6 DHAQ)/2,6 reduced DHAQ is stored in the negolyte tank. The electrolyte is circulated to the porous electrodes, which act as active sites for the electrochemical reactions, via peristaltic pumps during battery operation. An ion-selective membrane is placed between the electrodes, which allows only selected ions to flow through while preventing cross-contamination. The following electrochemical processes are taking place on the cathode and anode surfaces:

$$\textit{Positive}: \mathrm{Fe(CN)_6^{3-}} + \mathrm{e^-} \underset{\text{charge}}{\overset{\text{discharge}}{\rightleftharpoons}} \mathrm{Fe(CN)_6^{4-}} \tag{1}$$

$$\textit{Negative}: 2.6 - \mathrm{DHAQ} + 2\,\mathrm{e^-} \underset{\text{discharge}}{\overset{\text{charge}}{\rightleftharpoons}} 2.6 - \mathrm{reDHAQ} \tag{2}$$

A two-dimensional cell model has been developed for this work, which includes three domains: the cathode, the ion exchange membrane, and the anode. The key assumptions to reduce the complexity of the multiphysics behavior of the battery taken into consideration are:

- The model is stationary.
- Every property of the electrode and membrane is isotropic.
- It is assumed that the entire cell is isothermal.

- The electrolyte is considered incompressible.
- There is no modeling of parasitic reactions.
- Infinite dilute approximation is considered.
- The membrane only permits K^+ ions to pass through. All other ion crossover is disregarded.

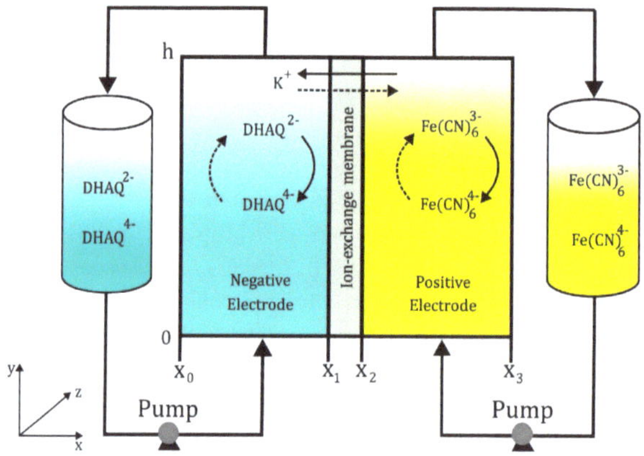

Figure 1. Schematic representation of the numerical model.

The concept of obstacle placement compared to a base case is shown in Figure 2. The obstacles are rectangular prisms.

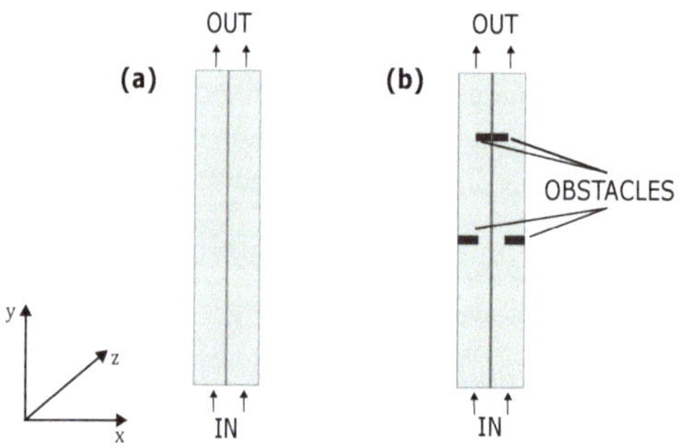

Figure 2. Schematic representation of the (**a**) typical no-obstacle flow-through configuration and (**b**) the proposed obstacle configuration.

2.1. Governing Equations

2.1.1. Mass Transport

The conservation of mass for each species i is expressed by Equation (3):

$$\frac{\partial}{\partial t}(\varepsilon c_i) + \nabla \cdot \mathbf{N_i} = -S_i \quad (3)$$

where c denotes the concentration of species i, ε denotes the electrode porosity, and S stands for the source term of each substance i, provided in Table 1.

Table 1. Source terms of species.

Species	Positive Electrode	Negative Electrode
$[Fe(CN_6)]^{4-}$	i_r/F	-
$[Fe(CN_6)]^{3-}$	$-i_r/F$	-
$DHAQ^{2-}$	-	$2i_r/F$
$DHAQ^{4-}$	-	$-2i_r/F$

The Nernst–Planck equation expresses the flux of charged species by diffusion, migration, and convection, represented as $\mathbf{N_i}$:

$$\mathbf{N_i} = -D_i^{eff}\nabla c_i - z_i u_i c_i F \nabla \varphi_l + \mathbf{u} c_i \quad (4)$$

where D^{eff} is the effective diffusivity, z stands for the charge of species, u indicates ionic mobility, F represents the Faraday constant, and φ_l is the liquid phase potential. $\mathbf{u}$ indicates the electrolyte velocity. Table 2 provides a summary of electrolyte characteristics.

Table 2. Properties of electrolyte.

Quantity	Symbol	Value	References
Diffusivity of $[Fe(CN_6)]^{4-}$	$D_{[Fe(CN)_6]^{4-}}$	8.2×10^{-10} m^2 s^{-1}	[36]
Diffusivity of $[Fe(CN_6)]^{3-}$	$D_{[Fe(CN)_6]^{3-}}$	8.6×10^{-10} m^2 s^{-1}	[36]
Diffusivity of $DHAQ^{2-}$	$D_{DHAQ^{2-}}$	4.8×10^{-10} m^2 s^{-1}	[17]
Diffusivity of $DHAQ^{4-}$	$D_{DHAQ^{4-}}$	4.8×10^{-10} m^2 s^{-1}	[17]
Diffusivity of K^+	D_{K^+}	1.96×10^{-9} m^2 s^{-1}	[37]
Diffusivity of OH^-	D_{OH^-}	5.2×10^{-9} m^2 s^{-1}	[37]

D_i^{eff} is obtained by means of the Bruggemann correlation:

$$D_i^{eff} = \varepsilon^{3/2} D_i \quad (5)$$

The Nernst–Einstein equation is used to calculate ionic mobility:

$$u_i = \frac{D_i^{eff}}{RT} \quad (6)$$

where R refers to the universal gas constant and T is the temperature.

Darcy's Law is used for the convection term in porous electrodes:

$$\mathbf{u} = -\frac{K}{\mu}\nabla p \quad (7)$$

where K stands for the porous electrode permeability, μ is the dynamic viscosity of electrolyte, and p refers to pressure. The following is the Kozeny–Carman equation for permeability:

$$K = \frac{d_f^2 \varepsilon^3}{16 k_{ck}(1-\varepsilon)^2} \quad (8)$$

where d_f stands for the electrode fiber diameter and k_{ck} is the Kozeny–Carman constant, listed with additional electrode parameters in Table 3.

Table 3. Electrode parameters.

Quantity	Symbol	Value	References
Non-compressed electrode porosity	ε	0.895	[38]
Non-compressed electrode specific surface area	a	3.5×10^4 m^2 m^{-3}	Fitted
Non-compressed electrode conductivity	σ_s	66.7 S m^{-1}	[39]
Kozeny–Carman constant	K_{ck}	4.28	[40]

Since only K^+ ions are permitted to pass through the membrane, the ion flux is determined by:

$$\mathbf{N_{K^+}} = -\frac{\sigma_{mem}}{F}\Delta\varphi_{mem} \quad (9)$$

where the potential and membrane conductivity are denoted by φ_{mem} and σ_{mem}, respectively.

2.1.2. Reaction Kinetics

The electrochemical reactions at the electrode surface are modeled using the Butler–Volmer model:

$$i_{r,pos} = i_{0,pos} a \left[exp\left(\frac{(1-\alpha_{pos})F\eta_+}{RT} \right) - exp\left(\frac{(-\alpha_{pos})F\eta_+}{RT} \right) \right] \quad (10)$$

$$i_{r,neg} = i_{0,neg} a \left[exp\left(\frac{(1-\alpha_{neg})F\eta_-}{RT} \right) - exp\left(\frac{(-\alpha_{neg})F\eta_-}{RT} \right) \right] \quad (11)$$

The positive and negative reaction rates are denoted by the terms $i_{r,pos}$ and $i_{r,neg}$, respectively. Specific surface area of the porous electrode is denoted by a, η_+ and η_- are the positive and negative overpotential, charge transfer coefficients for positive and negative reactions are denoted by α_{pos} and α_{neg}, and exchange current density $i_{0,pos}$ and $i_{0,neg}$ are determined by:

$$i_{0,pos} = Fk_{pos}(c_{[Fe(CN)_6]^{3-}})^{1-\alpha_{pos}}(c_{[Fe(CN)_6]^{4-}})^{\alpha_{pos}} \quad (12)$$

$$i_{0,neg} = Fk_{neg}(c_{DHAQ^{2-}})^{1-\alpha_{neg}}(c_{DHAQ^{4-}})^{\alpha_{neg}} \quad (13)$$

where the reaction rate constants for the positive and negative sides are denoted by k_{pos} and k_{neg}, respectively.

The following formulas can be used to determine the overpotential of the positive and negative electrodes:

$$\eta_+ = \varphi_s - \varphi_l - E_{eq,+} \quad (14)$$

$$\eta_- = \varphi_s - \varphi_l - E_{eq,-} \quad (15)$$

where φ_s represents the solid phase potential, and $E_{eq,+}$ and $E_{eq,-}$ denote open circuit potentials of positive and negative reactions, respectively. These potentials can be obtained using the Nernst equation:

$$E_{eq,+} = E'_{eq,+} + \frac{RT}{F} ln\left(\frac{c_{[Fe(CN)_6]^{3-}}}{c_{[Fe(CN)_6]^{4-}}} \right) \quad (16)$$

$$E_{eq,-} = E'_{eq,-} + \frac{RT}{2F} ln\left(\frac{c_{DHAQ^{2-}}}{c_{DHAQ^{4-}}} \right) \quad (17)$$

where $E'_{eq,+}$ and $E'_{eq,-}$ indicate the standard equilibrium potentials for the positive and negative side reactions, outlined in Table 4 alongside the other kinetic parameters.

Table 4. Kinetic parameters

Quantity	Symbol	Value	References
Standard equilibrium potential for positive reaction	$E'_{eq,+}$	0.33 V	[29]
Standard equilibrium potential for negative reaction	$E'_{eq,-}$	−0.71 V	[29]
Cathodic transfer coefficient	α_{pos}	0.5	[17]
Anodic transfer coefficient	α_{neg}	0.5	[17]
Rate constant for positive reaction	k_{pos}	6×10^{-7} m s^{-1}	[29]
Rate constant for negative reaction	k_{neg}	7×10^{-7} m s^{-1}	[29]

2.1.3. Charge Conservation

The principle of electroneutrality requires that the total charge in the electrolyte equals zero, ensuring electrical neutrality:

$$\sum_i z_i c_i = 0 \quad (18)$$

To satisfy charge conservation, current flow in the solid and liquid phases is coupled with electrochemical reactions through the following equation:

$$\nabla \cdot \mathbf{i_l} = -\nabla \cdot \mathbf{i_s} = \mathbf{i_r} \tag{19}$$

which implies that the current leaving the electrolyte, $\mathbf{i_l}$, matches the current entering the electrode, $\mathbf{i_s}$, and collectively equals the electrochemical reaction rate, $\mathbf{i_r}$. The solid and liquid phase currents are expressed as follows:

$$\mathbf{i_s} = -\sigma_s^{eff} \nabla \varphi_s \tag{20}$$

$$\mathbf{i_l} = F \sum_i z_i \mathbf{N_i} \tag{21}$$

where σ_s^{eff} represents the effective conductivity of the electrode, determined using the bulk conductivity of the electrode material, σ_s, and calculated as follows:

$$\sigma_s^{eff} = (1-\varepsilon)^{3/2} \sigma_s \tag{22}$$

2.1.4. Boundary Conditions

Boundary conditions are defined using the x and y coordinates shown in Figure 1. At the external boundary of the anode, located at $x = x_0$, the solid potential is set to zero by grounding this boundary:

$$\varphi_s = 0 \quad at \quad x = x_0 \tag{23}$$

At the cell's lower boundary, $y = 0$, the species inlet flux is defined based on the flow rate Q, electrode width w_e, and electrode thickness t_e:

$$\mathbf{n} \cdot \mathbf{u} = \frac{Q}{\varepsilon w_e t_e} \quad at \begin{cases} x_0 < x < x_1 \quad and \quad x_2 < x < x_3 \\ y = 0 \end{cases} \tag{24}$$

The species concentration at the electrode inlet remains constant.

$$c_i = c_i^{in} \quad at \begin{cases} x_0 < x < x_1 \quad and \quad x_2 < x < x_3 \\ y = 0 \end{cases} \tag{25}$$

At all other boundaries, the mass flux is zero, with the exception of K^+ at the membrane–electrode interface:

$$-\mathbf{n} \cdot \mathbf{N_i} = 0 \quad at \begin{cases} y = 0 < y < y = h \\ for \\ x = (x_0, x_1, x_2, x_3) \end{cases} \tag{26}$$

Here, h represents the height of the cell. At this location, a pressure outlet is defined:

$$p = p_{out} \quad at \begin{cases} x_0 < x < x_1 \quad and \quad x_2 < x < x_3 \\ y = h \end{cases} \tag{27}$$

and the diffusive flux of all species is neglected:

$$-D_i^{eff} \nabla c_i \cdot \mathbf{n} = 0 \quad at \begin{cases} x_0 < x < x_1 \quad and \quad x_2 < x < x_3 \\ y = h \end{cases} \tag{28}$$

At the membrane–electrode interface, charge continuity is maintained by defining the continuity of the electrolyte current density to the current density in the membrane:

$$\mathbf{n} \cdot \mathbf{i}_l = \mathbf{n} \cdot \mathbf{i}_{l,\mathbf{mem}} \qquad at \begin{cases} y = 0 < y < y = h \\ for \\ x = (x_1, x_2) \end{cases} \tag{29}$$

The top and bottom edges of both the membrane and electrodes are electrically insulated:

$$\begin{matrix} -\mathbf{n} \cdot \mathbf{i}_l = 0 \\ -\mathbf{n} \cdot \mathbf{i}_s = 0 \end{matrix} \begin{cases} x_0 < x < x_3 \\ y = 0 \quad and \quad y = h \end{cases} \tag{30}$$

A constant current density, controlled by the user, is applied to the external boundary of the positive electrode:

$$\mathbf{n} \cdot \mathbf{i}_s = i_{avg} \qquad at \qquad x = x_3 \tag{31}$$

All geometric and operational parameters are provided in Table 5.

Table 5. Geometrical and operation parameters.

Quantity	Symbol	Value	References
Electrode length	h	0.04 m	[38]
Non-compressed electrode thickness	t_e	0.006 m	[38]
Electrode width	w_e	0.04 m	[38]
Membrane thickness	t_m	0.000183 m	[38]
Flow rate	Q	60 mL min^{-1}	-
Temperature	T	298 K	-
Outlet pressure	p_{out}	0 Pa	-
State of Charge	SOC	50%	-

2.2. Battery Performance Parameters

The concentration of each species is determined based on the user-defined parameter SOC, with the total concentrations on the positive and negative sides denoted as c_1 and c_0, respectively:

$$c_{reDHAQ} = c_0 \cdot SOC \tag{32}$$

$$c_{DHAQ} = c_0 \cdot (1 - SOC) \tag{33}$$

$$c_{Fe(CN)_6^{4-}} = c_1 \cdot (1 - SOC) \tag{34}$$

$$c_{Fe(CN)_6^{3-}} = c_1 \cdot SOC \tag{35}$$

A crucial factor in assessing species distribution is the uniformity factor, which measures how evenly a species is spread throughout the system. It quantifies the concentration variation of species i and is calculated as follows:

$$U_i = 1 - \frac{1}{c_{i,\mathrm{avg}}} \sqrt{\frac{1}{V} \iiint \left(c_i - c_{i,\mathrm{avg}}\right)^2 dV} \tag{36}$$

where $c_{i,avg}$ denotes the average concentration of species i, and V represents the volume of the electrode.

2.3. Numerical Model

The simulation model was created using COMSOL Multiphysics 5.5. For the positive and negative electrodes, the Tertiary Current Distribution module was utilized, while the Secondary Current Distribution module was applied to the membrane. The fluid dynamics equations within the porous electrodes were addressed using the Brinkman equations, solved via a finite volume method. The PARDISO solver was selected, with a convergence criterion of 1×10^{-6} relative error to ensure high precision in the results. To optimize computational efficiency while maintaining accuracy, a mesh independence study was carried out, demonstrating that grid refinements beyond 3100 elements did not significantly alter the outcomes, as shown in Figure 3.

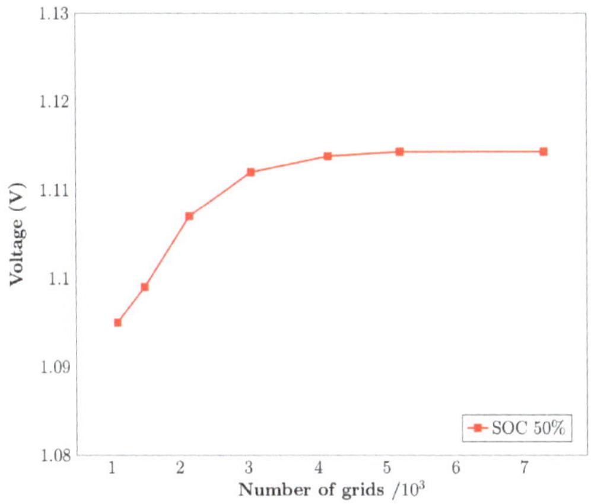

Figure 3. Grid dependency study.

3. Results

3.1. Experimental Validation

To validate the numerical model, the results were compared with experimental data from Chu et al. [29], using a rectangular electrode with dimensions of 20 mm × 20 mm × 3.5 mm. The flow rate of the circulation pump was set to 60 ml min^{-1}, with a current density of 40 mA cm^{-2}, and a Nafion 212 cation exchange membrane was used as the ion-exchange medium. In these tests, the concentration of the positive active material was 200 mol m^{-3}, while the concentration of the negative active material was 100 mol m^{-3}. Figure 4 presents the cell potential during charge and discharge modes for both experimental and numerical models. The maximum error, which occurs at the endpoints of the charge-discharge curves, remained below 3.6%, with an average error of 1.45%, confirming the accuracy and reliability of the numerical model for subsequent analysis.

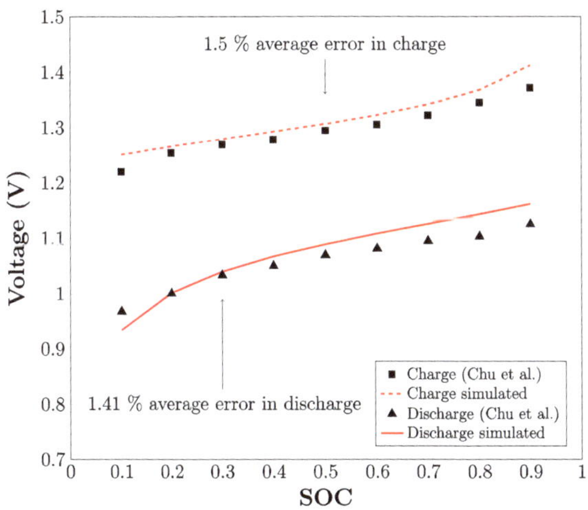

Figure 4. Results of experimental validation [29].

3.2. Case Studies

Two configurations were investigated to assess the influence of obstacle placement. Case 1 serves as the baseline, with no obstacles incorporated, providing a reference for evaluating the effects of the modified configuration. In Case 2, obstacles are introduced within the electrode structure. Two obstacles are introduced in each electrode, both obstructing 50% of the electrode thickness. One is located at the electrode outer boundary, and the second is placed at the electrode membrane interface. All obstacles have a dimension of 1 mm in electrolyte flow direction. The obstacle at the outer boundary is located at one-third of the electrode length, and the other one at two-thirds of the electrode length.

For both configurations, the concentration of active material in the positive and negative compartments is set to 400 and 200 mol m^{-3}, respectively. Additionally, the concentrations of K^+ and OH^- ions are maintained at 1000 mol m^{-3} on both sides.

3.3. Effects of SOC

Looking at Figure 5, discharge curves for Case 1 and 2 can be analyzed. The discharge curves reveal that Case 1 consistently exhibits a slightly higher voltage, 0.18% on average, compared to Case 2 across the entire range of state of charge. This behavior indicates that the baseline configuration benefits from a more direct flow path for the electrolyte, minimizing the hydraulic resistance and associated pressure drop. In contrast, the inclusion of obstacles in Case 2 leads to additional flow resistance and localized pressure gradients, which may impede uniform electrolyte penetration. However, when looking at the lower end of the SOC range, the difference between Case 1 and 2 decreases to 0.07%, see the zoom in Figure 5. While the obstacle configuration aims to enhance mass transport and mitigate concentration gradients by promoting better mixing, these potential advantages appear offset by the increased Ohmic polarization effects.

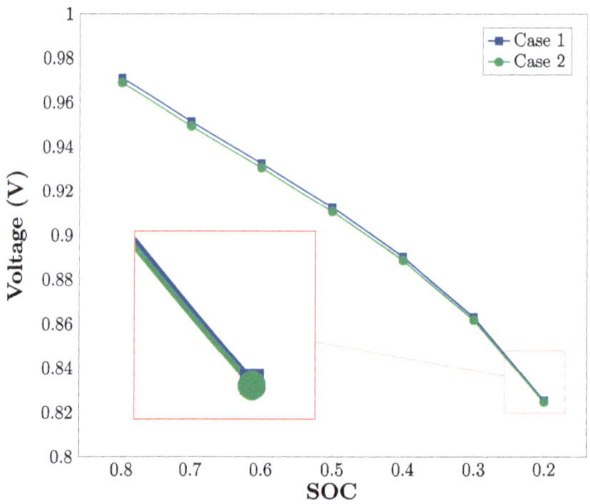

Figure 5. Discharge curves obtained for a flow rate of 60 mL min^{-1} and 50 mA cm^{-2}.

Despite the slightly lower discharge voltage observed in Case 2, this configuration demonstrates significant improvements in terms of concentration uniformity and minimum species concentration, as depicted in Figure 6a,b. Across the entire SOC range, Case 2 demonstrates a higher concentration uniformity factor compared to the baseline Case 1, which indicates enhanced distribution of active species within the electrode, see Figure 6a. For a SOC value of 0.2, Case 1 achieves a value of 88.7%, while for Case 2, this rises to 90.1%. The obstacles actually help spread electrolyte more evenly through the electrode, which prevents troublesome zones with low concentration. Though the advantage in concentration uniformity narrows as SOC increases, the design in Case 2 still provides better distribution than in Case 1, reinforcing stable electrochemical reactions across varying SOC conditions. One potential explanation for this trend is that at lower SOCs, the availability of reactants throughout the electrolyte is significantly limited, leading to pronounced concentration gradients across the cell. Under these conditions, reactant depletion near the electrode surface becomes a critical issue, particularly in regions where the flow is less effective at replenishing the consumed species. This is the operating point where obstacles can force portions of the electrolyte flow into regions that might otherwise receive minimal reactant supply, promoting a more uniform utilization of the electrode area. For higher SOCs, this becomes less critical and obstacles seem to be less effective. Furthermore, the higher minimum concentration observed in Figure 6b highlights the improved utilization of active species in Case 2. This suggests that the presence of obstacles mitigates stagnant regions and optimizes mass transport.

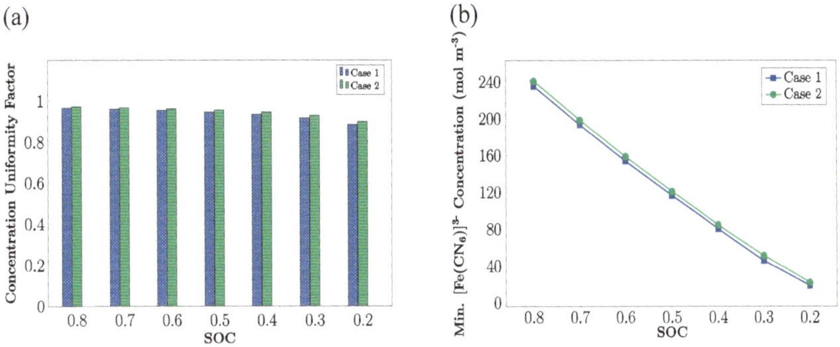

Figure 6. The (**a**) concentration uniformity factor and (**b**) minimum concentration of $[Fe(CN_6)]^{3-}$ in the positive electrode for a flow rate of 60 mL min^{-1} and 50 mA cm^{-2}.

Better electrolyte distribution translates into higher limiting current density, see Figure 7. This improvement can be attributed to effectively reducing concentration gradients and mitigating localized depletion of active species. The more uniform concentration profile in Case 2, as illustrated in Figure 6a, corroborates this observation, as a well-distributed reactant supply ensures that the electrode regions farther from the inlet remain reactive even at higher current densities, even at near discharged state.

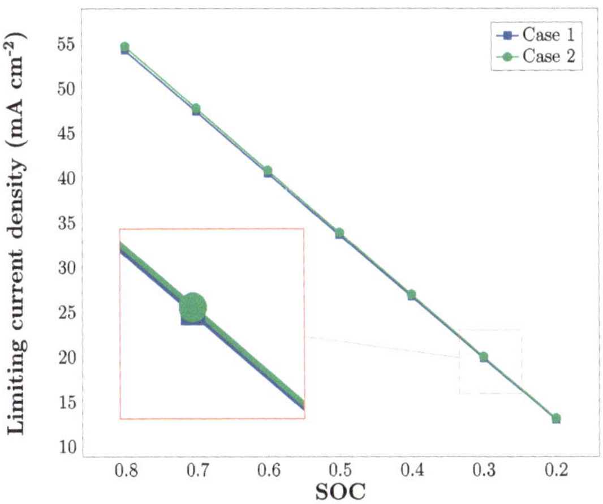

Figure 7. Limiting current density for a flow rate of 60 mL min^{-1} and 50 mA cm^{-2}.

3.4. Effects of Flow Rate

A key consideration in evaluating the effectiveness of electrode design modifications is the associated pressure drop, see Figure 8, as it directly impacts the pumping energy required for system operation. In the case of obstacle placement (Case 2), a higher pressure drop is observed compared to Case 1, by an average of 14%. The increased hydraulic resistance in Case 2 can be attributed to the obstacles interrupting the electrolyte flow path, creating regions of recirculation, while these flow dynamics improve electrolyte penetration and mixing, as seen in the higher concentration uniformity and elevated limiting current density, they simultaneously impose a penalty on pumping efficiency. This trade-off highlights a challenge on optimizing the electrode structure to enhance electrochemical performance while minimizing the energy costs associated with fluid transport.

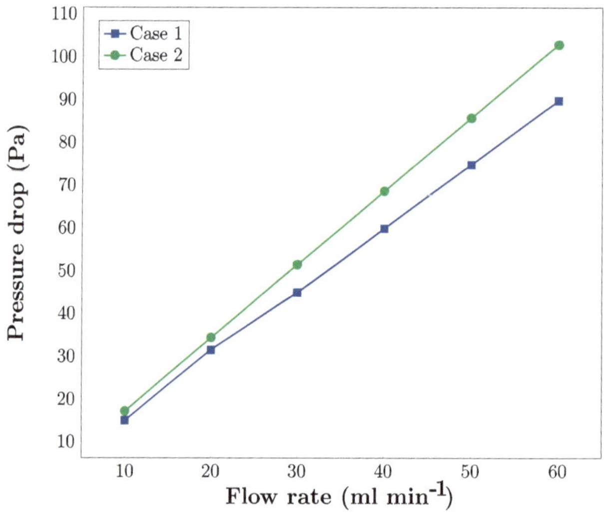

Figure 8. Pressure drop values for a State of Charge of 50% and 50 mA cm^{-2} in discharge mode.

Regardless of the increase in pressure drop for Case 2, improvements in other metrics are obtained. Figure 9a shows the relation between concentration uniformity factor and flow rate for both configurations. Both curves exhibit a rising trend as the flow rate increases, reflecting enhanced reactant distribution across the electrode. However, there are notable differences in their behavior. Case 2 demonstrates a 1.8% improvement in concentration uniformity for the lowest flow rate of 10 mL min^{-1}, and it maintains higher values across the studied range. However, this difference shrinks as the flow rate increases, down to 1.0% for 60 mL min^{-1}. Looking at Figure 9b, benefits in minimum concentration are observed. Case 2 obtains higher minimum concentration values across the studied range, averaging a 15% increase, which translates to reduced local reactant depletion. Thanks to reducing local reactant depletion, limiting current density is raised, see Figure 10, by an average of 1% over the whole range. It is worth noting that when zooming in on Figure 10, even at lower flow rates, Case 2 consistently outperforms Case 1.

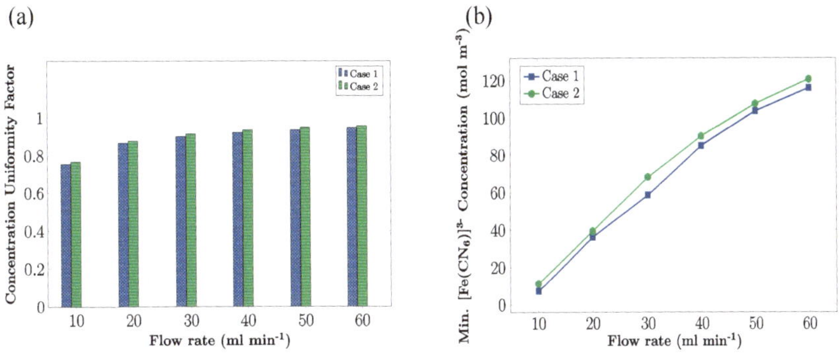

Figure 9. The (**a**) concentration uniformity factor and (**b**) minimum concentration of $[Fe(CN_6)]^{3-}$ in the positive electrode for a State of Charge of 50% and 50 mA cm^{-2}.

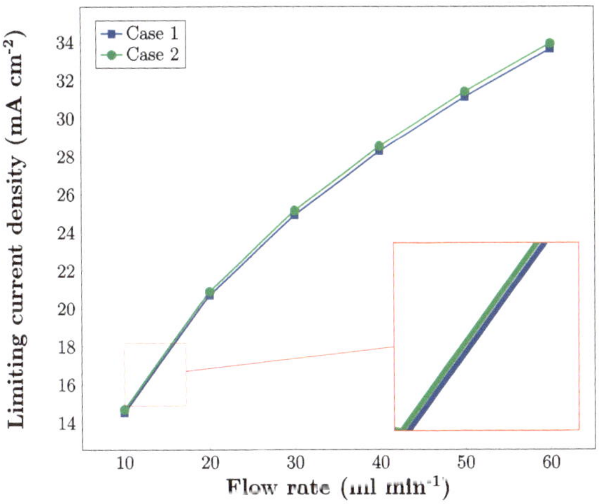

Figure 10. Limiting current density for a State of Charge of 50% and 50 mA cm^{-2}.

4. Conclusions

In this work, the effect of placing obstacles in the electrode structure of an organic redox flow battery on reactant distribution and local depletion is investigated. Two configurations were studied: Case 1 was designated as the control without obstacles, while Case 2 featured a design with obstacles to improve mass transport and uniformity.

Analyzing the discharge profiles, it is noted that Case 1 maintained a marginally higher voltage compared to Case 2 throughout the complete state-of-charge range; the average difference was about 0.18%. Such a voltage difference can be explained due to the fact that Case 1 had a simpler flow channel leading to lower hydraulic resistance and smaller pressure drops. Conversely, the obstacles in Case 2 introduced more resistance to the flow, which consequently generated localized pressure gradients. Additionally, obstacle placement means reducing electrode area, which translates to lower electronic conductivity and increased Ohmic polarization. However, this disadvantage was compensated at lower SOC values, where the gap in performance between the two configurations narrowed considerably. Although the discharge voltage was marginally lower in Case 2, there were marked improvements in the concentration uniformity and minimum concentration of active species with this configuration. The distribution of reactants became much more uniform with the obstacles placed in Case 2, especially at low levels of SOC, thereby preventing areas of low concentration that could damage the cell, while the benefit of concentration uniformity decreased with higher values of SOC, Case 2 still showed better distribution throughout the whole range.

The higher pressure drop of Case 2 is mainly due to the fact that the obstacles block electrolyte flow. This suggests a trade-off between the electrochemical performance and energy consumption: while the obstacles improved the mixing quality and reduced local depletion, the obstacles also created additional flow resistance that required more pumping energy.

Looking at the performance of the design with obstructions at different flow rates, Case 2 did better than Case 1, especially with lower flow rates. Although the advantage became smaller with an increase in flow rate, Case 2 still had better concentration uniformity and higher minimum concentration values within the studied range. Furthermore, having obstacles in Case 2 increased the limiting current density by 1% on average. Even at the lower flow rates, Case 2 showed better performance, indicating that the obstacles may enhance electrochemical efficiency at different operating conditions.

In conclusion, though the introduction of obstacles in the electrode structure increased the pressure drop, the enhanced concentration uniformity, limiting current density, and active species utilization bring huge benefits. It is indicated that obstacle placement could be a very promising strategy to optimize the performance of redox flow batteries. The introduction of strategically placed obstacles presents a simple yet scalable design modification that can be adapted to various operating conditions. Furthermore, the insights into the relationship between flow dynamics and electrochemical performance establish a foundation for future optimization efforts, including refining obstacle geometry and placement to balance improvements in mass transport with the associated pumping energy costs.

Author Contributions: Conceptualization, J.M.-L. and U.F.-G.; methodology, A.B.-A.; software, E.S.-D.; validation, J.M.-L. and U.F.-G.; investigation, U.F.-G.; resources, E.S.-D.; data curation, A.B.-A. and Í.O.-F.; Writing—Review and Editing, J.M.-L. and Í.O.-F.; visualization, A.B.-A.; supervision, E.S.-D.; project administration, U.F.-G.; funding acquisition, U.F.-G. All authors have read and agreed to the published version of the manuscript.

Funding: Government of the Basque Country: Elkartek CICe2022; KK-2022/00043 and ELKARTEK24/15, KK-2024/00035; CIEMAT: Energía eólica offshore para el ensayo y el desarrollo energético de energías renovables e hidrógeno verde; Government of the Basque Country: Mobility Lab Foundation; Government of the Basque Country: ITSAS-REM (IT1514-22).

Data Availability Statement: All data generated in the current study are available upon reasonable request to the corresponding author.

Acknowledgments: The authors are grateful for the support provided by SGIker of UPV/EHU. This research was developed under the frame of the Joint Research Laboratory on Offshore Renewable Energy (JRL-ORE).

Conflicts of Interest: Authors Eduardo Sánchez-Díez, Aitor Beloki-Arrondo and Íñigo Ortega-Fernández were employed by Basque Research and Technology Alliance (BRTA). The remaining authors declare that the research was conducted in the absence of any commercial or financial relationships that could be construed as a potential conflict of interest.

Nomenclature

List of symbols

a	Specific surface area
c	Concentration
D	Diffusion coefficient
d_f	Fiber diameter

E	Equilibrium potential
F	Faraday constant
h	Height
i_0	Exchange current density
i_r	Electrochemical reaction rate
K	Permeability
Kck	Kozeny–Carman constant
k	Reaction rate constant
N	Flux of charged species
p	Pressure
Q	Volumetric flow rate
R	Constant of ideal gases
S	Source term
T	Temperature
t	Time
u	Mobility
u	Velocity
w	Width
z	Species charge
Greek	
α	Charge transfer coefficient
ε	Electrode porosity
η	Overpotential
φ	Potential
σ	Conductivity
μ	Dynamic viscosity
Superscripts and subscripts	
$+$	Positive side
$-$	Negative side
$'$	Standard
avg	Average
e	Electrode
eff	Effective
ι	Species
l	Liquid
mem	Membrane
out	Outlet
s	Solid

Abbreviations

The following abbreviations are used in this manuscript:

CFD	Computational fluid dynamics
RFB	Redox flow battery
VRFB	Vanadium redox flow battery
ORFB	Organic redox flow battery

References

1. Nguyen, T.; Savinell, R.F. Flow Batteries. *Electrochem. Soc. Interface* **2010**, *19*, 54. [CrossRef]
2. Asiaban, S.; Kayedpour, N.; Ebneali Samani, A.; Bozalakov, D.; De Kooning, J.; Crevecoeur, G.; Vandevelde, L. Wind and solar intermittency and the associated integration challenges: A comprehensive review including the status in the Belgian power system. *Energies* **2021**, *14*, 630. [CrossRef]
3. Mutz, M.; Perovic, M.; Gümbel, P.; Königer, V.; Taranovskyy, A.; Li, Y.; Beran, L.; Käfer, T.; Dröder, K.; Knoblauch, V.; et al. Toward a Li-Ion Battery Ontology Covering Production and Material Structure. *Energy Technol.* **2022**, *11*, 2200681. [CrossRef]

4. Braff, W.A.; Mueller, J.M.; Trancik, J.E. Value of storage technologies for wind and solar energy. *Nat. Clim. Chang.* **2016**, *6*, 964–969. [CrossRef]
5. Yu, J.; Fan, L.; Yan, R.; Zhou, M.; Wang, Q. Redox Targeting-Based Aqueous Redox Flow Lithium Battery. *ACS Energy Lett.* **2018**, *3*, 2314–2320. [CrossRef]
6. Reed, D.; Thomsen, E.; Li, B.; Wang, W.; Nie, Z.; Koeppel, B.; Kizewski, J.; Sprenkle, V. Stack Developments in a kW Class All Vanadium Mixed Acid Redox Flow Battery at the Pacific Northwest National Laboratory. *J. Electrochem. Soc.* **2015**, *163*, A5211. [CrossRef]
7. Arevalo-Cid, P.; Dias, P.; Mendes, A.; Azevedo, J. Redox flow batteries: A new frontier on energy storage. *Sustain. Energy Fuels* **2021**, *5*, 5366–5419. [CrossRef]
8. Lourenssen, K.; Williams, J.; Ahmadpour, F.; Clemmer, R.; Tasnim, S. Vanadium redox flow batteries: A comprehensive review. *J. Energy Storage* **2019**, *25*, 100844. [CrossRef]
9. Liu, Y.; Xu, J.; Lu, S.; Xiang, Y. Titanium Nitride Nanorods Array-Decorated Graphite Felt as Highly Efficient Negative Electrode for Iron-Chromium Redox Flow Battery. *Small* **2023**, *19*, e2300943. [CrossRef]
10. Cunha, A.; Martins, J.; Rodrigues, N.; Brito, F. Vanadium redox flow batteries: A technology review. *Int. J. Energy Res.* **2014**, *39*, 889–918. [CrossRef]
11. Bhattarai, A.; Whitehead, A.; Schweiss, R.; Scherer, G.; Skyllas-Kazacos, M.; Wai, N.; Nguyen, T.; Ghimire, P.; Oo, M.; Hng, H. Anomalous Behavior of Anion Exchange Membrane during Operation of a Vanadium Redox Flow Battery. *ACS Appl. Energy Mater.* **2019**, *2*, 1712–1719. [CrossRef]
12. Skyllas-Kazacos, M.; Chakrabarti, B.; Hajimolana, Y.S.; Mjalli, F.; Saleem, M. Progress in Flow Battery Research and Development. *J. Electrochem. Soc.* **2011**, *158*, R55–R79. [CrossRef]
13. Minke, C.; Kunz, U.; Turek, T. Techno-economic assessment of novel vanadium redox flow batteries with large-area cells. *J. Power Sources* **2017**, *361*, 105–114. [CrossRef]
14. Leung, P.; Shah, A.A.; Sanz, L.; Flox, C.; Morante, J.R.; Xu, Q.; Mohamed, M.R.; Ponce de León, C.; Walsh, F.C. Recent developments in organic redox flow batteries: A critical review. *J. Power Sources* **2017**, *360*, 243–283. [CrossRef]
15. Yu, L.C.; Luo, Y.C.; Feng, W.; Zhang, S.; Zhang, X. Fluorinated TEMPO: A new redox-active catholyte material for aqueous Zn-anode hybrid flow batteries. *J. Mater. Chem. A* **2023**, *11*, 18911–18921. [CrossRef]
16. Sánchez-Díez, E.; Ventosa, E.; Guarnieri, M.; Trovò, A.; Flox, C.; Marcilla, R.; Soavi, F.; Mazur, P.; Aranzabe, E.; Ferret, R. Redox flow batteries: Status and perspective towards sustainable stationary energy storage. *J. Power Sources* **2021**, *481*, 228804. [CrossRef]
17. Lin, K.; Chen, Q.; Gerhardt, M.R.; Tong, L.; Kim, S.B.; Eisenach, L.; Valle, A.W.; Hardee, D.; Gordon, R.G.; Aziz, M.J.; et al. Alkaline quinone flow battery. *Science* **2015**, *349*, 1529–1532. [CrossRef] [PubMed]
18. Huang, S.; Zhang, H.; Salla, M.; Zhuang, J.; Zhi, Y.; Wang, X.; Wang, Q. Molecular engineering of dihydroxyanthraquinone-based electrolytes for high-capacity aqueous organic redox flow batteries. *Nat. Commun.* **2022**, *13*, 4746. [CrossRef] [PubMed]
19. Peng, K.; Li, Y.; Tang, G.; Liu, Y.; Yang, Z.; Xu, T. Solvation regulation to mitigate the decomposition of 2,6-dihydroxyanthraquinone in aqueous organic redox flow batteries. *Energy Environ. Sci.* **2023**, *16*, 430–437. [CrossRef]
20. Jin, S.; Fell, E.M.; Vina-Lopez, L.; Jing, Y.; Michalak, P.W.; Gordon, R.G.; Aziz, M.J. Near Neutral pH Redox Flow Battery with Low Permeability and Long-Lifetime Phosphonated Viologen Active Species. *Adv. Energy Mater.* **2020**, *10*, 2000100. [CrossRef]
21. Zou, H.; Xu, Z.; Xiong, L.; Wang, J.; Fu, H.; Cao, J.; Ding, M.; Wang, X.; Jia, C. An alkaline S/Fe redox flow battery endowed with high volumetric-capacity and long cycle-life. *J. Power Sources* **2024**, *591*, 233856. [CrossRef]
22. Shen, X.; Kellamis, C.; Tam, V.; Sinclair, N.; Wainright, J.; Savinell, R. An All-Soluble Fe/Mn-Based Alkaline Redox Flow Battery System. *ACS Appl. Mater. Interfaces* **2024**, *16*, 18686–18692. [CrossRef]
23. Aramendia, I.; Fernandez-Gamiz, U.; Martinez-San-Vicente, A.; Zulueta, E.; Lopez-Guede, J.M. Vanadium Redox Flow Batteries: A Review Oriented to Fluid-Dynamic Optimization. *Energies* **2021**, *14*, 176. [CrossRef]
24. Esan, O.C.; Shi, X.; Pan, Z.; Huo, X.; An, L.; Zhao, T. Modeling and Simulation of Flow Batteries. *Adv. Energy Mater.* **2020**, *10*, 2000758. [CrossRef]
25. Sun, J.; Guo, Z.; Pan, L.; Fan, X.; Wei, L.; Zhao, T. Redox flow batteries and their stack-scale flow fields. *Carbon Neutrality* **2023**, *2*, 30. [CrossRef]
26. Pan, L.; Xie, J.; Guo, J.; Wei, D.; Qi, H.; Rao, H.; Leung, P.; Zeng, L.; Zhao, T.; Wei, L. In-plane gradient design of flow fields enables enhanced convections for redox flow batteries. *Energy Adv.* **2023**, *2*, 2006–2017. [CrossRef]
27. Martinez Lopez, J.; Aramendia, I.; Fernandez-Gamiz, U.; Sanchez-Diez, E.; Beloki, A.; Kurt, E.; Lopez-Guede, J.M. Computational Modeling of a 2D Vanadium Redox Flow Battery Cell. *JOM* **2024**, *76*, 130–140. [CrossRef]
28. Xu, Q.; Zhao, T.S.; Leung, P.K. Numerical investigations of flow field designs for vanadium redox flow batteries. *Appl. Energy* **2013**, *105*, 47–56. [CrossRef]
29. Chu, F.; Su, M.; Xiao, G.; Tan, Z.; Yang, G. Analysis of Electrode Configuration Effects on Mass Transfer and Organic Redox Flow Battery Performance. *Ind. Eng. Chem. Res.* **2022**, *61*, 2915–2925. [CrossRef]

30. Aparicio-Mauricio, G.; Jiménez-Lima, L.; Rivero, E.P.; Cruz-Díaz, M.R. 3D modeling and simulation of an alkaline flow battery considering the tertiary current distribution on the electrodes: Anthraquinone-ferro/ferricyanide as organic electroactive species. *J. Power Sources* **2023**, *582*, 233533. [CrossRef]
31. Akuzum, B.; Alparslan, Y.C.; Robinson, N.C.; Agar, E.; Kumbur, E.C. Obstructed flow field designs for improved performance in vanadium redox flow batteries. *J. Appl. Electrochem.* **2019**, *49*, 551–561. [CrossRef]
32. Messaggi, M.; Gambaro, C.; Casalegno, A.; Zago, M. Development of innovative flow fields in a vanadium redox flow battery: Design of channel obstructions with the aid of 3D computational fluid dynamic model and experimental validation through locally-resolved polarization curves. *J. Power Sources* **2022**, *526*, 231155. [CrossRef]
33. Liu, Y.; Huang, Z.; Xie, X.; Liu, Y.; Wu, J.; Guo, Z.; Huang, Q. Redox flow battery:Flow field design based on bionic mechanism with different obstructions. *Chem. Eng. J.* **2024**, *498*, 155663. [CrossRef]
34. Gurieff, N.; Keogh, D.F.; Timchenko, V.; Menictas, C. Enhanced Reactant Distribution in Redox Flow Cells. *Molecules* **2019**, *24*, 3877. [CrossRef] [PubMed]
35. Gurieff, N.; Keogh, D.F.; Baldry, M.; Timchenko, V.; Green, D.; Koskinen, I.; Menictas, C. Mass Transport Optimization for Redox Flow Battery Design. *Appl. Sci.* **2020**, *10*, 2801. [CrossRef]
36. Dhillon, A. Multiphysics Modelling of an Alkaline All-Iron, All-Soluble Aqueous Redox Flow Battery. Master's Thesis, University of Waterloo, Waterloo, ON, Canada, 2021.
37. Chen, Z.; Yu, W.; Liu, Y.; Zeng, Y.; He, Q.; Tan, P.; Ni, M. Mathematical modeling and numerical analysis of alkaline zinc-iron flow batteries for energy storage applications. *Chem. Eng. J.* **2021**, *405*, 126684. [CrossRef]
38. Wang, Q.; Qu, Z.G.; Jiang, Z.Y.; Yang, W.W. Experimental study on the performance of a vanadium redox flow battery with non-uniformly compressed carbon felt electrode. *Appl. Energy* **2018**, *213*, 293–305. [CrossRef]
39. Knehr, K.W.; Agar, E.; Dennison, C.R.; Kalidindi, A.R.; Kumbur, E.C. A Transient Vanadium Flow Battery Model Incorporating Vanadium Crossover and Water Transport through the Membrane. *J. Electrochem. Soc.* **2012**, *159*, A1446. [CrossRef]
40. You, D.; Zhang, H.; Chen, J. A simple model for the vanadium redox battery. *Electrochim. Acta* **2009**, *54*, 6827–6836. [CrossRef]

batteries MDPI

Review

Sodium-Ion Batteries: Applications and Properties

Petr Bača, Jiří Libich *, Sára Gazdošová and Jaroslav Polkorab

Department of Electrotechnology, Faculty of Electrical Engineering and Communication, Brno University of Technology, Technická 10, 616 00 Brno, Czech Republic; baca@vut.cz (P.B.); 230305@vut.cz (J.P.)
* Correspondence: libich@vutbr.cz

Abstract: With the growing interest in reducing CO_2 emissions to combat climate change, humanity is turning to green or renewable sources of electricity. There are numerous issues associated with the development of these sources. One of the key aspects of renewable energy sources is their problematic controllability, namely the control of energy production over time. Renewable sources are also associated with issues of recycling, utilization in different geographical zones, environmental impact within the required area, and so on. One of the most discussed issues today, however, is the question of efficient use of the energy produced from these sources. There are several different approaches to storing renewable energy, e.g., supercapacitors, flywheels, batteries, PCMs, pumped-storage hydroelectricity, and flow batteries. In the commercial sector, however, mainly due to acquisition costs, these options are narrowed down to only one concept: storing energy using an electrochemical storage device—batteries. Nowadays, lithium-ion batteries (LIBs) are the most widespread battery type. Despite many advantages of LIB technology, the availability of materials needed for the production of these batteries and the associated costs must also be considered. Thus, this battery type is not very ideal for large-scale stationary energy storage applications. Sodium-ion batteries (SIBs) are considered one of the most promising alternatives to LIBs in the field of stationary battery storage, as sodium (Na) is the most abundant alkali metal in the Earth's crust, and the cell manufacturing process of SIBs is similar to that of LIBs. Unfortunately, considering the physical and electrochemical properties of Na, different electrode materials, electrolytes, and so on, are required. SIBs have come a long way since they were discovered. This review discusses the latest developments regarding the materials used in SIB technology.

Keywords: sodium-ion batteries; positive electrode; negative electrode; electrolyte; separator battery materials

1. Introduction

The collective effort to reduce greenhouse gas emissions, the replacement of fossil fuels, and the boom in a wide range of different applications, ranging from small portable devices to electric vehicles (EVs) and large stationary off-grid storage, has led to continuous developments in the field of battery storage. In the case of portable devices or, for example, drones, portability and an adequate size and weight are essential. At the same time, safety and cost are of primary importance for stationary storage, where the power output may be in the MW range [1].

Historically, the development of batteries began in about the year 1800 with non-cyclable cells, i.e., primary cells (an illustration of historical development can be seen in Figure 1). As these cells are non-rechargeable with lower energy density, they have been used in low-power consumer electronics. Present-day examples of these cells include, among others, zinc–carbon cells (Zn/MNO_2+C), alkaline cells (Zn/MNO_2+C with KOH electrolyte), silver-oxide cells (Zn-Ag_2O) for specific purposes, and lithium batteries (e.g., Li/MnO_2), with an operating life of several years [2,3].

The first battery of the secondary (i.e., rechargeable) type was invented in the middle of the 19th century. It is the lead/acid (Pb/acid) battery and it was invented by Gaston Planté [2]. Since then, the battery has undergone considerable development and its performance has been enhanced. Currently, Pb/acid batteries can reach an energy density of around 30 to 50 Wh/kg at a nominal voltage of 2.1 V. Their application in stationary renewable energy sources, however, is cost-ineffective and not environmentally friendly in terms of their capacity, size, life time, and lead content. Despite these drawbacks, this type of battery is widely used in the automotive industry. At the end of the 19th century, the nickel–cadmium (Ni-Cd) battery was invented as another secondary battery type and a potential competitor to Pb/acid. This battery achieves an energy density ranging from 50 to 75 Wh/kg, has a long life span, and has low self-discharge. However, due to disadvantages such as expensive processing and Cd storage, these batteries did not enter the commercial sphere until the mid-20th century. With respect to their environmental and safety properties, they contain highly toxic Cd and for this reason are subject to regulations in certain countries [2,4]. As a promising successor to Ni-Cd, nickel–metal hydride (Ni-MH) batteries were invented around 1975. This type of battery has a higher energy density in the range of 40 to 110 Wh/kg but has a lower cycling capacity in comparison to Ni-Cd [5].

Since SONY introduced the first LIB battery to the market in 1991, the share of Ni-MH has been on a decline. LIB research itself started in the 1970s to 1980s, and it was at this time that the first SIB research also appeared. However, more research on SIBs followed only in later years, due to the superior properties of Lithium (Li). LIBs have conquered the market of secondary cells due to their high capacity, energy density, nominal voltage, the possibility of faster charging, large operating temperature range, lifespan and nearly absent memory effect [6]. Their capacity has steadily increased over the years and now exceeds 200 Wh/kg and have become the most widespread energy storage devices in terms of installed capacity [7]. The main drawbacks of LIBs today are the Li content itself and, in the case of positive-electrode materials, the presence of cobalt (Co) [3–5,8]. Both of these materials have a very limited supply in the Earth's crust (Li availability is negligible compared to Na, which makes up 2.36% of the Earth's continental crust), and their extraction is associated with the unethical treatment of workers [1].

SIB batteries are seen as a possible replacement for LIBs. The properties of Na are very similar to those of Li, and its abundance in the Earth's crust is a significant advantage (as of 19 November 2024, sodium price is 15 times lower than that of lithium) [9,10]. Sodium mining is not nearly as environmentally demanding, and therefore, sodium appears to be an ideal competitor to lithium. The possibility of using current collectors made of more cost-effective aluminum is another practical advantage of SIBs. Unlike Li, Na does not form alloys with aluminum, so there is no need to use more expensive copper on the negative collector, as is the case with LIBs. This leads to a decrease in the overall battery production cost, as the price of copper is now at USD 9.1 per 1 kg, while a kilogram of aluminum now comes to USD 2.5 [11,12]. However, the different properties of Na come into play when it comes to the use of both positive- and negative-electrode materials. One of the most significant obstacles is the size of the Na ion (0.102 nm), which is 1.3 times larger compared to the Li-ion (0.076 nm). The dimensions of Na prevent the use of already known materials used for LIBs. For instance, when using a negative graphite electrode with a positive $LiCoO_2$ electrode, intercalation of Li ions into graphite occurs during discharge, forming Li_xC_6 with a final energy density of about 110 Wh/kg (first LIB) [3]. In 1988, it was found that unlike Li ions, Na ions are not capable of effecticely forming intercalating compounds with graphite. Thus, when the Li ions are substituted by Na ions, they do not form Na_xC_6, as this bond is unstable. NaC_{64} is formed instead, resulting in a reduction in the theoretical capacity to 35 mAh/g [13]. Using P2-$Na_{0.7}CoO_2$ and graphite as electrodes, Hasa et al. obtained an energy density of only 60 Wh/kg [14]. Another disadvantage is the size of the electrodes. Since Na ions are larger, more electrode material is needed in order to achieve the same capacity as in LIBs. As for the nominal voltage, it is lower in SIBs than in LIBs. This is due to the higher electrochemical potential of sodium (−2.71 V vs. Standard Hydrogen electrode, SHE) compared to lithium (−3.04 vs. SHE) [5,8,11–13,15,16]. When compared to LIBs, SIBs do not exhibit as high specific energy densities (Wh/kg) due to factors such as the sloped voltage profiles of layered oxide electrodes and narrow voltage windows of organic electrolytes, and due to different ion sizes [17,18].

The failure of graphite led to the development of alloying (materials forming an alloy with an alkali metal) and conversion (based on an anionic reaction between a transition metal and an alkali metal) materials until 2001 when the group of Dahn et al. demonstrated hard carbon and its properties. By combining Na with a hard-carbon electrode, a reversible capacity of up to 300 mAh/g can be achieved [19]. Since 2010, carbon-based materials have been the subject of increased interest in the field of research. Their simple manufacturing process and high theoretical capacity (up to 500 mAh/g) have been the subject of many studies to date [20,21].

In 2011, Johnson et al. demonstrated the use of amorphous TiO_2 nanotubes as a negative-electrode material for SIBs, with a theoretical capacity of 150 mAh/g. Sodium titanate $Na_2Ti_3O_7$, with a theoretical capacity of up to 200 mAh/g, was introduced by Tarascon et al. in that same year [22,23]. In 2012, the utilization of elements such as tin (Sn) and antimony (Sb) to form alloys in negative electrodes was reported. In the case of Sn, a theoretical capacity of 847 mAh/g can be achieved with the formation of the compound $Na_{15}Sn_4$; in the case of Sb, a value of 660 mAh/g can be reached with the formation of Na_3Sb [20,24,25]. During the same year, Kim and Park et al. presented a red phosphorus/carbon composite material with a reversible capacity of 1540 mAh/g in the formation of the Na_3P phase at a ca. 1C rate [26].

In 2014, the negative-electrode material α-MoO_3 was discovered by the group of Hariharan et al. The rocking chair $MoO_3/Na_3V_2(PO_4)_3$ cell exhibited a plateau voltage of 1.4 V and discharge capacity of 164 mAh/g depending on the weight of the negative electrode [26].

In addition, in 2014, the MXene (transition metal-based carbide, carboxide materials—2D-nanometer layers) material Ti_3C_2 with a reversible capacity of ~100 mAh/g was experimentally developed by the group of Kent and Gogotsi et al. [27]. In 2018, an LSG (laser-scribed graphene) material was discovered by the group of Alshareef et al. with a layer width of 3.8 nm in a 002 lattice arrangement, exhibiting high capacitances of up to 425 mAh/g [28].

There has been a steady increase in the number of published research studies on SIBs over the last decade, as depicted in Figure 2. China is the leading country in terms of the number of contributions in the materials field. A comparison of the number of publications over time and the subsequent distribution by individual countries can be seen in Figure 3.

In recent years, in the field of SIB material science, hard carbon-based materials have been the most promising negative-electrode materials thanks to their satisfactory electrochemical properties. In the context of positive-electrode materials, the trends point towards Prussian blue analogs or transition metal oxides (characterized by non-toxicity, long life/cyclability, safety, etc.). In terms of future SIB development, for example, low-temperature Na-S and Na-O (still rather theoretical) batteries are being discussed [29].

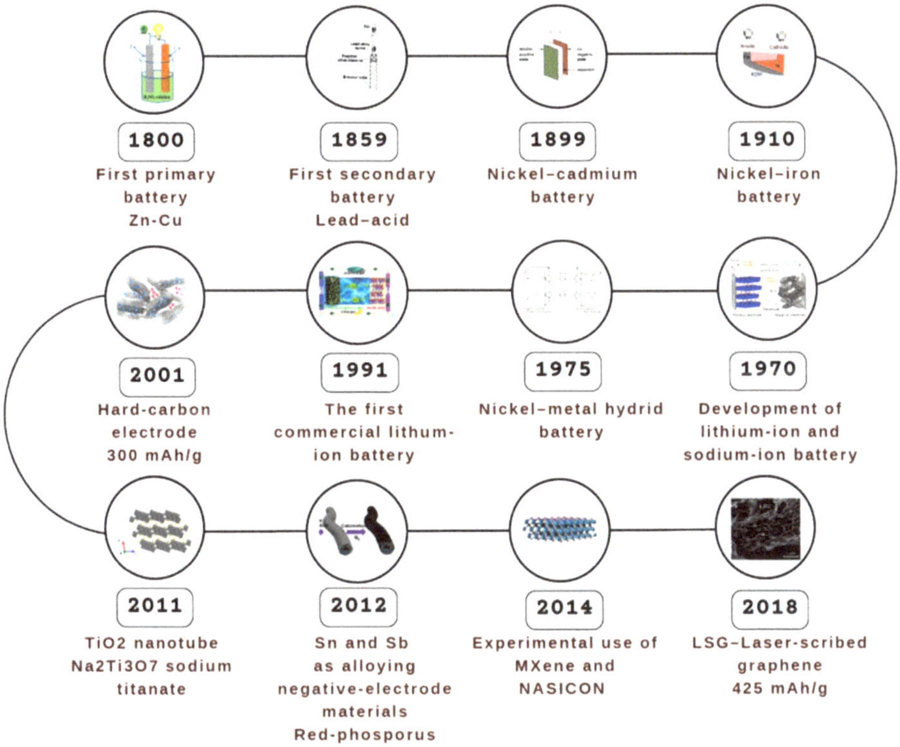

Figure 1. Brief history of battery development [2,28,30–35]. Parts of this figure were adapted with permission from [15,36,37]. Copyright 2020, 2014, 2015 American Chemical Society.

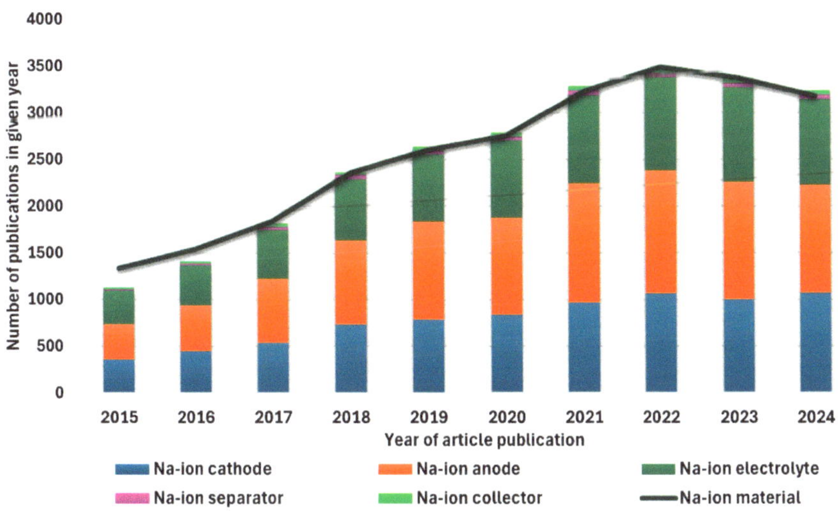

Figure 2. Number of article publications since 2015 according to Web of Science.

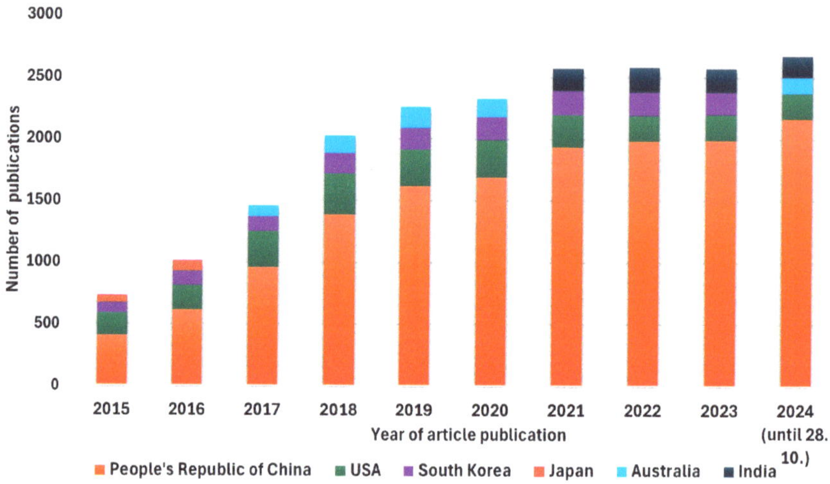

Figure 3. Number of released articles by leading country in each year according to Web of Science.

2. Sodium-Ion Battery Types and Operation Principles

Each part of the SIB cell plays a crucial role in the overall battery performance as the materials directly impact the battery's energy density, safety, lifespan, cost, and much more. The basic cell consists of a positive electrode and a negative electrode, an electrolyte, a separator, and current collectors, which can be seen in Figure 4. SIBs operate on a similar fundamental principle to LIBs, meaning they rely on the reversible movement of ions between the positive and the negative electrodes to either store or release energy. The main difference is that instead of Li ions, Na ions are used as the charge carriers [29]. During the discharge process, sodium ions at the negative electrode release an electron, making them carry a positive charge (Na^+). The Na^+ ions are then transferred to the positive electrode through the electrolyte. In the meantime, the released electrons travel through a current collector and an external circuit, creating a current capable of powering devices. During the charge, an external voltage is applied, forcing the electrons in the positive electrode to flow in the opposite direction toward the negative electrode through the external circuit. At the same time, the Na^+ ions travel through the electrolyte back to the negative electrode, and recombination with the electrons occurs [38].

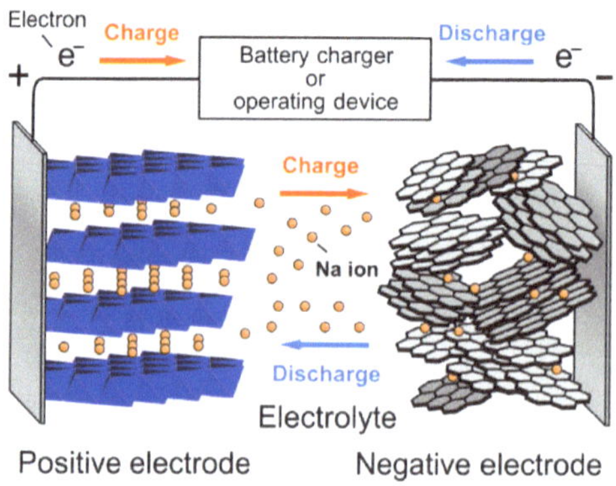

Figure 4. Principle of the sodium-ion battery. Adapted with permission from [15]. Copyright 2014 American Chemical Society.

2.1. Room-Temperature Batteries

According to their ideal operating temperature, batteries can be classified as room-temperature (RT), intermediate-temperature (IT), and high-temperature (HT). Each of these types has its specific properties related to the materials used, the cell construction, and the working principle. The research on SIBs primarily focuses on room-temperature applications due to their similar mechanism to that of LIBs [29].

This review has its main focus on RT SIB and the most used conventional materials and their combinations, which are discussed in more detail in Chapter 3. Among the SIBs capable of operating at room temperature are batteries using sodium metal negative electrodes. The usage of sodium metal as a negative-electrode materials is attractive thanks to their abundance of sodium, high theoretical specific capacity of 1166 mAh/g, and low redox potential. Their biggest limitation is their significant volume change while cycling, leading to the unstable formation of solid electrolyte interphase (SEI) and uncontrollable formation and growth of dendrites. Another disadvantage is the loss of active material due to the presence of an irreversible reaction, which leads to fast capacity fading. These shortcomings become even more serious when paired with a high-capacity sulfur (S) or oxygen (O) positive electrode. Stabilizing the SEI using chemical and physical methods is desirable for constructing a stable sodium metal negative-electrode SIB. A few strategies aiming to overcome this problem have arisen in recent times, such as the utilization of artificial SEI layers, modification of the electrolyte composition (salts, additives, solvents), or usage of 3D current collectors (e.g., copper nanowires) to promote a more homogenous distribution of sodium ions [39].

Na/O_2 batteries as well as Na/S batteries use sodium metal as the negative-electrode material. Na/O_2 is usually made with a liquid electrolyte (usually alkali metal salts in solvents) while using an external O_2 electrode. On the battery casing, openings allow air diffusion on the positive-electrode side, which is also occupied by a porous carbon matrix gas diffusion layer. Na/S batteries are similar to Na/O_2 batteries; however, they are sealed and make use of sulfur hosted in the porous structure of carbon in the positive electrode. The cell configurations can be seen in Figure 5. The role of carbon in this case is to accommodate the large volumetric changes in sulfur during cycling, but also to improve electron conductivity. Both Na/O_2 and Na/S batteries have the potential for fairly high energy densities; however, their development is in the early stages, and even though a great deal of research has been performed on these battery types, they are not yet commercially feasible [38].

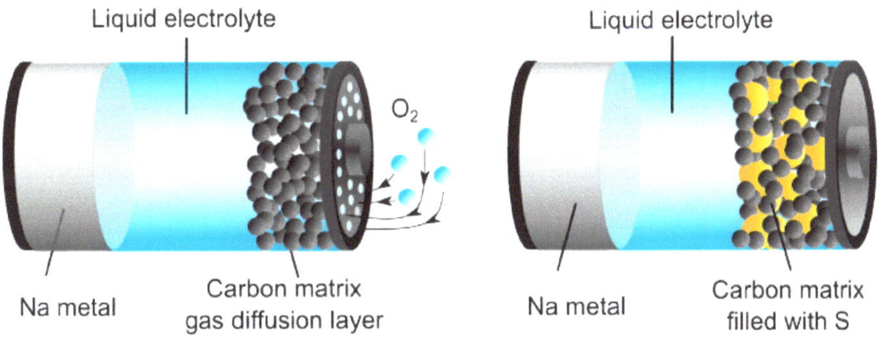

Figure 5. Schematic cell configuration of Na/O_2 and Na/S batteries [38].

2.2. High-Temperature Batteries

Na/S batteries, for example, can operate at high temperatures (typically 300 °C), in which case, both the sodium negative electrode and sulfur positive electrode are in a molten state. In this state, the materials possess high chemical reactivity as well as corrosivity. As these attributes cause serious safety hazards, inert current collectors, stable solid-state electrolytes, and proper sealing need to be used, which ultimately increases the costs of the battery [29,40]. The high reactivity and ion conductivity of molten sodium allow for high current densities (ca. 2.1×10^7 S/m) [41]. Due to high working temperatures, inorganic solid electrolytes (ISEs) are frequently employed, among which an inflexible ceramic material β''-Al_2O_3 (beta alumina) is included [40]. HT Na/S batteries have a life expectancy of over 10 years and have been implemented as test units for stationary energy storage and part of wind and solar installations in, for instance, Japan and the United States [42]. For transportation and stationary energy storage applications, Na/S batteries with organic-based electrolytes seem to be a more suitable option in terms of safety, manufacturing costs, and energy density [40].

MXene materials, such as $Ti_3C_2T_x$, have the potential to be used in elevated temperature conditions. Seredych et al. evaluated the thermal stability of this material, synthesized by selectively etching Al layers from the Ti_3AlC_2 MAX phase. The obtained $Ti_3C_2T_x$ remained stable until 800 °C, at which point it started to degrade, release CO, and transform into cubic titanium carbide (TiC). They also found that a lower concentration of hydrofluoric acid in the etching agent improves the material's thermal stability. Its thermal stability makes $Ti_3C_2T_x$ a promising material for various applications, such as ceramic and matrix composites requiring high-temperature processing [43].

3. Materials for Sodium-Ion Batteries

Specific capacity, cycling stability, and operation voltage are all crucial characteristics of a battery. Among the most important characteristics of the battery is its safety, mainly the thermal runaway issue. Unlike in LIBs, the Al current collectors in NIBs do not undergo dissolution at 0 V; therefore, no short-circuit-causing dendrites are formed, which allows safer storage. However, when discharged to 0 V, SEI in NIBs tends to dissolve and form a thicker layer when charged again, which can lead to sodium metal plating and rollover cell failure [44]. The battery characteristics are all determined by the intrinsic electrochemical properties of the electrode materials, which is why finding suitable electrode materials is essential [29].

3.1. Negative-Electrode Materials

Negative-electrode materials for SIBs can be categorized by the mechanism by which they store sodium ions. These are materials based on intercalation mechanisms, alloying reactions, and conversion reactions. Intercalation-based (also known as insertion-based) materials allow sodium ions to be repeatedly inserted into their crystal structure with minimal structural change. An example of such material is hard carbon, or titanium-based oxides, such as $Na_2Ti_3O_7$. Materials based on alloying reactions work on the principle of forming alloys with sodium during the charging process. These materials provide high theoretical capacities (at the cost of high volume changes) and include tin (Sn), antimony (Sb), silicon (Si), bismuth (Bi), and others [38]. Conversion-type materials undergo chemical conversion reactions during cycling between transition metal oxides (TMOs = Cu, Fe, Mn, Ti...) and alkali metals (Li, Na, K...) using anions (P, O, S, N). TMs include transition metal oxides, transition metal sulfides, transition metal nitrides, and transition metal phosphides. Various storage mechanisms of Na+ are shown in Figure 6. Similarly to alloying materials, they offer high theoretical capacity with the disadvantage of large volume changes. Another drawback is the presence of voltage hysteresis, i.e., large voltage differences between charging and discharging, which harms the battery's overall efficiency [45–47].

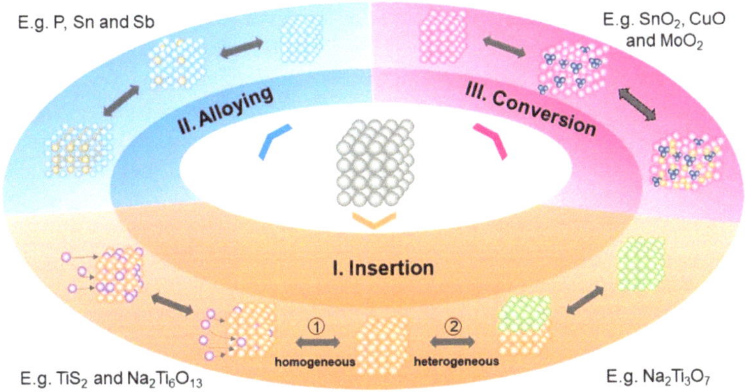

Figure 6. Na^+ storage mechanisms [48].

3.1.1. Insertion Materials

Hard carbon

Hard carbon and its modifications have recently been one of the most researched materials in the negative SIB and LIB electrode fields. Its high reversible capacity of up to 700 mAh/g (with certain modifications), simple production process, and availability make hard carbon a suitable candidate for SIBs [49]. Its development in recent years has mainly focused on biomass processing using the pyrolytic method. The pyrolysis method relies on heating the biomass precursor to temperatures of up to 1000 °C with no air access, which leads to the release of molecules such as H_2O and CO_2 from the material. This release results in high porosity (Brunauer–Emmett–Teller (BET) surface area increase) and surface inhomogeneity, resulting in low ICE (Initial Coulombic Efficiency) and significant loss of active material during SEI layer formation. A gradual increase in temperature beyond 1000 °C leads to material carbonization [50,51]. During carbonization, more stable molecules such as H_2 "evaporate" and form graphene nanodomains, which increase the number of closed HC pores, contributing to increased capacity. For increased hard-carbon purity, HCl, for example, can be used to remove inorganic compounds such as alkali salts or transition metals [52].

Wang et al. used this method in the formation of a negative electrode made from walnut shells, in which the sample carbonized after HCl treatment and firing at 1400 °C showed better electrochemical properties compared to the untreated sample—namely a reversible capacity of 342 mAh/g with a capacity retention of 91% after 100 cycles at a current density of 20 mA/g. This represents simple hard-carbon synthesis with high cyclability. Unfortunately, the use of HCl may be a disadvantage from an environmental point of view. As far as the electrochemical properties are concerned, a high decrease in reversible capacity at higher current densities can be observed, which could affect the utilization of this material in fast-charging batteries [53].

In addition, Kamiyama et al. created a hard-carbon material with MgO nanotubes using the freeze-dry synthesis of magnesium gluconate and glucose. This material shows a reversible capacity of up to 478 mAh/g (depending on the material ratios and firing method) at a current density of 25 mA/g with an ICE of 88% and a reversible capacity of 400 mAh/g at 250 mA/g [54].

In 2022, Zhou et al. created a rosewood-based material. With a specific chemical treatment, the material achieves a reversible capacity of 326 mAh/g at 20 mA/g. The material has the advantage of a reversible capacity of 230 mAh/g at 5000 mA/g. Moreover, it retains 85% of its capacity after 800 cycles at a current density of 500 mA/g [55]. Since these materials are relatively new, more extensive research is required in the future.

In order to mitigate the low ICE typical for hard carbon (typically around 60%), presodation of the negative-electrode material may be used. Chemical presodation, involving the immersion of hard carbon in ether solvents or reductive reagents, seems to provide the best compatibility with the existing SIB manufacturing process. Man et al. were able to improve the ICE of hard carbon by nearly 34% by using tetraethylene glycol dimethyl ether (TEGDME) presodation solution. After pairing the presodiated hard carbon with $Na_3V_2(PO_4)_3$, an ICE of 91.25% was achieved [56].

In the commercial sphere, Faradion Limited uses its own patented hard carbon, which has a specific capacity in excess of 330 mAh/g. This material is used in combination with a positive electrode composed of layered oxides with both O3- and P2-type structures, achieving a reversible specific capacity of 134.5 mAh/g and an ICE of over 90% [57].

Schütte et al. published a work presenting the first Na-ion battery full cell parametrization for use in a physico-chemical model (PCM). PCMs are widely used in battery research, as they can help identify the battery parameters, determine anode potential during fast charging, predict available battery power, model material degradation during cycling, and more. The parametrization was performed on a commercial 1.2 Ah 18,650 cell using hard carbon and layered oxide as electrodes. Their work utilizes already established analysis techniques for LIBs and acknowledges their transferability to SIB cell analyses. It emphasizes the importance of SoC-dependent parameters (such as the diffusion coefficient of hard carbon), thermal models, aging, and hysteresis models, providing guidance for future parameterizations and PCMs in the SIB field [58].

Soft carbon

Soft carbon is a type of carbon with arranged layers and a disordered structure characterized by tight spacing between layers. These short distances prevent Na^+ intercalation, as seen in the work of Cheng et al., where the precursor from pyrene and terephthaloyl chloride is prepared at a ratio of 1:0.5 and subsequently fired at 1300 °C, resulting in the formation of soft carbon with an ordered structure. The C-1:0.5 sample reached a capacity of only 181 mAh/g at a potential of 0.1 V vs. Na/Na^+. The BET area was 8.2 m^2/g, the percentage of closed pores equaled 0, and the d022 plane distance (narrowest interlayer) was 0.345 nm. A similar sample was formed at a C-1:10 ratio, which resulted in the formation of hard carbon featuring a capacity of 399 mAh/g at a potential of 0.1 V vs. Na/Na^+, a BET area of 6.4 m^2/g, 70% of closed pores, and a d022 layer distance of 0.382 nm [59]. The difference between the soft- and hard-carbon structures can be seen in Figure 7.

Pendashteh et al. conducted an experiment with soft carbon formed by pyrolysis at 500 °C of PVC (polyvinyl chloride), subsequent treatment with ball milling, and repeated pyrolysis at 800 °C. For this material, a reversible capacity of 207 mAh/g was achieved at a current density of 25 mA/g, a BET measurement area of 5.8 m^2/g, and an ICE of 82%. At a current density of 372 mA/g, this material maintains a capacity of 145 mAh/g, and this capacity does not drop below 100 mAh/g for the duration of 400 cycles [60].

Soft carbon, despite its less demanding production and good stability, offers low capacities. This is caused by the aforementioned interlayer distances. Referring to the Web of Science, lower interest in this material in connection to SIB can be observed (only 75 articles have been published in the last 6 years when searching for "Soft Carbon Na-ion") [61]. Thus, the use of soft carbon in SIB batteries and stationary storage systems is a question of future development, as today, there are more promising materials in terms of capacities and current load.

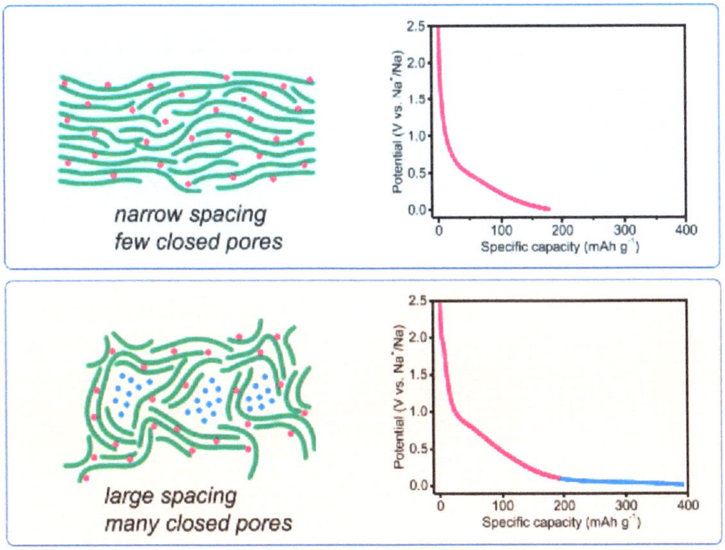

Figure 7. Na^+ storage mechanism in soft carbon and hard carbon, respectively [59].

Graphene/rGO

Graphene is the most stable allotrope of carbon, consisting of carbon atoms arranged in a hexagonal lattice which together form a layer with a one-atom thickness. The individual layers are attracted to each other by van der Waals forces. The interlayer distance of such layered graphene amounts to 3.35 nm [62].

Graphene exhibits very favorable properties in terms of mechanical strength, electrical conductivity, and high theoretical surface area. However, when only one layer of graphite is used in conjunction with Na^+, there is a lack of proper absorption of this ion, as it is energetically disadvantageous for graphite, resulting in a decrease in the theoretical capacity [62].

In order to use graphene as a negative-electrode material, it needs to be modified using defects. These defects are meant to disrupt the graphene lattice and create gaps in which Na+ can then intercalate. Graphene lattice defects can be divided into two groups. The first group is made up of intrinsic defects, among which are defects caused by vacancies, Thrower–Stone–Wales defects (change in the rotation of the π-bonds of the C=C carbon atoms), and other defects not caused by the presence of foreign atoms. The second group covers extrinsic defects, which are formed by foreign atoms [63,64].

However, these defects also pose the disadvantage of low ICE in the first few cycles. Ding et al. pointed out the importance of the preparation of graphene layers, where low firing temperatures of about 600 °C lead to the formation of graphene with rGO properties, while higher temperatures (1400 °C) lead to graphene with lower porosity and larger interlayer spacing. Despite the higher-quality graphene, the ICE still remains low at around 60% [65].

These defects were used by Wang et al. in the fabrication of a 3D-printed Au/rGO (rGO—reduced graphene oxide) anode by mixing GO with $HAuCl_4$ compound. The mixture was then washed in deionized water and centrifuged to form an Au/GO ink suitable for 3D printing. After printing, excess water was removed using the freeze-dry method, leading to the formation of the Au/rGO structure. Prior to use, Na deposition was performed on the material to obtain Na@Au/rGO. Combined with a 3D-printed NVP@C-rGO cathode, this material exhibited a reversible capacity of 91.2 mAh/g at a current load of 100 mA/g. Furthermore, the cell exhibited a reversible capacity of 84.95 mAh/g at a current density of 100 mA/g with high CE (up to 96%) after 205 cycles [66]. This work provides insight into a new method of 3D-printed materials, which is among the areas of interest in the production of electrode materials. Unfortunately, the demanding production of rGO as well as the price of $HAuCl_4$ may affect their large-scale use, for example, in stationary storage applications.

Sodium titanates

Sodium titanates are of interest due to their stability, material availability, and environmental friendliness. The most well-known sodium titanate is $Na_2Ti_3O_7$ based on the familiar and commercially used LIB negative-electrode material $Li_4Ti_5O_{12}$. The basic structure of $Na_2Ti_3O_7$ consists of a TiO_6 octahedral exhibiting a zigzag layer structure, as shown in Figure 8. The benefit of this material is the low Na^+ insertion potential (0.3 V vs. Na^+/Na), providing a higher SIB operating voltage. The drawback of this material is the slow diffusion of Na^+ ions attributed to the large bandgap voltage of 3.7–3.9 eV [67].

All of this results in low electrochemical activity at higher charging and discharging currents, which limits the use of $Na_2Ti_3O_7$ in high-power applications. This shortcoming has led to efforts to enhance ion diffusion using a range of doping techniques, nanostructures, fabrication using carbon-based composites, or surface modifications [67,68].

The group of Wang et al. also used nanostructure techniques. To create a negative electrode, they used carbon nanotubes with a diameter of 270 nm, which were then used to prepare $Na_2Ti_3O_7$ nanowires and nanobelts by synthesizing TiO_2 (titanium dioxide) in NaOH (sodium hydroxide). An example of the final structure can be seen in Figure 8. The material formed by the nanobelts exhibited a surface area of 38.68 m^2/g with a capacity of 239 mAh/g at a current density of 0.1 A/g. Moreover, this material achieved high cyclability, with over 5000 cycles at a current density of 10 A/g at 100% capacity retention. Nevertheless, the excellent properties of the negative electrode were diminished by the relatively complex and energy-intensive production of carbon nanofibers (CNF), as their synthesis is performed by electrospinning the material at 15 kV followed by firing at 1000 °C [67].

Zhong et al. used carbon MXene composite $Ti_3C_2T_x$ in the preparation of a negative electrode. Its fabrication utilized a hydrothermal reaction in H_2O_2 (hydrogen peroxide) and NaOH solution to form $Na_2Ti_3O_7$ materials, followed by calcination in Tris-buffer solution in order to create the $Na_2Ti_3O_7$@C material. The resulting structure can be observed in Figure 8. Using the BEL method, an area of 132.93 m^2/g was calculated. In addition, the material shows a voltage window ranging from 0.01 at charge to 3 V at discharge (vs Na/Na^+) at a scan rate of 0.1 mV/s on the CV curve. $Na_2Ti_3O_7$@C exhibits high cyclability at a current density of 2 A/g, reaching a reversible capacity of 119 mAh/g while maintaining 93.5% of its capacity after 200 cycles. While this material fails to achieve the same capacitance and stability as the material of Wang et al., $Na_2Ti_3O_7$@C has less energy and process-intensive production [69].

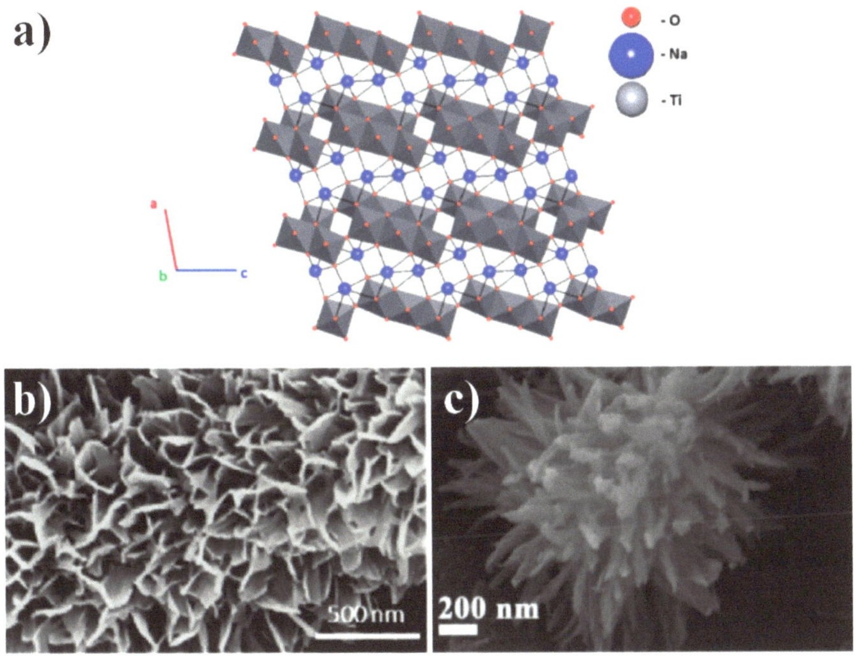

Figure 8. (**a**) $Na_2Ti_3O_7$ zigzag structure [70]. (**b**) SEM image of nanobelts/CNF [67]. (**c**) FESEM (fast emission scanning electron microscopy) image of $Na_2Ti_3O_7$@C [69].

3.1.2. Alloying and Conversion Materials

Sn-based materials

The use of Sn involves both alloy and conversion reactions for Na^+ deposition. Large volumetric changes occur in alloy materials during sodation and after desodation. This leads to a very rapid capacity loss due to cracking of the SEI layer and its eventual restoration, which reduces the active-electrode material. At the same time, the electrode material undergoes decomposition, manifested by the loss of contact between the electrode and the collector. This results in disconnection of the material from the potential, and its participation in current and capacitance reactions is hindered. Other issues with Sn-based materials are, for example, low ICE and low intrinsic conductivity. These disadvantages result in the inability to develop SIBs with a long lifetime and good safety, mainly due to the unstable SEI layer [71].

In order to limit undesirable volumetric changes, a number of material modifications have been employed, such as nanosizing (modification of and reduction in Sn material particles), combination with carbon materials (carbon nanotubes, graphene, etc.), defects (oxygen or sulfur vacancies), or structural modifications [46,71].

Li et al. used the nanosizing method in combination with carbon material to synthesize Sn-Ni@NC (NC = nitrogen-doped carbon) material. This material exhibited a very high initial capacity of 537.5 mAh/g when discharged and 319.7 mAh/g when charged with an ICE of only 59.47%, which was presumably due to the formation of the SEI layer. At a current load of 0.2 A/g, Sn-Ni@NC maintained a reversible capacity of 264.9 mAh/g compared to the Sn-Ni material, which only reached 141.8 mAh/g. Furthermore, the Sn-Ni@NC material exhibited a reversible capacity of 177.4 mAh/g at a current density of 1 A/g, maintaining a capacity of 148.1 mAh/g after 200 cycles [72].

Sb-based materials

Sb-based materials suffer from the same undesirable properties as Sn-based materials. They employ similar modifications to achieve higher capacities, lower volumetric expansion, and higher electrical conductivity. Yu et al. used intermetallic compounds with antimony ($TiSb_2$ and $NbSb_2$) to synthesize Sb@TiC@C and Sb@NbC@C materials.

The individual materials use amorphous carbon to compensate for volumetric changes. The different materials achieved very high volumetric capacities, with Sb@TiC@C reaching 746 mAh/cm^3 (at a mass of 1.86 g/cm^3, the gravimetric capacity reaches 401 mAh/g). Furthermore, a volumetric capacity of 726 mAh/cm^3 was achieved for Sb@NbC@C (at a mass of 1.93 g/cm^3, the gravimetric capacity amounts to 376 mAh/g). This capacity was retained by the materials even after 100 cycles at a current density of 50 mA/g [73].

Pb-based materials

Lead can be classified as a possible negative-electrode material for SIBs. Lead itself has a reversible capacity of up to 464 mAh/g, with 98.5% capacity retention after 50 cycles, at a current density of 13 mA/g. These properties were utilized in an experiment by Pandit et al. in the construction of a full cell made of $Na_3V_2(PO_4)_3$/C material as a positive electrode and Pb as a negative electrode. The full cell achieved a reversible capacity of 233 mAh/g at a current density of 0.1 C with an excellent energy density of 170 Wh/kg [74]. Recycled Pb, which can be extracted quite successfully from older Pb–acid batteries, could be employed in negative electrodes. The high toxicity and high density of Pb still represent considerable disadvantages [75].

MXene-based materials

MXenes are inorganic materials forming 2D structures from transition metals (commonly carbides, nitrides, and carbonitrides). Some of the characteristic properties of MXene include excellent electrical conductivity, an intrinsic layered structure (aiding in the fast movement of ions), and the possibility of adjusting interlayer distances (allowing for the deposition of different ions or molecules) [76,77].

Tang et al. focused on the confinement of SnP (Tin monophosphide) into the $Ti_3C_2T_x$ MXene structure. By etching the Ti_3AlC_2 material with highly concentrated hydrofluoric acid, $Ti_3C_2T_x$ MXene was prepared (T_x denotes terminal groups H, OH, and F on the surface and edges). Afterwards, positive NH_4^+ ions were introduced into the intergrid spaces of MXene by electrostatic intercalation, resulting in stretching of the interlayer distances from 0.96 nm to 1.21 nm. Next, SnO_2 nanocrystals were formed in the interlayer spaces of MXene using a hydrothermal method. Next, the MXene mixture was manually crushed together with NaH_2PO_2 and fired in an argon atmosphere, resulting in the formation of the SnP phase, and enabling the assembly of the M-SnP-in material. This material achieved a very strong reversible capacity (584.5 mAh/g) at a current density of 0.2 A/g over 200 cycles and 438.2 mAh/g at a current density of 15 A/g. Furthermore, the material exhibited excellent cyclability—maintaining a capacity of 436.6 mAh/g at a current density of 2 A/g for 1500 cycles (the SnP_x alloy material lasted for only 600 cycles, with less than half of its original capacity remaining after 300 cycles). Finally, a full cell was constructed in combination with the cathode material $Na_3V_2(PO_4)_3$, yielding an energy density of 265.4 Wh/kg [76].

3.2. Positive-Electrode Materials

In the last decade, numerous materials have been studied as potential positive-electrode materials for SIBs. The options for positive-electrode materials include layered transition metal oxides, polyanionic compounds (phosphates, fluorophosphates...), Prussian blue derivatives, conversion materials (transition metal fluorides, oxyfluorides, sulfides, and selenides), and organic compounds [29]. Among these options, the first three listed have been shown to be the most promising [78].

Layered Transition Metal Oxides

Due to their high structural compatibility for Na^+ insertion and extraction, simple structure, ease of synthesis, and high operating potential, layered transition metal oxides (TMOs) are considered to be among the most promising candidates among all considered materials for positive SIB electrodes. Depending on how the sodium ion and transition metal layers are stacked, layered transition metal oxides can have a different structures. P2-type (having higher rate performance and capacity retention) and O3-type phases (able to achieve high capacities) have been shown to be the most interesting in the SIB field. The biggest drawbacks of layered transition oxides are large volume changes caused by phase transitions. Many synthesis methods are used to form layered metal oxides, such as sol–gel, hydrothermal, solid-state, and coprecipitation [29,47]. For example, the sol–gel method is very promising, as it offers the desired structural, morphological, and electrochemical properties, as the homogeneity and purity are easily controlled by selecting suitable precursors. It consists of dissolved Na in transition metal salts in a solvent with an added gel-forming chelating agent. Hydrothermal synthesis uses quite low temperatures, resulting in low energy consumption and ease of synthesis, while solid-state synthesis has the advantage of ease in achieving the right stoichiometry [79]. Along with the quite simple coprecipitation method, the synthesis methods in general are not too complicated and readily attainable [78].

The layered structures are made of edge-sharing TMO_6 (M = Fe, Mn, Ni, Co, Cr, Ti, V, and their combinations) octahedra, forming repeating layers between which Na^+ ions are positioned in the octahedral (O) oxygen environment, which is called O-type stacking. On top of that, in the case of Na-ion oxides, P-type stacking can occur, where the P-type refers to prismatic Na-ion coordination. The most studied layered stacking configurations are the P2 phase (ABBA oxygen stacking) and the O3 phase (ABCABC oxygen stacking), as seen in Figure 9. P2-type usually provides higher rate performance and good cycling stability when compared to O3-type analogs due to its greater Na ion conductivity and better structural integrity; however, its first charge storage capacity is limited due to the lower initial Na content. It is typical that upon Na ion intercalation and deintercalation during cycling, the structure shifts between O- and P-types, commonly degrading its cycle stability. Metal oxides consisting of more elements (especially those including Mn) tend to provide higher energy densities than those with a single element [80].

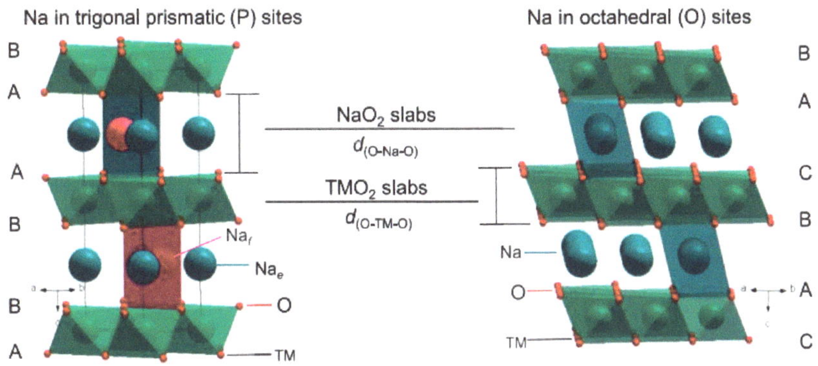

Figure 9. Illustration of crystal representative P2-type and O3-type layered oxides [80].

P2 type $Na_{2/3}Ni_{1/3}Mn_{2/3}O_2$, for instance, stands out for its high capacity of 160 mAh g^{-1}; however, its capacity quickly fades after reaching 4.2 V due to the P2-to-O3 phase transition. A layer of the $NaPO_3$ surface is able to help improve stabilization for long-term cyclability, as it is able to suppress oxygen release in the highly desodinated state of the battery. O3-type $NaNi_{0.5}Mn_{0.5}O_2$ is able to provide capacities in the range of 105–125 mAh/g at 2.2–3.8 V. Zhao et al. prepared a high-entropy O3-type $NaNi_{0.12}Cu_{0.12}Mg_{0.12}Fe_{0.15}Co_{0.15}Mn_{0.1}Ti_{0.1}Sn_{0.1}Sb_{0.04}O_2$ using a solid-state synthesis method, achieving a reversible capacity of about 110 mAh/g. This material retained 83% of its capacity after 500 cycles and ~80% capacity retention at a 5C rate, operating in the voltage range of 2.0–3.9 V, versus Na^+/Na [78]. O3-type materials could contribute to higher energy densities, while P-2 exhibits higher capacity retention and stability. Both types are promising candidates for low-cost SIB energy storage systems application in the near future, although further research is needed to enhance the performance of these materials [47].

Extensive research has been performed on both O3- and P2-type layered structures. In 2013, $Na_{0.950}Ni_{0.317}Mn_{0.317}Mg_{0.158}Ti_{0.208}$-$O_2$ material with a typical O3 phase provided high energy and a good cycle life; however, no further developments were made at the material level. Eventually, a mixed-phase material was developed, averaging the stoichiometries of O3- and P2-type materials. The stoichiometries used were $Na_{0.66}7Ni_{0.300}Mn_{0.600}Mg_{0.033}Ti_{0.067}O_2$ for the P-type and $NaNi_{0.333}Mn_{0.033}Mg_{0.16}7Ti_{0.167}O_2$ for the O3-type (an image of the stacked morphologies can be seen in Figure 10). The material delivers a capacity of 156 mAh/g at 0.2C in the 4.35–2 V window. It also retains 80% of its capacity after 3000 cycles and allows for charging at 4C without capacity drops when paired with hard carbon in a pouch cell [57].

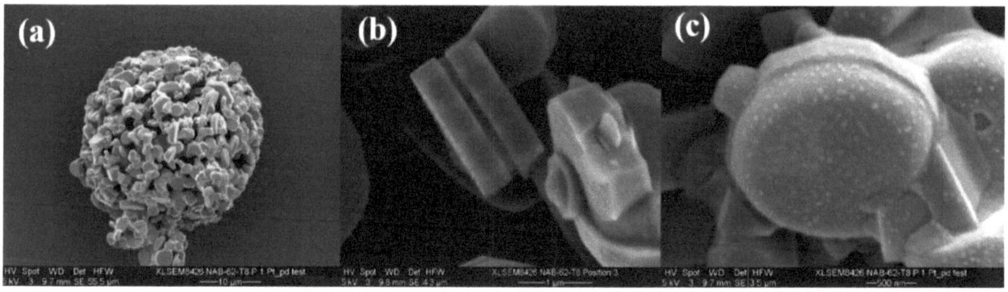

Figure 10. FESEM images of the combined stoichiometries of the Faradion positive-electrode material. (**a**) A zoomed-out image. (**b**,**c**) Magnified images showing the stacked O3/P2 phase morphologies of the primary particles [57].

Polyanion-type materials

Owing to the stability of their crystal structure, thermal stability, adjustable voltage, and high ionic conductivity, polyanion-type materials seem to be a strong candidate for positive-electrode materials used in SIBs. The main obstacles with these materials are low electron conductivity, unsatisfactory theoretical specific capacity, and low energy density. The structure of these materials is composed of MO_x polyhedral (M = metal) and anion groups $(XO_4)^{n-}$ (X = B, C, S, P, Si, As, etc.) or their derivatives connected by a covalent bond [81]. The most commonly adopted material synthesis methods are the sol–gel and the solid–solid method. The sol–gel method allows for the synthesis of materials with a high surface area and allows for properties such as homogeneity and purity to be easily controlled, but controlling the pore structure and morphology is quite problematic. The solid-state technique provides many benefits, such as low cost and simple equipment requirements; however, it provides minimal control over the morphology and shape of the material, influencing the final electrochemical performance [82].

Among polyanionic compounds, NASICONs (Na Super Ionic Conductors) are a subject of interest thanks to their strong 3D framework, high rate capability, long cyclability, and satisfactory kinetics of sodium-ion mobility, making them suitable for high-power applications. NASICONs used as positive electrodes are generally phosphates combined with various transition metals, such as vanadium (V), iron (Fe), and manganese (Mn). Combinations of these metals may be used as well [29]. $Na_3V_2(PO_4)_3$ (NVP) is a typical example of a NASICON structure, with a specific capacity of 117 mAh/g and a voltage plateau at ~3.5 V. Zhao et al. analyzed the $Na_4MnCr(PO_4)_3$ structure, which retained 85% of its capacity after 500 cycles at 100 mA/g at a cutoff potential of 4.3 V, meaning the energy density at the highest potential is over 400 Wh/kg [83]. Hu et al., for instance, constructed NASICON-type $Na_{3.2}MnTi_{0.8}V_{0.2}(PO_4)_3$, with five redox couples ranging from 2.1 to 4.1 V, showing a capacity of 172.5 mAh/g and quite a high energy density of 527.2 Wh/g [84].

Fluorophosphates, such as $Na_3V_2(PO_4)_2F_3$ (NVPF), are the subject of extended studies. They offer a theoretical capacity of roughly 128 mAh/g and a working voltage of ~3.9 V, achieving an energy density of ~506 Wh/kg. However, they suffer from capacity fading at high rates due to low conductivity and insufficient ion diffusivity [39,85]. Gu et al. prepared a $Na_3V_{1.98}Mn_{0.02}(PO_4)_2F_3$ electrode material using a hydrothermal synthesis method, which offers a specific capacity of 123.8 mAh/g at 0.1C in the voltage range from 2.0 to 4.3 Vversushigher than Na^+/Na, and power density around 400 Wh/kg [85,86].

Plewa et al. synthesized $Na_2FeM(SO_4)_3$ (M = Fe, Mn, Ni) nanometric size grain materials through the uncomplicated dissolution of Na_2SO_4 and transition metal sulfates in a liquid solution, subsequent solvent evaporation, and annealing. All of the tested $Na_2FeM(SO_4)_3$/C cells represented systems of high voltage >3.6 V with a capacity of 110 mAh/g (and therefore an energy density of over 396 Wh/kg) and capacity retention of 90% after 50 cycles [87]. While polyanion-type materials offer advantages in terms of voltage, thermal stability, and cycle life, their lower specific capacity might not be suitable for energy-dense applications. NVPs are, for instance, suitable for high-power batteries (rather than energy-dense ones) as they exhibit fast electrochemical kinetics at the cost of quite low specific capacities [45].

Prussian Blue Materials

Prussian blue analogs (PBAs) represent high-entropy electrode materials from the family of metal–organic frameworks (MOFs). Prussian blue suffers from low Coulombic efficiency and capacity fading, but offers a high theoretical capacity, and its framework provides enough space for Na^+ insertion and extraction. Thanks to its high rate capability and satisfactory cycling stability, it can be suitable for high-power applications [29,88]. PBAs exhibit face-centered cubic structures and are described by the molecular formula $Na_xM[Fe(CN)_6]_{1-y}\cdot{*}_y\cdot mH_2O$, where M stands for transition metal elements (Fe, Co, Ni, and Mn). Here, * denotes [Fe(CN)$_6$] defects, where the range of x is 0 < x < 2, and the range of y is 0 < y < 1. M-based PBs (apart from Ni) show a theoretical capacity of up to 170 mAh/g; however, the [Fe(CN)$_6$] defects generated during the synthesis process alongside crystalline water cause lattice distortion and structural collapse during the cycling process, resulting in a reduction in capacity and cycling performance [89]. In terms of theoretical capacity, this material can be compared to commercially used positive-electrode LIB materials, where $LiCoO_2$ has a theoretical capacity of 274 mAh/g (practical discharge capacity of 173 mAh/g with acceptable reversibility), $LiFePO_4$ shows a theoretical capacity of 154 mAh/g, and $LiMn_2O_4$ achieves 148 mAh/g [90–92]. In general, PBAs are a promising class of materials for energy storage applications, with Fe-based PBAs standing out as the most environmentally friendly option owing to their use of abundant, non-toxic, and sustainable elements [89].

Hydrothermal, coprecipitation, and electrodeposition methods are the most popular synthesis techniques employed in electrochemical applications. In the widely used hydrothermal method, the morphology of PBA is determined by the used precursor structure. The coprecipitation approach is used by many because of its simplicity and low cost; however, it does not allow for adjustment of the crystal shape and size as well as the hydrothermal method. As for the electrodeposition method, the utilized substrate greatly influences the morphology of the final electrode material [93].

To address the issues of $Fe(CN)_6$ defects, Wang et al. developed a hollow layered Fe-PB composite using the hydrothermal synthesis method. 1,3,5-benzene tricarboxylic acid (BTA) was utilized as a chelating and etching agent as its carboxyl groups (-COOH) can chelate with Fe_2^+ ions, so high-quality cubic Fe-PB can be produced by controlling the reaction rate of Fe-PB formation. The hollow layered structure was shown to significantly reduce the diffusion path of Na^+ ions and limit the volume changes during Na-ion insertion/extraction. The electrode's initial discharge capacity equaled 95.9 mAh/g and remained at 73.1 mAh/g after 500 cycles in the voltage range of 2.0–4.2 V. The full cell paired with hard carbon provided a considerable energy density of 312.2 Wh/kg [89].

There are many PBA modification methods, among which metal-ion doping (doped at the M-site or Na-site) is currently one of the most effective modification methods, with the goal of reducing $Fe(CN)_6$ vacancies. Chen et al. directly prepared a Cu-doped Fe-PBA using the coprecipitation synthesis method. The resulting material contained fewer $Fe(CN)_6$ vacancies and low crystal water content. The added Cu contributed to structure stability, allowing the material to achieve an initial capacity of 127.4 mAh/g (while undoped Fe-PBA achieved 107.9 mAh/g) at 100 mA/g between 2.5 and 4.2 V. Cu doping is therefore an effective modification method, improving the overall performance of PBA positive-electrode materials [94].

PBA-based SIBs can be put between supercapacitors and other batteries (Ni-Cd, lead–acid, LIBs) in terms of cycle life and power density (up to 1250 W/kg). As they have the highest cycle number (>40,000 cycles) among all batteries and use abundant, non-toxic raw materials, they could be an alternative for energy storage in high-power applications, including backup power for data centers and regenerative braking. An example of the commercial use of these batteries is the PBA-based pluggable sodium-ion module (four-cell stacks in series connection) developed by ABB and Natron. The utilized iron-based PB electrode material can be seen in Figure 11. These cells were subjected by He et al. to UL 9540A thermal runaway tests, comprising external short-circuit tests, external heating, and mechanically induced events (e.g., nail penetration). None of these tests triggered thermal runaway, demonstrating the SIBs' safety. Their energy density sits quite low (23 Wh/kg at the cell level), but they show high rate capability and very satisfactory capacity retention [95]. Another example is SIB prototype cells by Novasis Energies, Inc., using a PBA positive electrode, commercial hard-carbon negative electrode, and a nonaqueous (e.g., salts dissolved in non-water-based solvents) electrolyte. This company was able to construct 0.5–5 Ah capacity high-safety pouch cells with noteworthy rate capabilities and cycle life [96].

Figure 11. SEM image of Natron iron-based positive-electrode material $Na_xMn_yFe(CN)_6{\cdot}nH_2O$ [95].

Organic materials

Organic-based positive-electrode materials have attracted considerable attention, as they are made up of weak intermolecular interactions (e.g., hydrogen bonds and van der Waals forces), which results in a flexible structure capable of accommodating Na^+ ions without significant volume change while allowing their smooth transport. They can be divided into two groups—carbonyl small molecules (ketones and quinones, anhydride compounds, imide compounds, and carboxylate compounds) and organic polymers (carbonyl and conjugated conductive polymers, covalent organic frameworks (COFs), organometallic compounds, and organic radial polymers). Organic compounds have the advantages of low toxicity, structural diversity, ease of synthesis, and material abundance tied to sustainability. They suffer from electrolyte dissolution, low stability, and poor electronic conductivity, preventing their utilization in practice [97].

To tackle these issues, Kuan et al. reported the use of nitrogen- and carbonyl-rich highly extended p-conjugated small organic molecule hexaazatrianthranylene (HATA) embedded quinone (HATAQ). HATAQ molecules create supramolecular graphite-like 2D layered arrangements in the solid state, promoting charge transfer and helping to improve structural stability. The material, when combined with Na metal negative electrodes in a coin cell, demonstrated a 460 mAh/g capacity at 500 mA/g (1C) and retained 99% of its capacity (138 mA) after 5000 cycles at 60 A/g, which is one of the top reported values for organic small molecules so far. The estimated gravimetric energy density is estimated to be 920 Wh/kg, based on the specific capacity obtained at 1C and an average discharge potential of 2.0 V [97].

Metal–organic polymers (MOPs) have been subjected to many studies on conjugated carbonyl-derived linkers, as they offer multiple redox centers and are low-cost. Wang et al. synthesized a flower-like p–d conjugated MOP (copper–tetramino-benzoquinone, Cu-TABQ) where TABQ serves as an organic ligand and $Cu_2{}^+$ as a transition metal node. Both elements act as dual redox centers, allowing the achievement of a reversible capacity of 322.9 mAh/g at 50 mA/g in the voltage range of 1.0–3.0 V (meaning the average power density revolves around 645.8 Wh/kg). With an ICE of 75.6%, the electrode material retained 98.7% of its capacity after 50 cycles. Owing to the stable coordination between $Cu_2{}^+$ and TABQ, the positive-electrode electronic conductivity and Na^+ redox reaction were significantly improved, so the capacity reached 198.8 mAh/g even at a high current density of 4000 mA/g [98].

Due to their structure, organic materials have the potential to be used in applications where light weight, portability, and flexibility are crucial. If the mentioned limitations can be effectively mitigated, organic materials could emerge as promising electrode materials, especially in renewable and environmentally responsible organic batteries [3].

3.3. Electrolytes and Salts

Electrolyte research is focused on non-flammable electrolytes with applications in a large temperature range. As for stationary SIB applications, finding electrolytes with properties that would help reduce additional costs regarding the heating, fire safety, etc., of the area where the battery is stored is needed. In addition, such materials must not significantly increase the battery's cost [29].

SIBs can be used with liquid or solid electrolytes. The liquid electrolyte is the medium used to transfer the charge (Na^+ ions) from the negative to the positive electrode and vice versa. The most commonly used liquid electrolytes are electrolytes based on salts dissolved in organic aprotic solvents (mainly based on ethers, esters, and carbonates). As for solid-state electrolytes (SSE), polymeric bonds with sodium salts are the most common type. Furthermore, there are also inorganic solid electrolytes (ISEs) consisting of oxides, phosphates, sulfates, and hydrates. ISEs are hard, inflexible materials that are mainly used in high-temperature Na-S systems [29].

Individual electrolytes may differ in terms of their working temperature range, solubility of sodium salts, and Na^+ transfer rate. Du et al. compared ether-based non-flammable electrolytes (1 M $NaBF_4$ in tetraglyme) with carbonate-based electrolytes (1 M $NaClO_4$ in 1:1 ethylene carbonate (EC)/propylene carbonate (PC)) in HC half-cells in a voltage range of 0.001–1.5 V. At 0.2 C, the HC achieved a high initial charge capacity of 273 mAh/g, ICE of 87%, and 99% capacity retention after 50 cycles using 1 M $NaBF_4$ in tetraglyme. The carbonate-based, 1 M $NaClO_4$ in EC/PC electrolyte only reached 178 mAh/g, and exhibited lower ICE of 69% and capacity retention of 84%. It is worth noting that the ether-based electrolyte achieved a considerable charge capacity of 223 mAh/g at a 2C rate. For this material, Differential Scanning Calorimetry (DSC) showed higher onset temperatures for the first exothermic peak (SEI cracking), which implies higher thermal stability. The non-flammable nature of the ether-based electrolyte can reduce the need for extensive safety measures in battery system storage. Along with the electrolyte's contribution to the battery's long lifespan, the overall costs may be lowered, which is a critical aspect for battery systems in stationary applications [99].

The development of aqueous SIBs is likely to be limited by their low energy density and unsatisfying output potential due to the low electrochemical stability window of the electrolyte (only 1.23 V under thermodynamic conditions). In addition, aqueous electrolytes suffer from decomposition with hydrogen and oxygen evolution, resulting in poor cycling stability [100].

SSEs are a non-flammable type of electrolyte, which eliminates the risk of fire hazards and enhances the overall safety of the battery system. This is due to the high reactivity of sodium, which could lead to a severe reaction with organic liquid electrolytes. SSEs exhibit high thermal stability, wide electrochemical stability windows, and exceptional mechanical properties [1,38]. This means they can withstand higher voltages without decomposing, leading to increased energy density without compromising safety. Due to their solid nature, SSEs can act as a more effective barrier against dendrites than porous separators in liquid electrolytes, as they can be pierced causing a short-circuit in the cell [40]. Although SSEs are more costly in terms of manufacturing and integration than liquid electrolytes, their extended lifetime and low maintenance could reduce costs in the long run. This, among other characteristics, could be advantageous, especially in stationary applications where the size and weight of the battery are not as critical compared to other fields of application [1]. SSEs can be divided into multiple types, such as solid polymer electrolytes

(SPEs), composite solid polymer electrolytes (CSPEs), and inorganic solid electrolytes (ISEs). All of these types have their unique characteristics and are under active development. In the present day, ISEs (among which β''-Al_2O_3 (beta alumina) and NASICON $Na_3Zr_2Si_2PO_{12}$ are the most used) are recognized for their high ionic conductivity and are frequently employed in high-temperature Na/S batteries [29].

3.4. Separators

A wide operating temperature range and non-flammability are among the key qualities of separators. In addition, the porosity and wettability of separators are equally as important, as they affect the internal resistivity of the battery and affect its general performance [101].

The most widely used separator materials in SIB applications include polyolefin-based materials. This group contains, for example, polypropylene (PP), polyethylene (PE), and glass fiber (GF). These materials are mainly used for their chemical stability and inexpensive large-scale production. The disadvantages of PP and PE include poor wettability and insufficient electrolyte retention when used with carbon-based electrolytes, e.g., propylene carbonate (PC), ethylene carbonate (EC), and dimethyl carbonate (DMC). Another drawback is the low temperature range of PP and PE (their melting points are at 165 °C and 132 °C, respectively). The greatest potential danger to the battery is the material's expansion—at higher temperatures, PP and PE tend to shrink, leading to internal short-circuits and other critical issues. As far as GF separators are concerned, the thickness of the material and its arrangement are not ideal. Due to the high porosity of GF, a larger amount of electrolyte is needed for filling and poses a risk in the case of dendrite formation [102–105]. Separators based on cellulose can also be found, as it is the most widely used biopolymer (plant biomass consists of 40–45 wt% of cellulosic content). It represents a very promising material in the field of renewable materials, with numerous applications, including energy storage systems. The downsides of cellulose-based separators include their flammability and moisture content (depending on the manufacturing process); however, they are highly available and affordable [106].

Aquion Energy, Inc., a company developing aqueous SIBs as an alternative to LIBs and other systems for grid storage, uses a fire-safe seawater solution as an electrolyte and a synthetic cotton separator for their batteries. The output potential of each cell would be limited, however, as cells utilizing aqueous electrolytes suffer from a low electrochemical stability window [107]. Faradion Limited uses low-viscosity nonaqueous liquid electrolytes along with commercially widely used polyolenic separators to stay consistent with established Li-ion cell manufacturing practices [99].

Unfortunately, the development of separators for SIBs has not received nearly as much attention as the development of separators for LIBs. Although LIBs and SIBs work on the same principle, different materials are used, as Na^+ exhibits different kinetic properties compared to Li^+. It may be beneficial to investigate the suitability of LIB separators in SIBs in the future, as they operate in similar temperature conditions and may utilize electrolytes based on similar principles (e.g., salts in solvent—$LiPF_6$ for Li-ion, $NaPF_6$ for Na-ion), etc.

4. Conclusions

SIBs have come a long way since the beginning of their development, and are still being advanced by introducing new materials, improving existing materials, and reducing their environmental impact. Their impact is related both to the use of materials associated with unethical mining practices (e.g., Co) and to materials used in electrodes, such as PVDF binders (requiring the use of the toxic volatile solvent methylpyrrolidone (NMP) [108]. Fluorine contained in, e.g., PVDF and electrolyte salts poses a significant risk, especially at elevated temperatures, where toxic gases, particularly highly toxic hydrogen fluoride, can be released when the battery catches on fire. For this reason, alternatives to this element are being actively explored, not limited to the battery industry [109]. The production of SIBs is cost-effective due to the use of Na, which is the most abundant alkaline element in the Earth's crust (2.36% of the continental crust), and also because of the possibility of substituting Cu used in LIBs ($9.1/kg) with Al (USD 2.5/kg) for current collectors [11,12].

However, there is a limited range of usable materials due to the difficulty of production and the possibility of scaling up. For example, the 3D-printed Au/rGO negative-electrode material is very appealing from an electrochemical point of view, but it is not suitable for mass production. Among the negative-electrode materials with the greatest scale-up potential, HC can be considered due to its simple pyrolytic production from biological materials (shells, algae, wood). PB belongs to the category of very promising positive-electrode materials, with the possibility of hydrothermal synthesis using available materials, such as Fe, Mn, and Ni. In the commercial sphere, the current development of Na-Ion batteries is moving towards the use of widely studied HC and PB electrode materials owing to their high capacities, cycling stability, and environmental friendliness. Regarding electrolytes, great efforts have been made towards the development of SSEs that allow for the application of sodium metal electrode materials. To date, however, the most used types of electrolytes are aprotic solvents (EC/PC or EC/DMC) combined with sodium salts.

The availability and ease of production of electrode and electrolyte materials, along with the ability to avoid less ideal materials like Co and Au, could pave the way for cheaper production and the commercial implementation of sodium-ion batteries (SIBs) in stationary storage facilities, ranging from small- to large-scale. These batteries would help ensure the stability of power grids while reducing environmental impact. However, this is still a theoretical near-future development of SIBs (initial prototypes and applications already exist), as challenges such as the operating temperature ranges, safety, cyclability, and current loads of battery systems have to be addressed.

Author Contributions: Conceptualization, P.B. and J.L.; Methodology, P.B. and J.L.; Validation, P.B. and J.L.; Formal Analysis, S.G. and J.P.; Investigation, S.G. and J.P.; Resources, S.G. and J.P.; Data Curation, S.G. and J.P.; Writing—Original Draft Preparation, S.G. and J.P.; Writing—Review and Editing, P.B., J.L., S.G. and J.P.; Visualization, S.G. and J.P.; Supervision, P.B. and J.L.; Project Administration, P.B. and J.L. All authors have read and agreed to the published version of the manuscript.

Funding: This research was supported by the specific graduate research of the Brno University of Technology No. FEKT-S-23-8286.

Data Availability Statement: No new data were created or analyzed in this study. Data sharing is not applicable to this article.

Conflicts of Interest: The authors declare no conflicts of interest.

References

1. Hirsh, H.S.; Li, Y.; Tan, D.H.S.; Zhang, M.; Zhao, E.; Meng, Y.S. Sodium-Ion Batteries Paving the Way for Grid Energy Storage. *Adv. Energy Mater.* **2020**, *10*, 2001274. [CrossRef]
2. Dell, R.M. Batteries: Fifty years of materials development. *Solid State Ion.* **2000**, *134*, 139–158. [CrossRef]
3. Yang, X.; Rogach, A.L. Anodes and Sodium-Free Cathodes in Sodium Ion Batteries. *Adv. Energy Mater.* **2020**, *10*, 2000888. [CrossRef]
4. Armand, M.; Tarascon, J.-M. Building better batteries. *Nature* **2008**, *451*, 652–657. [CrossRef] [PubMed]
5. Olabi, A.G.; Abbas, Q.; Shinde, P.A.; Abdelkareem, M.A. Rechargeable batteries: Technological advancement, challenges, current and emerging applications. *Energy* **2023**, *266*, 126408. [CrossRef]
6. Huang, Y.; Li, J. Key Challenges for Grid-Scale Lithium-Ion Battery Energy Storage. *Adv. Energy Mater.* **2022**, *12*, 2202197. [CrossRef]
7. DOE Global Energy Storage Database. Available online: https://gesdb.sandia.gov/statistics.html (accessed on 9 January 2025).

8. Usiskin, R.; Lu, Y.; Popovic, J.; Law, M.; Balaya, P.; Hu, Y.-S.; Maier, J. Fundamentals, status and promise of sodium-based batteries. *Nat. Rev. Mater.* **2021**, *6*, 1020–1035. [CrossRef]
9. Battery Grade Sodium Carbonate Price. Available online: https://www.metal.com/Sodium-ion%20Battery/202404290001 (accessed on 19 November 2024).
10. Lithium Carbonate Price. Available online: https://www.metal.com/Lithium/201102250059 (accessed on 19 November 2024).
11. Copper Price. Available online: https://markets.businessinsider.com/commodities/copper-price (accessed on 19 November 2024).
12. Aluminum Price. Available online: https://markets.businessinsider.com/commodities/aluminum-price (accessed on 19 November 2024).
13. Ge, P.; Fouletier, M. Electrochemical intercalation of sodium in graphite. *Solid State Ion.* **1988**, *28–30*, 1172–1175. [CrossRef]
14. Hasa, I.; Dou, X.; Buchholz, D.; Shao-Horn, Y.; Hassoun, J.; Passerini, S.; Scrosati, B. A sodium-ion battery exploiting layered oxide cathode, graphite anode and glyme-based electrolyte. *J. Power Sources* **2016**, *310*, 26–31. [CrossRef]
15. Yabuuchi, N.; Kubota, K.; Dahbi, M.; Komaba, S. Research Development on Sodium-Ion Batteries. *Chem. Rev.* **2014**, *114*, 11636–11682. [CrossRef]
16. Zhang, W.; Zhang, F.; Ming, F.; Alshareef, H.N. Sodium-ion battery anodes: Status and future trends. *EnergyChem* **2019**, *1*, 100012. [CrossRef]
17. Wu, Y.; Shuang, W.; Wang, Y.; Chen, F.; Tang, S.; Wu, X.-L.; Bai, Z.; Yang, L.; Zhang, J. Recent Progress in Sodium-Ion Batteries: Advanced Materials, Reaction Mechanisms and Energy Applications. *Electrochem. Energy Rev.* **2024**, *7*, 17. [CrossRef]
18. Kim, H. Sodium-Ion Battery: Can It Compete with Li-Ion? *ACS Mater. Au* **2023**, *3*, 571–575. [CrossRef] [PubMed]
19. Stevens, D.A.; Dahn, J.R. High Capacity Anode Materials for Rechargeable Sodium-Ion Batteries. *J. Electrochem. Soc.* **2000**, *147*, 1271. [CrossRef]
20. Zhu, Y.; Han, X.; Xu, Y.; Liu, Y.; Zheng, S.; Xu, K.; Hu, L.; Wang, C. Electrospun Sb/C Fibers for a Stable and Fast Sodium-Ion Battery Anode. *ACS Nano* **2013**, *7*, 6378–6386. [CrossRef] [PubMed]
21. Kim, Y.; Park, Y.; Choi, A.; Choi, N.-S.; Kim, J.; Lee, J.; Ryu, J.H.; Oh, S.M.; Lee, K.T. An Amorphous Red Phosphorus/Carbon Composite as a Promising Anode Material for Sodium Ion Batteries. *Adv. Mater.* **2013**, *25*, 3045–3049. [CrossRef] [PubMed]
22. Xiong, H.; Slater, M.D.; Balasubramanian, M.; Johnson, C.S.; Rajh, T. Amorphous TiO_2 Nanotube Anode for Rechargeable Sodium Ion Batteries. *J. Phys. Chem. Lett.* **2011**, *2*, 2560–2565. [CrossRef]
23. Senguttuvan, P.; Rousse, G.; Seznec, V.; Tarascon, J.-M.; Palacín, M.R. $Na_2Ti_3O_7$: Lowest Voltage Ever Reported Oxide Insertion Electrode for Sodium Ion Batteries. *Chem. Mater.* **2011**, *23*, 4109–4111. [CrossRef]
24. Ellis, L.D.; Hatchard, T.D.; Obrovac, M.N. Reversible Insertion of Sodium in Tin. *J. Electrochem. Soc.* **2012**, *159*, A1801–A1805. [CrossRef]
25. Wang, W., Liu, X.H.; Mao, S.X.; Huang, J.Y. Microstructural Evolution of Tin Nanoparticles during In Situ Sodium Insertion and Extraction. *Nano Lett.* **2012**, *12*, 5897–5902. [CrossRef] [PubMed]
26. Hariharan, S.; Saravanan, K.; Balaya, P. A-MoO_3: A high performance anode material for sodium-ion batteries. *Electrochem. Commun.* **2013**, *31*, 5–9. [CrossRef]
27. Xie, Y.; Dall'Agnese, Y.; Naguib, M.; Gogotsi, Y.; Barsoum, M.W.; Zhuang, H.L.; Kent, P.R.C. Prediction and Characterization of MXene Nanosheet Anodes for Non-Lithium-Ion Batteries. *ACS Nano* **2014**, *8*, 9606–9615. [CrossRef] [PubMed]
28. Zhang, F.; Alhajji, E.; Lei, Y.; Kurra, N.; Alshareef, H.N. Highly Doped 3D Graphene Na-Ion Battery Anode by Laser Scribing Polyimide Films in Nitrogen Ambient. *Adv. Energy Mater.* **2018**, *8*, 1800353. [CrossRef]
29. Goikolea, E.; Palomares, V.; Wang, S.; de Larramendi, I.R.; Guo, X.; Wang, G.; Rojo, T. Na-Ion Batteries—Approaching Old and New Challenges. *Adv. Energy Mater.* **2020**, *10*, 2002055. [CrossRef]
30. Xu, X.; Zhang, Y.; Sun, H.; Zhou, J.; Yang, F.; Li, H.; Chen, H.; Chen, Y.; Liu, Z.; Qiu, Z.; et al. Progress and Perspective: MXene and MXene-Based Nanomaterials for High-Performance Energy Storage Devices. *Adv. Electron. Mater.* **2021**, *7*, 2000967. [CrossRef]
31. Linden, D.; Reddy, T.B. (Eds.) *Handbook of Batteries*, 3rd ed.; McGraw-Hill: New York, NY, USA, 2002.
32. Zhang, W.; Cao, P.; Li, L.; Yang, K.; Wang, K.; Liu, S.; Yu, Z. Carbon-encapsulated 1D SnO2/NiO heterojunction hollow nanotubes as high-performance anodes for sodium-ion batteries. *Chem. Eng. J.* **2018**, *348*, 599–607. [CrossRef]
33. May, G.J.; Davidson, A.; Monahov, B. Lead batteries for utility energy storage: A review. *J. Energy Storage* **2018**, *15*, 145–157. [CrossRef]
34. Zhu, Q.; Cheng, M.; Zhang, B.; Jin, K.; Chen, S.; Ren, Z.; Yu, Y. Realizing a Rechargeable High-Performance Cu-Zn Battery by Adjusting the Solubility of Cu^{2+}. *Adv. Funct. Mater.* **2019**, *29*, 1905979. [CrossRef]
35. Posada, J.O.G.; Rennie, A.J.R.; Villar, S.P.; Martins, V.L.; Marinaccio, J.; Barnes, A.; Glover, C.F.; Worsley, D.A.; Hall, P.J. Aqueous batteries as grid scale energy storage solutions. *Renew. Sustain. Energy Rev.* **2017**, *68*, 1174–1182. [CrossRef]
36. Zhang, Y.; Mu, Z.; Lai, J.; Chao, Y.; Yang, Y.; Zhou, P.; Li, Y.; Yang, W.; Xia, Z.; Guo, S. MXene/Si@SiOx@C Layer-by-Layer Superstructure with Autoadjustable Function for Superior Stable Lithium Storage. *ACS Nano* **2019**, *13*, 2167–2175.

37. Bommier, C.; Surta, T.W.; Dolgos, M.; Ji, X. New Mechanistic Insights on Na-Ion Storage in Nongraphitizable Carbon. *Nano Lett.* **2015**, *15*, 5888–5892. [CrossRef] [PubMed]
38. Hasa, I.; Mariyappan, S.; Saurel, D.; Adelhelm, P.; Koposov, A.Y.; Masquelier, C.; Croguennec, L.; Casas-Cabanas, M. Challenges of today for Na-based batteries of the future: From materials to cell metrics. *J. Power Sources* **2021**, *482*, 228872. [CrossRef]
39. Fang, H.; Gao, S.; Zhu, Z.; Ren, M.; Wu, Q.; Li, H.; Li, F. Recent Progress and Perspectives of Sodium Metal Anodes for Rechargeable Batteries. *Chem. Res. Chin. Univ.* **2021**, *37*, 189–199. [CrossRef]
40. Gebert, F.; Knott, J.; Gorkin, R.; Chou, S.-L.; Dou, S.-X. Polymer electrolytes for sodium-ion batteries. *Energy Storage Mater.* **2021**, *36*, 10–30. [CrossRef]
41. Nikiforidis, G.; van de Sanden, M.C.M.; Tsampas, M.N. High and intermediate temperature sodium–sulfur batteries for energy storage: Development, challenges and perspectives. *RSC Adv.* **2019**, *9*, 5649–5673. [CrossRef] [PubMed]
42. Li, M.; Lu, X.; Zhan, X.; Engelhard, M.H.; Bonnett, J.F.; Polikarpov, E.; Jung, K.; Reed, D.M.; Sprenkle, V.L.; Li, G. High performance sodium-sulfur batteries at low temperature enabled by superior molten Na wettability. *Chem. Commun.* **2021**, *57*, 45–48. [CrossRef] [PubMed]
43. Seredych, M.; Shuck, C.E.; Pinto, D.; Alhabeb, M.; Precetti, E.; Deysher, G.; Anasori, B.; Kurra, N.; Gogotsi, Y. High-Temperature Behavior and Surface Chemistry of Carbide MXenes Studied by Thermal Analysis. *Chem. Mater.* **2019**, *31*, 3324–3332. [CrossRef]
44. Desai, P.; Huang, J.; Foix, D.; Tarascon, J.-M.; Mariyappan, S. Zero volt storage of Na-ion batteries: Performance dependence on cell chemistry! *J. Power Sources* **2022**, *551*, 232177. [CrossRef]
45. Pu, X.; Wang, H.; Zhao, D.; Yang, H.; Ai, X.; Cao, S.; Chen, Z.; Cao, Y. Recent Progress in Rechargeable Sodium-Ion Batteries: Toward High-Power Applications. *Small* **2019**, *15*, 1805427. [CrossRef]
46. Wu, C.; Zhu, G.; Wang, Q.; Wu, M.; Zhang, H. Sn-based nanomaterials: From composition and structural design to their electrochemical performances for Li- and Na ion batteries. *Energy Storage Mater.* **2021**, *43*, 430–462. [CrossRef]
47. Liu, Q.; Hu, Z.; Chen, M.; Zou, C.; Jin, H.; Wang, S.; Chou, S.-L.; Dou, S.-X. Recent Progress of Layered Transition Metal Oxide Cathodes for Sodium-Ion Batteries. *Small* **2019**, *15*, 1805381. [CrossRef] [PubMed]
48. Dong, S.; Lv, N.; Wu, Y.; Zhang, Y.; Zhu, G.; Dong, X. Titanates for sodium-ion storage. *Nano Today* **2022**, *42*, 101349. [CrossRef]
49. Yin, J.; Zhang, Y.S.; Liang, H.; Zhang, W.; Zhu, Y. Synthesis strategies of hard carbon anodes for sodium-ion batteries. *Mater. Rep. Energy* **2024**, *4*, 100268. [CrossRef]
50. Dou, X.; Hasa, I.; Saurel, D.; Vaalma, C.; Wu, L.; Buchholz, D.; Bresser, D.; Komaba, S.; Passerini, S. Hard carbons for sodium-ion batteries: Structure, analysis, sustainability, and electrochemistry. *Mater. Today* **2019**, *23*, 87–104. [CrossRef]
51. Xiao, L.; Lu, H.; Fang, Y.; Sushko, M.L.; Cao, Y.; Ai, X.; Yang, H.; Liu, J. Low-Defect and Low-Porosity Hard Carbon with High Coulombic Efficiency and High Capacity for Practical Sodium Ion Battery Anode. *Adv. Energy Mater.* **2018**, *8*, 1703238. [CrossRef]
52. Senthil, C.; Lee, C.W. Biomass-derived biochar materials as sustainable energy sources for electrochemical energy storage devices. *Renew. Sustain. Energy Rev.* **2021**, *137*, 110464. [CrossRef]
53. Wang, J.; Zhao, J.; He, X.; Qiao, Y.; Li, L.; Chou, S.-L. Hard carbon derived from hazelnut shell with facile HCl treatment as high-initial-coulombic-efficiency anode for sodium ion batteries. *Sustain. Mater. Technol.* **2022**, *33*, e00446. [CrossRef]
54. Kamiyama, A.; Kubota, K.; Igarashi, D.; Youn, Y.; Tateyama, Y.; Ando, H.; Gotoh, K.; Komaba, S. MgO-Template Synthesis of Extremely High Capacity Hard Carbon for Na-Ion Battery. *Angew. Chem. Int. Ed.* **2021**, *60*, 5114–5120. [CrossRef]
55. Zhou, S.; Tang, Z.; Pan, Z.; Huang, Y.; Zhao, L.; Zhang, X.; Sun, D.; Tang, Y.; Dhmees, A.S.; Wang, H. Regulating closed pore structure enables significantly improved sodium storage for hard carbon pyrolyzing at relatively low temperature. *SusMat* **2022**, *2*, 357–367. [CrossRef]
56. Man, Q.; Wei, C.; Tian, K.; Shen, H.; Zhang, X.; Bai, X.; Xi, B.; Xiong, S.; Feng, J. Molecular-Level Design of High Flash Point Solvents Enables High-Safety and Dual-Function Chemical Presodiation of Hard Carbon and Alloy Anodes for High-Performance Sodium-Ion Batteries. *Adv. Energy Mater.* **2024**, *14*, 2401016. [CrossRef]
57. Rudola, A.; Rennie, A.J.R.; Heap, R.; Meysami, S.S.; Lowbridge, A.; Mazzali, F.; Sayers, R.; Wright, C.J.; Barker, J. Commercialisation of high energy density sodium-ion batteries: Faradion's journey and outlook. *J. Mater. Chem. A* **2021**, *9*, 8279–8302. [CrossRef]
58. Schütte, M.; Laufen, H.; Luder, D.; Ditler, H.; Kern, J.; Klick, S.; Junker, M.; Stahl, G.; Frie, F.; Sauer, D.U. First full cell parameterization of a commercial layered oxide/hard carbon sodium-ion 18650 battery cell for a physico-chemical model. *J. Energy Storage* **2025**, *107*, 114931. [CrossRef]
59. Cheng, D.; Zhou, X.; Hu, H.; Li, Z.; Chen, J.; Miao, L.; Ye, X.; Zhang, H. Electrochemical storage mechanism of sodium in carbon materials: A study from soft carbon to hard carbon. *Carbon* **2021**, *182*, 758–769. [CrossRef]
60. Pendashteh, A.; Orayech, B.; Suhard, H.; Jauregui, M.; Ajuria, J.; Silván, B.; Clarke, S.; Bonilla, F.; Saurel, D. Boosting the performance of soft carbon negative electrode for high power Na-ion batteries and Li-ion capacitors through a rational strategy of structural and morphological manipulation. *Energy Storage Mater.* **2022**, *46*, 417–430. [CrossRef]
61. Web of Science. Available online: https://www.webofscience.com/wos/woscc/summary/868023bc-3c7a-4dab-bfc3-92c8f01b13f7-0136cef274/relevance/1 (accessed on 9 January 2024).

62. Zhang, L.; Wang, W.; Lu, S.; Xiang, Y. Carbon Anode Materials: A Detailed Comparison between Na-ion and K-ion Batteries. *Adv. Energy Mater.* **2021**, *11*, 2003640. [CrossRef]
63. Tiwari, S.K.; Pandey, S.K.; Pandey, R.; Wang, N.; Bystrzejewski, M.; Mishra, Y.K.; Zhu, Y. Stone–Wales Defect in Graphene. *Small* **2023**, *19*, 2303340. [CrossRef] [PubMed]
64. Lin, Q.; Zhang, J.; Lv, W.; Ma, J.; He, Y.; Kang, F.; Yang, Q.-H. A Functionalized Carbon Surface for High-Performance Sodium-Ion Storage. *Small* **2020**, *16*, 1902603. [CrossRef]
65. Ding, J.; Wang, H.; Li, Z.; Kohandehghan, A.; Cui, K.; Xu, Z.; Zahiri, B.; Tan, X.; Lotfabad, E.M.; Olsen, B.C.; et al. Carbon Nanosheet Frameworks Derived from Peat Moss as High Performance Sodium Ion Battery Anodes. *ACS Nano* **2013**, *7*, 11004–11015. [CrossRef]
66. Wang, H.; Bai, W.; Wang, H.; Kong, D.; Xu, T.; Zhang, Z.; Zang, J.; Wang, X.; Zhang, S.; Tian, Y.; et al. 3D printed Au/rGO microlattice host for dendrite-free sodium metal anode. *Energy Storage Mater.* **2023**, *55*, 631–641. [CrossRef]
67. Wang, H.; Xu, D.; Qiu, R.; Tang, S.; Li, S.; Wang, R.; He, B.; Gong, Y.; Fan, H.J. Aligned Arrays of Na 2 Ti 3 O 7 Nanobelts and Nanowires on Carbon Nanofiber as High-Rate and Long-Cycling Anodes for Sodium-Ion Hybrid Capacitors. *Small Struct.* **2021**, *2*, 2000073. [CrossRef]
68. Costa, S.I.R.; Choi, Y.-S.; Fielding, A.J.; Naylor, A.J.; Griffin, J.M.; Sofer, Z.; Scanlon, D.O.; Tapia-Ruiz, N. Surface Engineering Strategy Using Urea to Improve the Rate Performance of $Na_2Ti_3O_7$ in Na-Ion Batteries. *Chem. Eur. J.* **2021**, *27*, 3875–3886. [CrossRef] [PubMed]
69. Zhong, W.; Tao, M.; Tang, W.; Gao, W.; Yang, T.; Zhang, Y.; Zhan, R.; Bao, S.-J.; Xu, M. MXene-derivative pompon-like $Na_2Ti_3O_7$@C anode material for advanced sodium ion batteries. *Chem. Eng. J.* **2019**, *378*, 122209. [CrossRef]
70. Rudola, A.; Saravanan, K.; Mason, C.W.; Balaya, P. $Na_2Ti_3O_7$: An intercalation based anode for sodium-ion battery applications. *J. Mater. Chem. A* **2013**, *1*, 2653–2662. [CrossRef]
71. Tomboc, G.M.; Wang, Y.; Wang, H.; Li, J.; Lee, K. Sn-based metal oxides and sulfides anode materials for Na ion battery. *Energy Storage Mater.* **2021**, *39*, 21–44. [CrossRef]
72. Li, H.; He, Y.; Li, X.; Yu, J.; Sun, X.; Gao, T.; Zhou, G. Pomegranate-like Sn-Ni nanoalloys@N-doped carbon nanocomposites as high-performance anode materials for Li-ion and Na-ion batteries. *Appl. Surf. Sci.* **2023**, *611*, 430–462. [CrossRef]
73. Yu, D.-K.; Park, C.-M. Sb-based intermetallics and nanocomposites as stable and fast Na-ion battery anodes. *Chem. Eng. J.* **2021**, *409*, 127380. [CrossRef]
74. Pandit, B.; Sougrati, M.T.; Fraisse, B.; Monconduit, L. Exploration of a $Na_3V_2(PO_4)_3$/C–Pb full cell Na-ion prototype. *Nano Energy* **2022**, *95*, 107010. [CrossRef]
75. Bai, Y.; Muralidharan, N.; Sun, Y.-K.; Passerini, S.; Whittingham, S.; Belharouak, I. Energy and environmental aspects in recycling lithium-ion batteries: Concept of Battery Identity Global Passport. *Mater. Today* **2020**, *41*, 304–315. [CrossRef]
76. Tang, J.; Peng, X.; Lin, T.; Huang, X.; Luo, B.; Wang, L. Confining ultrafine tin monophosphide in Ti3C2Tx interlayers for rapid and stable sodium ion storage. *EScience* **2021**, *1*, 203–211. [CrossRef]
77. Wang, T.; Yao, K.; Hua, Y.; Shankar, E.G.; Shanthappa, R.; Yu, J.S. Rational design of MXene-MoS2 heterostructure with rapid ion transport rate as an advanced anode for sodium-ion batteries. *Chem. Eng. J.* **2023**, *457*, 141363. [CrossRef]
78. Zhao, C.; Ding, F.; Lu, Y.; Chen, L.; Hu, Y.-S. High-Entropy Layered Oxide Cathodes for Sodium-Ion Batteries. *Angew. Chem. Int. Ed.* **2020**, *59*, 264–269. [CrossRef] [PubMed]
79. Mhaske, V.P.; Jilkar, S.; Yadav, M.D. Minireview on Layered Transition Metal Oxides Synthesis Using Coprecipitation for Sodium Ion Batteries Cathode Material: Advances and Perspectives. *Energy Fuels* **2023**, *37*, 16221–16244. [CrossRef]
80. Zhao, C.; Wang, Q.; Yao, Z.; Wang, J.; Sánchez-Lengeling, B.; Ding, F.; Qi, X.; Lu, Y.; Bai, X.; Li, B.; et al. Rational design of layered oxide materials for sodium-ion batteries. *Science* **2020**, *370*, 708–711. [CrossRef] [PubMed]
81. Guo, J.-Z.; Gu, Z.-Y.; Du, M.; Zhao, X.-X.; Wang, X.-T.; Wu, X.-L. Emerging characterization techniques for delving polyanion-type cathode materials of sodium-ion batteries. *Mater. Today* **2023**, *66*, 221–244. [CrossRef]
82. Kate, R.S.; Jadhav, H.S.; Chothe, U.P.; Bhattacharjee, K.; Kulkarni, M.V.; Deokate, R.J.; Kale, B.B.; Kalubarme, R.S. Critical review of the recent progress and challenges of polyanion Na_3V_2 $(PO_4)_3$ cathode materials in rechargeable sodium-ion batteries. *J. Mater. Chem. A* **2024**, *12*, 7418–7451. [CrossRef]
83. Zhao, Y.; Gao, X.; Gao, H.; Dolocan, A.; Goodenough, J.B. Elevating Energy Density for Sodium-Ion Batteries through Multielectron Reactions. *Nano Lett.* **2021**, *21*, 2281–2287. [CrossRef]
84. Hu, P.; Zhu, T.; Cai, C.; Wang, X.; Zhang, L.; Mai, L.; Zhou, L. A High-Energy NASICON-Type $Na_{3.2}MnTi_{0.8}V_{0.2}(PO_4)_3$ Cathode Material with Reversible 3.2-Electron Redox Reaction for Sodium-Ion Batteries. *Angew. Chem. Int. Ed.* **2023**, *62*, e202219304. [CrossRef]
85. Gu, Z.-Y.; Guo, J.-Z.; Sun, Z.-H.; Zhao, X.-X.; Wang, X.-T.; Liang, H.-J.; Zhao, B.; Li, W.-H.; Pan, X.-M.; Wu, X.-L. Aliovalent-Ion-Induced Lattice Regulation Based on Charge Balance Theory: Advanced Fluorophosphate Cathode for Sodium-Ion Full Batteries. *Small* **2021**, *17*, 2102010. [CrossRef]

86. Zhuang, S.-H.; Yang, C.-C.; Zheng, M.; Rengapillai, S.; Marimuthu, S.; Chiang, Y.-S.; Chang, B.K.; Huang, C.-H.; Liu, W.-R. A combined first principles and experimental study on Al-doped $Na_3V_2(PO_4)_2F_3$ cathode for rechargeable Na batteries. *Surf. Coat. Technol.* **2022**, *434*, 128184. [CrossRef]
87. Plewa, A.; Kulka, A.; Hanc, E.; Zając, W.; Sun, J.; Lu, L.; Molenda, J. Facile aqueous synthesis of high performance $Na_2FeM(SO_4)_3$ (M = Fe, Mn, Ni) alluaudites for low cost Na-ion batteries. *J. Mater. Chem. A* **2020**, *8*, 2728–2740. [CrossRef]
88. Du, G.; Tao, M.; Li, J.; Yang, T.; Gao, W.; Deng, J.; Qi, Y.; Bao, S.-J.; Xu, M. Low-Operating Temperature, High-Rate and Durable Solid-State Sodium-Ion Battery Based on Polymer Electrolyte and Prussian Blue Cathode. *Adv. Energy Mater.* **2020**, *10*, 1903351. [CrossRef]
89. Wang, C.-C.; Zhang, L.-L.; Fu, X.-Y.; Sun, H.-B.; Yang, X.-L. Hollow Layered Iron-Based Prussian Blue Cathode with Reduced Defects for High-Performance Sodium-Ion Batteries. *ACS Appl. Mater. Interfaces* **2024**, *16*, 18959–18970. [CrossRef]
90. Zhang, J.-N.; Li, Q.; Ouyang, C.; Yu, X.; Ge, M.; Huang, X.; Hu, E.; Ma, C.; Li, S.; Xiao, R.; et al. Trace doping of multiple elements enables stable battery cycling of $LiCoO_2$ at 4.6 V. *Nat. Energy* **2019**, *4*, 594–603. [CrossRef]
91. Altundag, S.; Altin, S.; Yaşar, S.; Altin, E. Improved performance of the $NaFePO_4$/Hardcarbon sodium-ion full cell. *Vacuum* **2023**, *210*, 111853. [CrossRef]
92. Xu, J.; Zhu, S.; Xu, Z.; Zhu, H. The basic physical properties of Li_2MnO_3 and $LiMn_2O_4$ cathode materials. *Comput. Mater. Sci.* **2023**, *229*, 112426. [CrossRef]
93. Zhao, D.; Lu, Y.; Ma, D. Effects of Structure and Constituent of Prussian Blue Analogs on Their Application in Oxygen Evolution Reaction. *Molecules* **2020**, *25*, 2304. [CrossRef]
94. Chen, Z.-Y.; Fu, X.-Y.; Zhang, L.-L.; Yan, B.; Yang, X.-L. High-Performance Fe-Based Prussian Blue Cathode Material for Enhancing the Activity of Low-Spin Fe by Cu Doping. *ACS Appl. Mater. Interfaces* **2022**, *14*, 5506–5513. [CrossRef] [PubMed]
95. He, M.; Davis, R.; Chartouni, D.; Johnson, M.; Abplanalp, M.; Troendle, P.; Suetterlin, R.-P. Assessment of the first commercial Prussian blue based sodium-ion battery. *J. Power Sources* **2022**, *548*, 232036. [CrossRef]
96. Bauer, A.; Song, J.; Vail, S.; Pan, W.; Barker, J.; Lu, Y. The Scale-up and Commercialization of Nonaqueous Na-Ion Battery Technologies. *Adv. Energy Mater.* **2018**, *8*, 1702869. [CrossRef]
97. Kuan, H.-C.; Luu, N.T.H.; Ivanov, A.S.; Chen, T.-H.; Popovs, I.; Lee, J.-C.; Kaveevivitchai, W. A nitrogen- and carbonyl-rich conjugated small-molecule organic cathode for high-performance sodium-ion batteries. *J. Mater. Chem. A* **2022**, *10*, 16249–16257. [CrossRef]
98. Wang, L.; Liu, N.; Zhao, X.; Wang, X.; Zhang, T.; Luo, Z.; Li, F. Copper and conjugated carbonyls of metal–organic polymers as dual redox centers for Na storage. *Chem. Sci.* **2024**, *15*, 2133–2140. [CrossRef] [PubMed]
99. Du, K.; Wang, C.; Subasinghe, L.U.; Gajella, S.R.; Law, M.; Rudola, A.; Balaya, P. A comprehensive study on the electrolyte, anode and cathode for developing commercial type non-flammable sodium-ion battery. *Energy Storage Mater.* **2020**, *29*, 287–299. [CrossRef]
100. Rong, J.-Z.; Cai, T.-X.; Bai, Y.-Z.; Zhao, X.; Wu, T.; Wu, Y.-K.; Zhao, W.; Dong, W.-J.; Xu, S.-M.; Chen, J.; et al. A free-sealed high-voltage aqueous polymeric sodium battery enabling operation at −25 °C. *Cell Rep. Phys. Sci.* **2022**, *3*, 100805. [CrossRef]
101. Janakiraman, S.; Khalifa, M.; Biswal, R.; Ghosh, S.; Anandhan, S.; Venimadhav, A. High performance electrospun nanofiber coated polypropylene membrane as a separator for sodium ion batteries. *J. Power Sources* **2020**, *460*, 228060. [CrossRef]
102. Zhang, L.; Li, X.; Yang, M.; Chen, W. High-safety separators for lithium-ion batteries and sodium-ion batteries: Advances and perspective. *Energy Storage Mater.* **2021**, *41*, 522–545. [CrossRef]
103. Liu, H.; Zheng, X.; Du, Y.; Borrás, M.C.; Wu, K.; Konstantinov, K.; Pang, W.K.; Chou, S.; Liu, H.; Dou, S.; et al. Multifunctional Separator Enables High-Performance Sodium Metal Batteries in Carbonate-Based Electrolytes. *Adv. Mater.* **2024**, *36*, 2307645. [CrossRef]
104. Song, Y.; Sheng, L.; Wang, L.; Xu, H.; He, X. From separator to membrane: Separators can function more in lithium ion batteries. *Electrochem. Commun.* **2021**, *124*, 106948. [CrossRef]
105. Zhou, D.; Tang, X.; Guo, X.; Li, P.; Shanmukaraj, D.; Liu, H.; Gao, X.; Wang, Y.; Rojo, T.; Armand, M.; et al. Polyolefin-Based Janus Separator for Rechargeable Sodium Batteries. *Angew. Chem. Int. Ed.* **2020**, *59*, 16725–16734. [CrossRef]
106. Muddasar, M.; Beaucamp, A.; Culebras, M.; Collins, M.N. Cellulose: Characteristics and applications for rechargeable batteries. *Int. J. Biol. Macromol.* **2022**, *219*, 788–803. [CrossRef] [PubMed]
107. Saltwater Battery: The Safest Long-Term Storage Solution. Available online: https://www.aquionenergy.com/technology/aqueous-hybrid-ion-ahi/ (accessed on 9 January 2024).
108. Li, R.-R.; Yang, Z.; He, X.-X.; Liu, X.-H.; Zhang, H.; Gao, Y.; Qiao, Y.; Li, L.; Chou, S.-L. Binders for sodium-ion batteries: Progress, challenges and strategies. *Chem. Commun.* **2021**, *57*, 12406–12416. [CrossRef] [PubMed]
109. Larsson, F.; Andersson, P.; Blomqvist, P.; Mellander, B.-E. Toxic fluoride gas emissions from lithium-ion battery fires. *Sci. Rep.* **2017**, *7*, 10018. [CrossRef]

batteries

MDPI

Article

Comprehensive Investigation of the Durability of Lithium-Ion Batteries Under Frequency Regulation Conditions

Yuxin Tian [1,2,3], Liye Wang [1,2,3,*], Chenglin Liao [1,2,3] and Guifu Yan [1,2,3]

1 Key Laboratory of Power Electronics and Electric Drives, Institute of Electrical Engineering, Chinese Academy of Sciences, Beijing 100190, China; tianyuxin@mail.iee.ac.cn (Y.T.); liaocl@mail.iee.ac.cn (C.L.); yanguifu@mail.iee.ac.cn (G.Y.)
2 Institute of Electrical Engineering, Chinese Academy of Sciences, Beijing 100190, China
3 University of Chinese Academy of Sciences, Beijing 101408, China
* Correspondence: wangliye@mail.iee.ac.cn

Abstract: Due to the large-scale use of renewable energy generation and its lack of inertia, the frequency of the grid is extremely unstable. At the same time, with the vigorous development of new energy vehicles, large-scale power batteries have huge potential for renewable energy consumption. In this context, the Vehicle-to-Grid (V2G) method is proposed. Electric vehicles are used as energy storage systems to provide frequency regulation services as flexible power grid resources. However, when electric vehicles are invested in large-scale frequency regulation, their own power battery durability will also be affected. Based on this problem, the pseudo-two-dimensions (P2D) model of the battery was established in this paper, and the effects of temperature, state of charge (SOC), reported power, and frequency regulation conditions on battery capacity attenuation and negative potential distribution were explored through experiments and simulations.

Keywords: pseudo-two-dimensions (P2D) model; state of charge (SOC); negative potential distribution; battery durability

1. Introduction

The large-scale investment in renewable energy makes the modern power system shift from the system based on rotary generators to the system based on inverters. Although this is very beneficial for the collection and utilization of renewable energy, because the power generation through the inverter does not provide enough mechanical inertia, it leads to insufficient inertia in the power system [1]. Because the moment of inertia has the ability to hinder the oscillation of the system and the frequency change speed is related to the moment of inertia, when the inertia of the system is insufficient, no matter what degree of power disturbance, it will lead to significant frequency fading and finally the frequency instability problem. At present, the main method to solve this problem is to add some virtual inertia into the system and deploy it appropriately to provide enough inertial response so as to enhance the stability of the system.

When the new energy power generation system is put into the industry on a large scale, considering the volatility characteristics of photovoltaic and wind power generation, the battery energy storage system will play a crucial role in the microgrid to maintain the continuous power supply for electric vehicle charging [2]. If the electric vehicle is charged without order in the power grid, it will cause the power grid load fluctuation to increase; that is, the peak value of the power grid fluctuation curve will increase. If the electric vehicle is charged orderly in the grid, that is, the V2G method, it will make the grid load "cut peak and fill valley". This method is widely used to adjust the frequency stability of the power grid, that is, through the power grid to regulate the charging and discharging power of electric vehicles, to provide auxiliary services for the power grid such as peak regulation, frequency regulation, and promotion of new energy consumption.

Although the power battery is beneficial to the stability of the power grid in the process of frequency modulation, its own life will also be affected to a certain extent. In the large-scale application of V2G technology, the study of battery durability is the core scientific issue of system design and economic evaluation. Because the V2G bidirectional charge and discharge mode can significantly change the operating boundary conditions of the battery, the asymmetric cycle condition caused by it will cause the electrochemical system to be subjected to multi-dimensional stress coupling. The reciprocating insertion/withdrawal of lithium ions between positive and negative electrodes will aggravate the lattice distortion of electrode materials and the dynamic reconstruction of the solid electrolyte interface (SEI) film. Meanwhile, frequent shallow charging and discharge cycles may lead to continuous loss of active lithium in the local state of charge (SOC) region, accelerating capacity decay. The random power fluctuation introduced by the grid dispatching demand will deteriorate the thermodynamic state of the battery, resulting in the accumulation of mechanical stress induced by temperature gradient. These multi-physics coupling degradation mechanisms will not only shorten the battery service life but also directly affect the energy handling capacity of the V2G system and the reliability of grid auxiliary services. Therefore, to establish a battery durability prediction model based on a cyclic attenuation mechanism, it is necessary to focus on quantifying the nonlinear relationship between capacity fade rate and internal resistance rise under dynamic working conditions. By constructing a state of health (SOH) prediction framework considering the synergistic effect of calendar aging and cycle aging, the life optimization design of the V2G charging and discharging strategy is realized. These will provide theoretical support for V2G's participation in the power market for battery residual value assessment, energy storage capacity allocation optimization, and life-cycle carbon footprint calculation, and ultimately achieve the dual goals of sustainable utilization of battery assets and improvement of grid elasticity in V2G technology. At present, determining the aging and degradation mechanism of the battery is the most important and most challenging goal, and the interaction of multiple factors in the use environment or use mode makes the battery produce different aging effects; that is, the capacity loss and the increase in internal resistance do not depend on the same variables, so these processes are not single but complex. This makes understanding battery aging a difficult task [3]. Over the years, many studies have attempted to explore the effects of battery aging more deeply.

Li Y. et al. [4] took electric vehicles as reactive power compensation devices to participate in voltage regulation of the power grid. The research results show that with the increase in the number of electric vehicles connected to the grid, electric vehicles can assist or even replace traditional reactive power compensation devices and obtain better voltage control performance. In order to alleviate the huge impact of disorderly charging of large-scale electric vehicles on the distribution network. Considering that the reduction in battery life will directly affect the economic benefits and sustainability of the V2G system, the battery capacity protection research usually quantifies the battery aging generated during the V2G process by different means and introduces the objective function. There are also some studies that limit battery aging to an acceptable range by adding constraints. Li S. et al. [5] pointed out that traditional methods for quantifying battery aging, such as electrochemical models and artificial intelligence algorithms, can only quantify battery degradation on a large time scale. Ebrahimi M. et al. [6] considered both cycle aging and calendar aging of batteries in the V2G process, but they believed that calendar aging only had a linear relationship with time.

Because lithium-ion batteries contain a variety of different chemical materials, the internal structure is very complex, and these materials will produce a variety of complex side reactions during the battery charging and discharging process, resulting in diversified aging mechanisms. Different aging modes will affect each other, and the same aging mechanism may lead to multiple aging modes. Battery temperature, SOC, charge/discharge ratio, depth of discharge, and cycle times are the main factors affecting battery aging [7]. Therefore, in order to study the durability of batteries, it is necessary to establish a perfect battery aging model. There are many kinds of battery aging models, and the accuracy, calculation amount, and application range of different models are different. At present, electrochemical models, empirical models, and semi-empirical models are commonly used in mainstream research. In the 1990s, Newman's group tried to estimate battery performance based on the Butler–Volmer equation and the theory of porous electrodes. Darling and Newman [8] for the first time tried to simulate the aging process in an electrochemical model and realized a simple solvent oxidation reaction, obtained good results, and gave a good conclusion on the selection of solvents. Dalverny et al. [9] determined the morphology of the $LiCoO_2$ electrode by ab initio calculation. In addition, the phase transition process of lithium iron phosphate (LFP) active particles was also theoretically studied. Ab initio calculations can explain battery aging by estimating the energy associated with solvent decomposition or dissolution of lithium components. Tasaki et al. [10] have linked the dissolution of lithium salts near the negative solid electrolyte interface (SEI) to calendar aging. As for empirical models, the aging model based on Arrhenius is the most widely used. Based on the empirical formula of the relationship between chemical reaction rate constant and temperature, the model can model cyclic aging and calendar aging, respectively, when various stress factors are comprehensively considered [7]. In recent years, with the development of artificial intelligence technology, the battery aging model based on a neural network can accurately reflect the relationship between various factors and capacity decay without establishing complicated mathematical expressions and updating network parameters through training [11]. The semi-empirical aging model not only has the simplicity and ease of use of the empirical model but also makes full use of the basic knowledge of the physical and chemical characteristics of the battery, which is very suitable for offline analysis. The Arrhenius formula is used to correct the growth of the SEI film and electrolyte loss during the initial use of the battery in the electrochemical model, which improved the accuracy of the model under different environments [12]. Based on the hypothesis that SEI film growth leads to battery degradation, Ecker M. et al. [13] combined the electrothermal model with empirical mathematical expressions to establish a semi-empirical aging model that can analyze battery degradation under different electrothermal environments. The lithium evolution behavior of lithium-ion batteries is one of the key factors leading to battery life decay, and the lithium evolution situation of batteries is mainly reflected by the negative electrode position of the battery, so this study will mainly focus on the analysis of the negative electrode potential distribution of lithium-ion batteries under the frequency modulation condition.

Through the above findings, the existing models, mostly based on laboratory-accelerated aging data, fail to effectively reflect the irregular charge–discharge characteristics brought by the grid demand response in the actual V2G working conditions. Secondly, the current economic analysis generally adopts a linear attenuation model, ignoring key nonlinear factors such as temperature change, charge and discharge rate, SOC, discharge depth heterogeneity, and calendar aging coupling effect. Third, most studies focus on battery degradation but do not establish a dynamic optimization model between battery degradation and power grid frequency fluctuations.

The goal of this study is to minimize the attenuation of lithium-ion batteries and maximize the benefits of the grid while controlling the frequency stability of the grid. Firstly, the typical frequency regulation conditions are obtained in the American PJM market. Secondly, a P2D model related to lithium-ion batteries was built by studying the power performance and durability mechanism of lithium-ion batteries, and the above conditions were input into the model to investigate the distribution of the model's negative potential under different SOC, temperature, and reported power, and the results were analyzed to find out the influence of different factors on the battery. The research route is shown in Figure 1. In the future, the frequency regulation model will be built according to the relevant knowledge of the power system, and the P2D model will be built according to the above results, and the two will be coupled. The coupling model is optimized to minimize the negative impact on the negative battery potential. At the same time, it is also necessary to consider that electric vehicles participate in the frequency regulation of the grid, and the grid gains the most benefits.

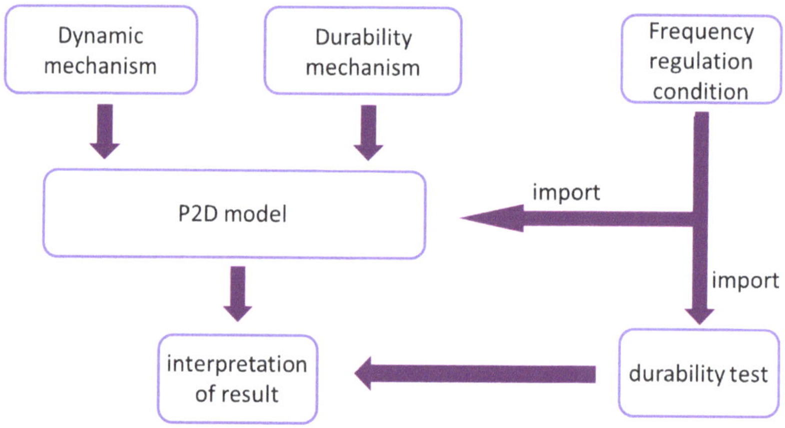

Figure 1. Research framework.

2. P2D Electrochemical Model Development

Based on the theory of porous electrodes, the battery P2D model describes the mass transfer, dynamics, thermodynamics, and other processes inside the battery through a set of partial differential equations. On the basis of the basic composition, the P2D model has the following three core assumptions: the electrode material is composed of spherical particles; the double-layer effect is not considered; and the positive and negative collector conductivities are very high, so the collector does not change significantly in the y- and z-axes; in other words, the electrochemical reaction dynamics only work in the x-axis. These three core assumptions simplify the battery model building process at three levels. The regular porous structure of spherical particles avoids the complex structure and particle distribution of active substances in practice, which is the most basic physical simplification method. Secondly, avoiding the double-layer effect can greatly simplify the distribution of ions on the electrolyte and electrode surface. Limiting the electrochemical dynamics to the x-axis further facilitates the mathematical processing. The internal structure of the battery is roughly shown in Figure 2. To study a battery from a mechanism point of view, it is necessary to investigate its internal material distribution and potential distribution, corresponding to the material transfer and charge transfer within the battery, respectively.

Therefore, in order to quantitatively describe the P2D model, it is necessary to establish the solid–liquid phase potential distribution, solid–liquid-phase diffusion process, electrochemical reaction model of interfacial reaction, and charge conservation as constraints. The following process is shown.

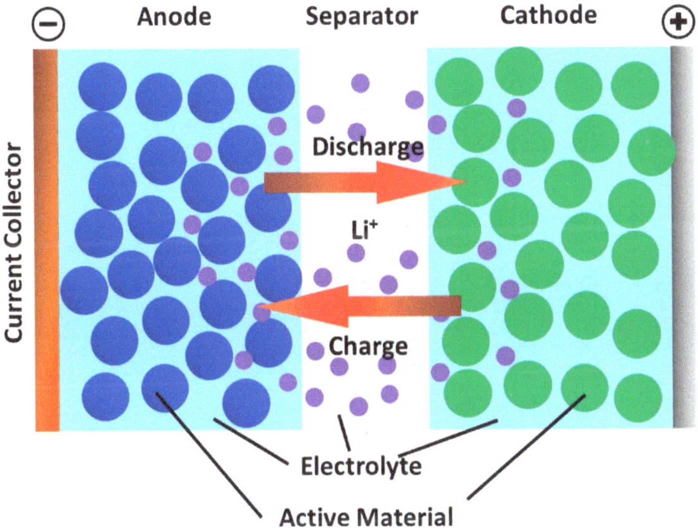

Figure 2. Battery internal structure diagram.

Solid-phase potential distribution: According to the conductivity of the electrode material, the relationship between current and potential gradient can be expressed (Equation (1)) as follows:

$$-\sigma_s^{eff}\frac{\partial\varphi_s}{\partial x}=i_s \tag{1}$$

where σ_s^{eff} is a solid-phase effective conductivity; φ_s is the potential for solid phase; and i_s is the current density. The minus sign in the formula is because the x direction takes the negative extreme as the origin during modeling, and the field strength and current direction are opposite; that is, the current inside the battery flows from the negative electrode to the positive electrode, while the field strength points from the positive electrode to the negative electrode. The boundary conditions for this equation are expressed as Equations (2) and (3).

$$\sigma_s^{eff}\frac{\partial\varphi_s}{\partial x}\bigg|_{x=0}=-\sigma_s^{eff}\frac{\partial\varphi_s}{\partial x}\bigg|_{x=L_{an}+L_{sep}+L_{ca}}=i_{app} \tag{2}$$

$$\frac{\partial\varphi_s}{\partial x}\bigg|_{x=L_{an}}=\frac{\partial\varphi_s}{\partial x}\bigg|_{x=L_{an}+L_{sep}}=0 \tag{3}$$

Liquid-phase potential distribution: According to the liquid-phase conductivity and particle-specific surface area, the relationship between the potential gradient and lithium-ion concentration, reaction rate, temperature, and current can be expressed through Equation (4).

$$\frac{\partial}{\partial x}\left(\sigma_e^{eff}\frac{\partial\varphi_e}{\partial x}\right)=-aFj_n+\frac{2RT\left(1-t_+^0\right)}{F}\frac{\partial}{\partial x}\left(\sigma_e^{eff}\frac{\partial \ln c_e}{\partial x}\right) \tag{4}$$

where σ_e^{eff} is the liquid effective conductivity; φ_e is the potential for the liquid phase; j_n is the solid–liquid junction chemical reaction rate per unit area; a is the particle-specific surface area; F is the Faraday constant; t_+^0 is the lithium-ion transference number; R is the state of the ideal gas constant; T is the battery temperature; and c_e is the liquid lithium-ion concentration. Integrating x on both sides of this equation yields the following equation (Equation (6)), where the boundary condition is shown.

$$\sigma_e^{eff}\frac{\partial \varphi_e}{\partial x} = -i_s + \frac{2RT\left(1 - t_+^0\right)}{F}\sigma_e^{eff}\frac{\partial \ln c_e}{\partial x} \tag{5}$$

$$\left.\frac{\partial \varphi_e}{\partial x}\right|_{x=0} = \left.\frac{\partial \varphi_e}{\partial x}\right|_{x=L_{an}+L_{sep}+L_{ca}} = 0 \tag{6}$$

Solid-phase ion diffusion: According to the solid-phase diffusion coefficient, the distribution of lithium-ion concentration over time in the polar diameter direction can be obtained via the following equation:

$$\frac{\partial c_s}{\partial t} = \frac{1}{r^2}\frac{\partial}{\partial r}\left(D_s r^2 \frac{\partial c_s}{\partial r}\right) \tag{7}$$

where c_s is the solid lithium-ion concentration; D_s is the solid-phase diffusion coefficient; t is time; and r is the pole size. This equation is a polar form of Fick's second law and represents the diffusion process of matter driven only by concentration. In the P2D model, the lithium ions of the porous electrode are removed or generated from the surface of active particles, so the lithium ions inside active particles will appear as a concentration gradient and will only be driven by it; therefore, the diffusion process of lithium ions in the solid particles can be described by this model. The boundary conditions are shown as Equations (8) and (9).

$$-D_s \left.\frac{\partial c_s}{\partial r}\right|_{r=0} = 0 \tag{8}$$

$$-D_s \left.\frac{\partial c_s}{\partial r}\right|_{r=R_s} = j_n \tag{9}$$

where R_s is the grain boundary radius, and j_n is the molar flow rate for ions.

Liquid-phase ion diffusion: Based on the diffusion coefficient and porosity of the liquid phase, and combined with the reaction process, the distribution of lithium-ion concentration in the liquid phase with time can be obtained via the following expression:

$$\varepsilon_e \frac{\partial c_e}{\partial t} = \frac{\partial}{\partial x}\left(D_e^{eff}\frac{\partial c_e}{\partial x}\right) + \left(1 - t_+^0\right)aj_n \tag{10}$$

where c_e is the liquid lithium-ion concentration; D_e^{eff} is the liquid-phase diffusion coefficient; ε_e is the electrode porosity; t is time; t_+^0 is the lithium-ion transference number; j_n is the solid–liquid junction chemical reaction rate per unit area; and a is the particle-specific surface area. This is a revision of Fick's second law, where lithium ions migrate from the negative active particles into the electrolyte diffusion, and eventually become embedded in the positive active particles. There is a lithium-ion concentration gradient in the battery internal electrolyte; this concentration gradient is driven by lithium-ion diffusion, and, at the same time, by the internal electric field and migration. The boundary conditions are shown in Equation (11).

$$\left.\frac{\partial c_e}{\partial t}\right|_{x=0} = \left.\frac{\partial c_e}{\partial t}\right|_{x=L_{an}+L_{sep}+L_{ca}} = 0 \tag{11}$$

Solid-charge conservation: As a constraint to the P2D model, it describes that the number of disembedded or generated lithium ions during the reaction is equal to the number of charge transfer, as shown in Equation (12).

$$-\frac{\partial i_s}{\partial x} = \sigma_s^{eff}\frac{\partial^2 \varphi_s}{\partial x^2} = aFj_n \tag{12}$$

where j_n is the solid–liquid junction chemical reaction rate per unit area; a is the particle-specific surface area; F is the Faraday constant; σ_s^{eff} is the solid-phase effective conductivity; φ_s is the potential for solid phase; and i_s is the current density. Boundary conditions are expressed via Equation (13).

$$\left.\frac{\partial \varphi_s}{\partial x}\right|_{x=L_{an}} = \left.\frac{\partial \varphi_s}{\partial x}\right|_{x=L_{an}+L_{sep}} = 0 \tag{13}$$

Solid–liquid interface reaction: According to the Bulter–Volmer equation of the electrochemical reaction of lithium ions during deintercalation and formation at the solid–liquid interface, changes in the current of the electrode and changes in the electrode potential can be described through Equation (14).

$$j_n = i_0\left[\exp\left(\frac{\alpha_{an}F}{RT}\eta_{act}\right) - \exp\left(-\frac{\alpha_{ca}F}{RT}\eta_{act}\right)\right] \tag{14}$$

where j_n is the current density; i_0 is the exchange-current density; α_{an} and α_{ca} are migration coefficients of anions and cations, respectively; η_{act} is the electrode potential; R is the state of the ideal gas constant; and T is the battery temperature.

After the mathematical foundation of the P2D model is completed, parameter identification of the research object is needed to further calibrate the parameters required by the P2D model. Finally, a model was built in the COMSOL Multiphysics® ver5.6 simulation software. The model is shown in Figure 3. To achieve subsequent measurements, 20 probes were selected at an interval of 1×10^{-5} m and numbered 120 from left to right.

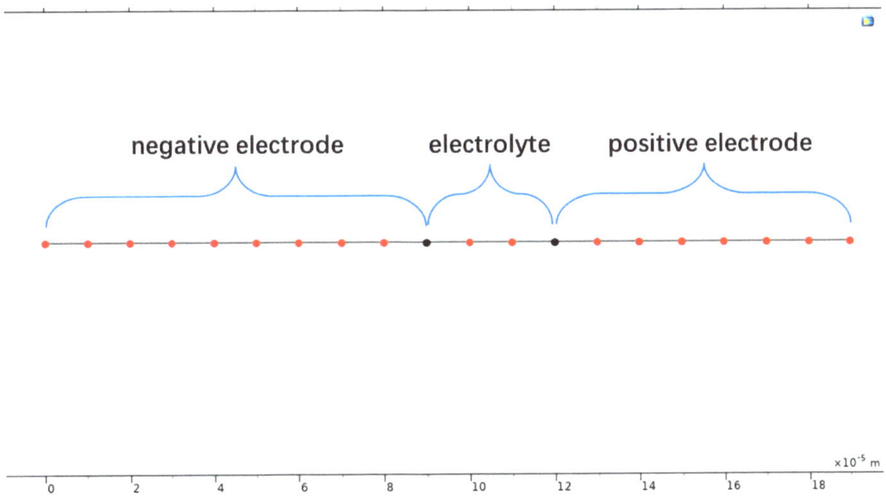

Figure 3. P2D model in COMSOL.

3. Experimental Setup

3.1. Durability Test Design

In this study, a 51 Ah ternary aluminum shell power lithium battery is used, and its nominal voltage is 3.7 V. The battery is suitable for a charging operating temperature of 20 °C to 50 °C, a discharge operating temperature from −20 °C to 50 °C, a charging limit voltage of 4.2 V, a discharge cut-off voltage of 2.7 V, and a main size of 148 × 27 × 92 mm.

In the battery durability test, the calendar aging of the battery is often ignored, and only cyclic aging is considered; that is, the battery is continuously charged and discharged, and the loss of battery capacity is investigated when a certain number of cycles are reached. In this study, the durability of the battery is explored under the frequency modulation condition, which can be regarded as the superposition of different pulse conditions, as shown in Figure 4. The following four test variables were selected for this test: SOC, half-pulse period, power, and temperature. The corresponding relationship between specific variables and levels is shown in Table 1.

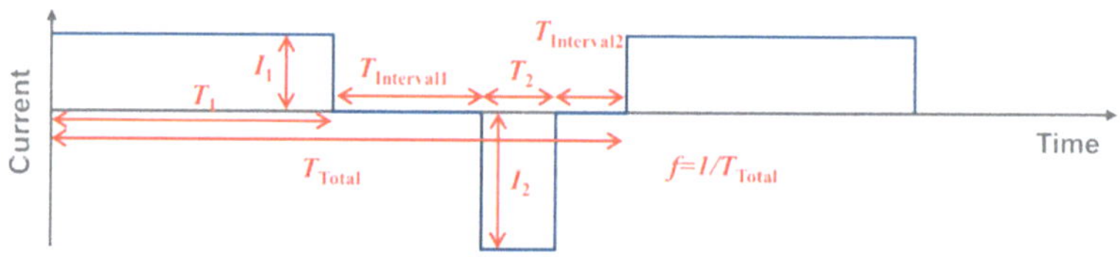

Figure 4. The superposition of different pulse conditions.

Table 1. Durability test variables.

Level	Factor 1	Factor 2	Factor 3	Factor 4
	SOC	Half-Pulse Period (s)	Power	Temperature (°C)
1	0.2	2	0.1 C*V	0
2	0.5	150	0.5 C*V	25
3	0.8	300	1 C*V	45

According to the factors and levels given in this table, a total of 81 sets of tests are required, covering all factor combinations. However, the durability test is a time-consuming test with limited equipment. In order to solve the problem of excessive number of tests and to cover the influence of various factors on the test results, the orthogonal test method was adopted in this study to simplify the number of experimental groups. This method has the characteristics of uniform dispersion and neat comparability. The final orthogonal test factor combination is shown in Table 2, with a total of 9 groups of tests. In this study, 9 batteries were tested and numbered, the same as the orthogonal test numbers; each cell corresponds to a set of factor combinations, and its capacity attenuation was measured after several cycles.

Each durability test consists of approximately 18 h of capacity testing and 50 h of bidirectional pulse tests. The capacity test mainly includes the following processes: the battery is placed in a 25 °C incubator Ito fully charge; constant discharge is set at 17 A to 2.5 V, let stand; charge is set at 17 A; constant current is set to 4.2 V to fully charge and stand; and repeat the process 3 to 5 times. The current capacity of lithium-ion battery can be obtained by calculating the average of multiple discharge capacities by the ampere–hour integration method. After the battery capacity is measured, the battery is tested in a bidirectional pulse condition according to the combination of test factors; each test lasts about 50 h, and the capacity test is carried out after each bidirectional pulse condition test, i.e., a durability test. Three rounds of durability tests were carried out in this study.

Table 2. Durability test orthogonal table.

No.	Factor 1	Factor 2	Factor 3	Factor 4
1	1	1	1	1
2	1	2	3	2
3	1	3	2	3
4	2	1	3	3
5	2	2	2	1
6	2	3	1	2
7	3	1	2	2
8	3	2	1	3
9	3	3	3	1

3.2. *Negative Potential Simulation Test Design*

In order to explore the influence of multiple factors on the negative battery potential, the following four factors were selected: SOC, temperature, reported power, and frequency regulation condition. Each factor is selected according to its characteristics that meet the actual needs of several levels. Reported power refers to the maximum capacity of the battery allowed to participate in frequency regulation, calculated by the product of the maximum current and the average voltage. The average voltage refers to the highest cut-off voltage and the lowest cut-off voltage, so the reported power can also reflect the full rate value of the battery participating in frequency modulation charge and discharge current.

As for frequency regulation condition, based on the PJM history instruction frequency regulation 2020, selected has the following characteristics of six kinds of typical working conditions: (1) the maximum frequency of full-rate charge and discharge; (2) the minimum frequency of full-rate charge and discharge; (3) the maximum frequency regulation mileage; (4) the minimum frequency regulation range; (5) the maximum integral value of current with respect to time in charge and discharge process; and (6) the minimum integral value of current with respect to time in charge and discharge process. The corresponding relationship between specific factors and levels is shown in Table 3. Six points of the frequency regulation condition correspond to levels 1 to 6 of factor 4, as seen in Table 3.

Table 3. Simulation test variables.

Level	Factor 1	Factor 2	Factor 3	Factor 4
	SOC	Temperature (°C)	Power	Frequency Regulation Condition
1	0.2	−10	0.1 C*V	p_1_max
2	0.4	0	0.5 C*V	p_1_min
3	0.6	25	1 C*V	m_max
4	0.8	45	2 C*V	m_min
5			3 C*V	int_max
6				int_min

Among them, frequency regulation mileage refers to the absolute value of the difference between the output value at the end of each control command and the output value of the command response after it is issued. In this study, it is the absolute value of the difference between adjacent two-second frequency regulation instructions, and the total frequency regulation mileage is the sum of frequency regulation mileage in the selected time interval. In order to make the selected frequency regulation conditions more universal and consider the complexity of the selection process, this study first selected four typical days in the whole year—March 1, June 1, September 1, and December 1—that is, one day in each of the four seasons. Then, in these four days, with one minute as the step length, Matlab was used to calculate the full rate charge and discharge frequency. Frequency regulation mileage and the integral value of the current with respect to time in the charge and discharge process every 5 min were compared in order to obtain six five-minute frequency regulation conditions.

The same as the durability test, this part adopts the orthogonal test method and obtains 49 groups of simulation tests according to the above factor levels. The specific factor colocation is shown in Table 4.

Table 4. Simulation test orthogonal table.

No.	Factor 1	Factor 2	Factor 3	Factor 4
1	1	1	1	1
2	1	2	3	4
3	1	3	5	6
4	1	4	5	3
5	1	4	2	6
6	1	4	4	2
7	1	4	5	5
8	2	1	5	5
9	2	2	2	2
10	2	3	4	5
11	2	4	5	1
12	2	4	1	4
13	2	4	3	6
14	2	4	5	3
15	3	1	5	4
16	3	2	1	6
17	3	3	3	3
18	3	4	5	6
19	3	4	5	2
20	3	4	2	5
21	3	4	4	1
22	4	1	3	5
23	4	2	5	5
24	4	3	2	1
25	4	4	4	4
26	4	4	5	6
27	4	4	1	3
28	4	4	3	6
29	4	1	2	6
30	4	2	1	3
31	4	3	1	6
32	4	4	3	2
33	4	3	4	5
34	4	2	3	2
35	3	1	2	4
36	3	1	3	5
37	3	2	3	1
38	3	3	5	4
39	3	4	2	6
40	2	3	4	3
41	2	2	5	5
42	2	1	1	2
43	2	1	2	3
44	2	2	4	6
45	1	3	5	1
46	1	4	1	5
47	1	3	3	1
48	1	2	5	4
49	1	1	4	5

4. Result and Discussion

4.1. Durability Test Result

The capacity loss data obtained from the three-round durability test were recorded, respectively, and the corresponding data of the combination of test factors and capacity loss for each group were obtained, as shown in Table 5. Next, the range analysis method is used to analyze the capacity loss results. The range analysis has the advantages of simplicity and intuition and is suitable for screening results with low accuracy requirements. However, it cannot estimate the size of the error or accurately estimate the importance of the influence of various factors on the results. The results of three rounds of durability analysis are shown in Figure 5. The K_{avg} in this figure is the average value of the test data applied to a certain level of a factor. In order to obtain the comparison among levels within each factor, we can compare the impact of each level on capacity loss by comparing the K_{avg} of each level.

Table 5. Durability orthogonal test results.

No.	Factor 1	Factor 2	Factor 3	Factor 4	Capacity Loss (Ah)		
					The First Round	The Second Round	The Third Round
1	1	1	1	1	0.002646	0.03314	0.201544
2	1	2	3	2	0.00789	0.029742	0.126122
3	1	3	2	3	0.35454	0.38947	0.601268
4	2	1	3	3	0.405462	0.446374	0.66037
5	2	2	2	1	0.233194	0.192172	0.23746
6	2	3	1	2	0.031688	0.008698	0.145326
7	3	1	2	2	0.112026	0.182386	0.210076
8	3	2	1	3	0.12343	0.22568	0.289484
9	3	3	3	1	0.852186	1.10946	0.882492

Compared with factor 1, it can be seen that the impact of each round of SOC on capacity loss presents a positive correlation trend; that is, the capacity loss increases with the increase in SOC, and the capacity loss is the smallest when SOC is 0.2. This is because the battery SOC and the battery voltage are highly correlated, and the battery voltage can be derived when the battery SOC and current are known. For the battery, the higher the SOC, the higher the terminal voltage; that is, the lower the negative potential, which increases the rate of negative side reactions and SEI film thickening, the faster the battery aging rate and capacity loss.

The decrease in the pulse period indicates that the frequency of charge and discharge conversion is increasing; that is, the direction of movement of lithium ions in the electrolyte is constantly changing and the change rate is accelerating, which will lead to the acceleration of the diffusion rate of lithium ions in the electrolyte. Moreover, this is conducive to solving the lithium deposition structure, but not conducive to the occurrence of the lithium evolution reaction, ultimately resulting in reduced capacity loss. When the pulse period is too small, i.e., the frequency is too high, the internal reaction rate becomes too fast, resulting in a rise in battery temperature, which is not conducive to reducing capacity loss. This can be explained by the fact that factor 2 has the smallest capacity loss when the half-pulse period is 150 s.

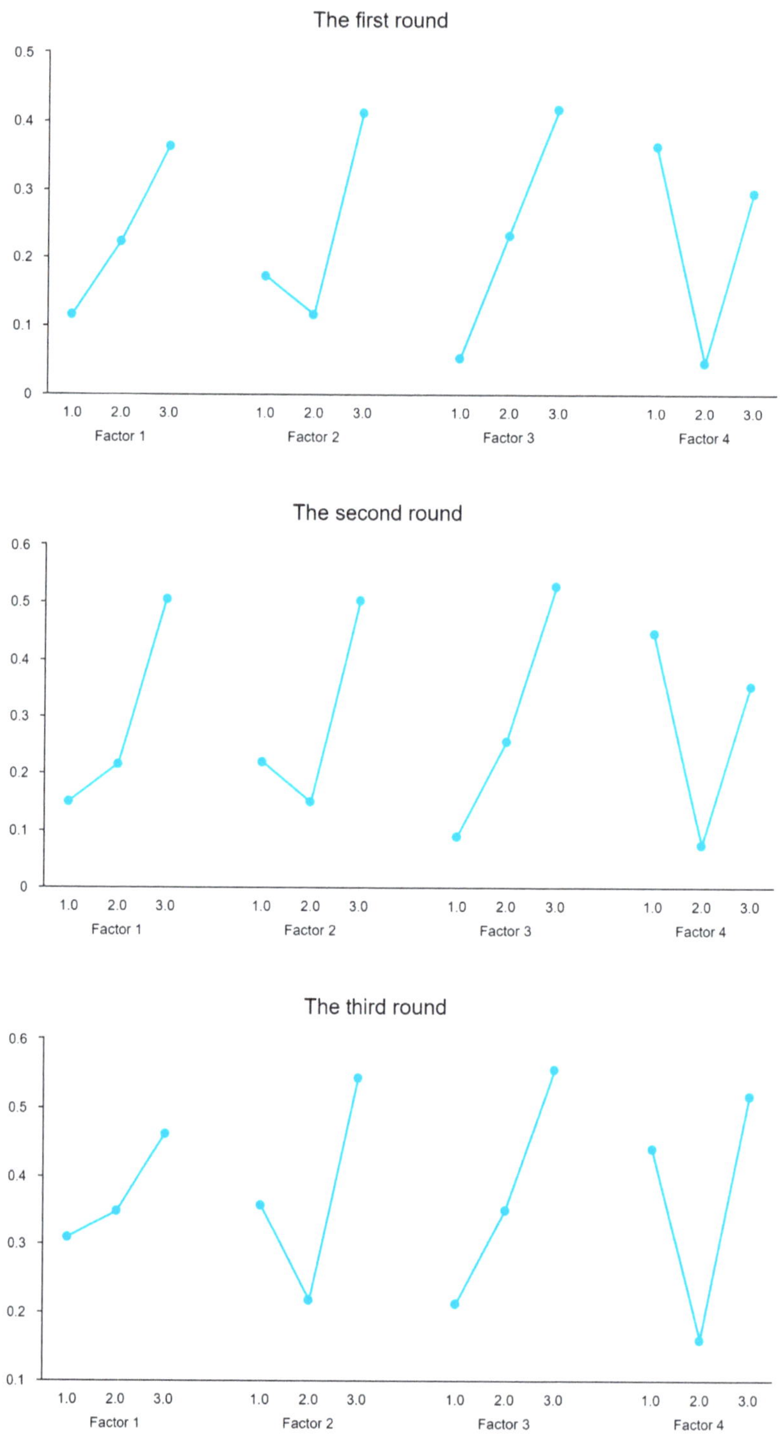

Figure 5. The K_{avg} of each level of the factor within first, second, and third rounds.

The reported power is related to the maximum charge and discharge currents—the greater the reported power, the greater the maximum charge and discharge current and the current during the charge and discharge process. Although this will accelerate the main reaction rate, it will also lead to the acceleration of the side reaction rate, which is conducive to the occurrence of lithium evolution reaction and SEI film thickening, making the loss of available lithium result in an increased capacity loss (factor 3 at 0.1 C*V) and minimum loss of battery capacity.

Compared with factor 4, it can be seen that each round of temperature has the smallest impact on capacity loss at 25 °C; that is, low and high temperatures bring greater capacity loss, which is in line with the appropriate operating temperature of general commercial batteries, i.e., 10–35 °C. This is because the temperature will affect the internal reaction rate of the battery. A temperature that is too high will speed up the reaction rate and the side reaction rate, resulting in an increased capacity loss. A temperature that is too low will lead to an increase in internal resistance, resulting in an intensified polarization phenomenon, which will lead to additional side reactions, especially at low temperatures. Charging may lead to the occurrence of lithium evolution, where the available lithium content is reduced so that the battery is rapidly degraded; the brittleness of the material at low temperatures will also affect battery life.

4.2. Negative Potential Simulation Test Results

The negative electrode potential refers to the solid–liquid-phase potential difference on the surface of the negative particle. When it is greater than the equilibrium potential of the lithium evolution reaction, the lithium evolution phenomenon will occur in the negative electrode of the battery. The phenomenon of lithium evolution leads to the appearance of lithium dendrites on the negative electrode surface, further leading to the loss of available lithium, increased polarization, and thus reduced battery capacity. Moreover, the longitudinal existence of a lithium-ion concentration gradient causes an uneven distribution of lithium dendrites, which will even puncture the diaphragm, resulting in a short circuit inside the battery. Therefore, the lithium evolution criterion can be expressed by Equation (15).

$$\varphi_s - \varphi_l < U_{e,2} \tag{15}$$

where φ_s is the solid-phase potential; φ_l is the liquid-phase potential; and $U_{e,2}$ is the lithium reaction equilibrium potential, usually considered to be 0 V (relative to Li/Li^{+}).

In this paper, the 15th group of test results is selected for display, as shown in Figure 6. Here, the abscissa of the thickness of the battery is present, the units are meters, and the diagram is the negative part of the selection, so the range is $\left[0.9 \times 10^{-5}\right]$ m. The ordinate is simulation time and the units are seconds, with a total of 300 s. The color diagram shows the negative electrode potential at the corresponding position and time, and the units are mV. To facilitate the detection of negative electrode potential, the color diagram interval is adjusted to $[-100, 50]$ mV.

It can be seen from the above-mentioned negative potential distribution results that some simulation groups have negative electrode potentials during the simulation process. Through the factor colocation of these groups, a qualitative conclusion can be obtained—under different working conditions, high SOC, low temperature, and high charging rate, negative electrode potential will appear negative. In addition, it can also be found that along the positive direction of the x-axis, the negative electrode potential has a decreasing trend; that is, the closer the electrode and the electrolyte interface, the smaller the negative electrode potential. Therefore, it is not only necessary to consider the influence of the time dimension on the negative potential, but also to take the x-axis direction into account. In order to better obtain the degree of influence of various factors on this phenomenon, statistical means are used for quantitative analysis. Therefore, in this study, the part of the negative electrode potential is a double integral for time and the *x*-axis direction, which can be briefly written as $\varphi_{neg} \times t \times l$. The obtained integral value was analyzed. The specific results are shown in Table 6. In order to express the results easily, the simulation groups without negative electrode potentials are not included in Table 6; that is, the integral value of these groups is 0. Although these groups do not appear in Table 6, they must be included in the data analysis process.

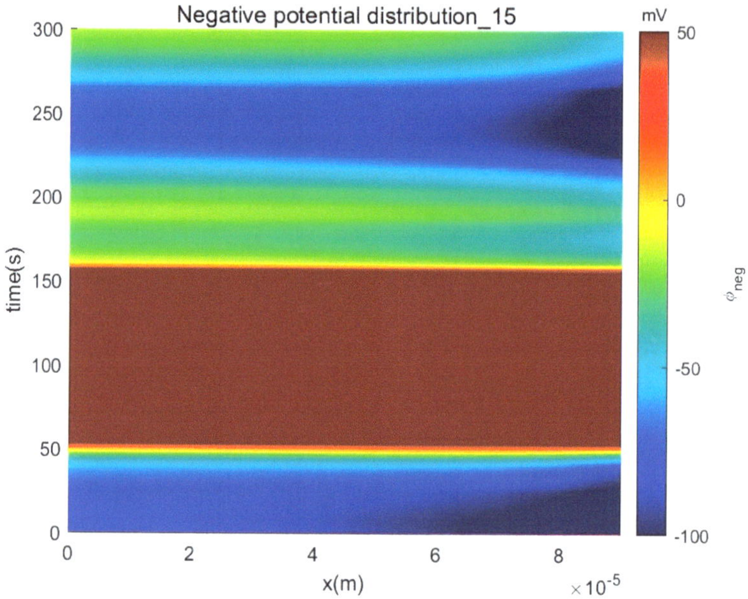

Figure 6. Negative potential distribution results.

Table 6. Negative potential negative integral result.

No.	Factor 1	Factor 2	Factor 3	Factor 4	$\varphi_{neg} \times t \times l$
2	1	2	3	4	0.00067558
8	2	1	5	5	0.4674
15	3	1	5	4	0.9619
22	4	1	3	5	0.3432
23	4	2	5	5	0.3573
35	3	1	2	4	0.1467
36	3	1	3	5	0.2518
37	3	2	3	1	0.8525
41	2	2	5	5	0.1939
43	2	1	2	3	0.7514
48	1	2	5	4	0.2517
49	1	1	4	5	0.1856

Multiple linear regression analysis among factor 1, factor 2, factor 3 and factor 4 were used as independent variables and dependent variables for multiple linear regression analyses. The R-square value of the model was 0.355, meaning that factor 1, factor 2, factor 3, and factor 4 could explain 35.5% of the variation of $\varphi_{neg} \times t \times l$. During the F-test of the model, it was found that the model passed the F-test (F = 6.049, $p = 0.001 < 0.05$), indicating that at least one of the factors (factor 1, factor 2, factor 3, and factor 4) would have an impact on $\varphi_{neg} \times t \times l$. In addition, by testing the multicollinearity of the model, it is found that the variance inflation factor (VIF) values in the model are all less than 5, which means that there is no collinearity problem, and the D-W value is near the number 2, which indicates that there is no autocorrelation in the model and there is no correlation between the sample data. The final analysis results showed that the regression coefficients corresponding to factor 1, factor 2, factor 3, and factor 4 were 0.021 (t = 0.852, $p = 0.399 > 0.05$), −0.110 (t = −4.687, $p = 0.000 < 0.01$), 0.040 (t = 2.098, $p = 0.042 < 0.05$), and −0.008 (t = −0.512, $p = 0.611 > 0.05$), respectively.

It means that factor 2 has a significant negative correlation with $\varphi_{neg} \times t \times l$; that is, it has the greatest impact on the result at −10 °C. Factor 3 has a significant positive correlation with $\varphi_{neg} \times t \times l$; that is, it has the greatest impact on the result when the maximum charging rate is 3 C. Factor 1 and factor 4 have no significant correlation with $\varphi_{neg} \times t \times l$.

Multiple linear regression analysis in factors: In order to further analyze the internal influences of factor 1 and factor 4 on $\varphi_{neg} \times t \times l$, internal multiple linear regression analysis was used, respectively. Firstly, factor 1 is analyzed, and level 1 is taken as a reference, and the linear regression coefficients of the four levels are 0.036, 0.081, 0.148, and 0.017, respectively. Compared with the linear regression coefficients of the other three levels, it can be seen that the coefficient at level 3 is the largest, so it can be considered that the SOC of 0.6 in factor 1 has the greatest impact on $\varphi_{neg} \times t \times l$. Next, factor 4 is analyzed, and level 1 is taken as a reference. The linear regression coefficients of the six levels are 0.122, −0.122, 0.073, −0.014, −0.122, and 0.04. Comparing the linear regression coefficients of the other five levels, it can be seen that between level 1 and level 2, level 1 is larger; between level 3 and level 4, level 3 is larger; and between level 5 and level 6, level 6 is larger. Therefore, it can be considered that in factor 4, the maximum integral value of current with respect to time in the charge and discharge process, the maximum frequency of the full rate charge and discharge, and the minimum frequency regulation range have the greatest influence on $\varphi_{neg} \times t \times l$.

To summarize, the analysis of temperature and reported power aligns with expectations, indicating that a lower temperature and higher charging rate have a greater impact on negative potential below 0. However, regarding SOC, this study reveals that the greatest influence is not observed at SOC of 0.8. This discrepancy may be attributed to the fact that the orthogonal simulation test conducted in this study did not encompass all possible factor combinations, resulting in some results being influenced by factor coupling and thereby affecting the isolated effect of individual factors. Furthermore, improvements can be made to the test methodology by selecting frequency regulation conditions for quantitative evaluation across six levels before analyzing the joint influence of three other factors on the target under investigation.

4.3. Theoretical Analysis

For temperature, there are two main mechanisms to affect the electrochemical system of the battery, i.e., one is the Arrhenius Equation (16) and the other is the Stokes–Einstein Equation (17).

$$k = A \cdot e^{-\frac{E_a}{(RT)}} \quad (16)$$

$$D = \frac{k_B T}{6\pi\eta r} \quad (17)$$

where A refers to the pre-factor; E_a is the activation energy; R is the gas constant; T is the absolute temperature; D represents the diffusion coefficient of the molecule; k_B is the Boltzmann constant; η is the viscosity of the solvent; and r is the radius of the molecule. It can be seen that the increase in electrolyte viscosity at low temperature reduces the diffusion coefficient of lithium ions, leading to the increase in charge transfer impedance and ohmic polarization. Furthermore, the kinetics of lithium embedding into negative graphite are hindered, and the concentration gradient of lithium ions on the negative surface increases, which may reduce the local potential to the lithium metal precipitation potential, form lithium dendrites, consume active lithium, and accelerate the SEI film rupture. At high temperatures, the side reaction accelerates, the dissolution or regeneration of the SEI film intensifies, the active lithium becomes continuously consumed, and the positive transition metal dissolves and migrates to the negative electrode to destroy the SEI film. Therefore, at low temperatures, the negative polarization increases, the potential distribution becomes more uneven, and the lithium threshold becomes easily reachable in local areas. At high temperatures, the SEI impedance decreases, the potential distribution becomes uniform, and the overall lithium loss accelerates. However, whether it is a high or low temperature, it will lead to a reduction in the effective lithium inventory, while the active material loss.

State of charge (SOC) directly affects the negative equilibrium potential, according to the Nernst Equation (18).

$$\varphi_{eq} = \varphi_0 + \frac{RT}{F}\ln\left(\frac{c_{Li^+}^{surf}}{c_{Li}^{max} - c_{Li^+}^{surf}}\right) \tag{18}$$

At high SOC, $c_{Li^+}^{surf}$ approaches c_{Li}^{max}, causing φ_{eq} to approach the lithium metal deposition threshold. At this time, the local current density distribution presents significant non-uniformity. The Butler–Volmer Equation (19) is expressed as follows:

$$j = j_0\left[e^{\alpha_a F\eta/RT} - e^{\alpha_c F\eta/RT}\right] \tag{19}$$

When $\eta > 0$ (charging process), the high SOC region preferentially triggers the lithium evolution side reaction, resulting in irreversible capacity loss.

As for charge and discharge ratio, high ratio will cause third-order polarization coupling. One is the ohmic polarization, i.e., $\Delta\varphi_\Omega = I \cdot R_\Omega$, which is related to collector and electrolyte conductivity. The second is the electrochemical polarization, controlled by the Butler–Volmer kinetics. The third is the concentration polarization, which follows Fick's second law. The porous electrode theory shows that the local current density distribution satisfies Equation (20).

$$\nabla \cdot \left(\sigma_{eff}\nabla\phi_s\right) = a_s j \tag{20}$$

It can be seen that a significant concentration gradient is formed inside the pore of the electrode at a high power rate, leading to lithium deposition, preferentially in the edge region.

In addition, the half-pulse period affects the dynamic response characteristics of the electrode, and the characteristic time constant τ_diff is determined by the diffusion process Equation (21).

$$\tau_{diff} = \frac{L^2}{D} \tag{21}$$

where L is the diffusion characteristic length. When $\tau < \tau_{diff}$, the pulse process causes an unsteady-state concentration fluctuation, ultimately resulting in an "insufficient relaxation" effect. When $\tau > \tau_{diff}$, lithium ions can be fully redistributed. The high-frequency pulse leads to periodic potential oscillation on the surface of the negative electrode, which aggravates the mechanical stress damage of the SEI film.

5. Conclusions

In this paper, a large number of research studies on power grid frequency regulation, virtual inertia, and battery durability are investigated. The virtual inertia system and battery life estimation models are introduced, and the advantages and disadvantages of each model are given. Based on the above-mentioned problems, the effects of temperature, SOC, half-pulse period, and reported power on battery capacity loss were investigated experimentally. The range analysis method is used to compare the difference in the influence of four factors on capacity loss; the internal influence of each factor is compared. In addition, in order to simplify the number of experimental groups, the orthogonal test method was also introduced.

Since the lithium evolution phenomenon of the battery will lead to the capacity loss of the battery, and the criterion that directly leads to the lithium evolution phenomenon states that the negative potential is less than zero, this study explored the influence of temperature, SOC, reported power, and frequency regulation conditions on the negative potential distribution of the battery and the degree of lithium evolution from the perspective of simulation. The P2D model was built according to the dynamic and durability mechanisms of lithium-ion batteries. The orthogonal test was also used to simplify the number of simulation groups; the negative electrode potential distribution diagram under each group of simulations and the qualitative results were obtained. Then, the battery thickness and simulation time are integrated with the negative electrode potential, and the integral value obtained is analyzed to obtain certain quantitative results.

Author Contributions: Conceptualization, Y.T. and L.W.; methodology, Y.T.; software, Y.T.; validation, Y.T.; formal analysis, Y.T.; investigation, Y.T.; resources, L.W.; data curation, L.W.; writing—original draft preparation, Y.T.; writing—review and editing, G.Y.; visualization, Y.T.; supervision, C.L.; project administration, L.W.; funding acquisition, L.W. All authors have read and agreed to the published version of the manuscript.

Funding: This research is supported by supported by the National Key Research and Development Program of China (Grant No.2023YFB2504100), the National Natural Science Foundation of China (52277228), the National Natural Science Foundation of China (52077208), and the Beijing Natural Science Foundation (L243021).

Data Availability Statement: The original contributions presented in this study are included within this article. Further inquiries can be directed to the corresponding author.

Conflicts of Interest: The authors declare no conflicts of interest. The funders had no role in the design of this study, in the collection, analysis, or interpretation of data, in the writing of the manuscript, or in the decision to publish the results.

Abbreviations

The following abbreviations are used in this manuscript:

SOC	state of charge
P2D	pseudo-two-dimensions
V2G	Vehicle-to-Grid
LFP	lithium iron phosphate
SEI	solid electrolyte interface

References

1. Caulfield, B.; Furszyfer, D.; Stefaniec, A.; Foley, A. Measuring the equity impacts of government subsidies for electric vehicles. *Energy* **2022**, *248*, 123588. [CrossRef]
2. Zou, W.; Li, J.; Yang, Q.; Wan, X.; He, Y.; Lan, H. A real-time energy management approach with fuel cell and battery competition-synergy control for the fuel cell vehicle. *Appl. Energy* **2023**, *334*, 120667. [CrossRef]
3. Capacity Degradation Minimization Oriented Optimization for the Pulse Preheating of Lithium-Ion Batteries Under Low Temperature—ScienceDirect. Available online: https://www.sciencedirect.com/science/article/abs/pii/S2352152X20315838 (accessed on 18 January 2025).

4. Li, Y.; Li, L.; Peng, C.; Zou, J. An MPC based optimized control approach for EV-based voltage regulation in distribution grid. *Electr. Power Syst. Res.* **2019**, *172*, 152–160. [CrossRef]
5. Optimization of Bi-Directional V2G Behavior with Active Battery Anti-Aging Scheduling. Available online: https://ieeexplore.ieee.org/document/8951120 (accessed on 18 January 2025).
6. Ebrahimi, M.; Rastegar, M.; Mohammadi, M.; Palomino, A.; Parvania, M. Stochastic Charging Optimization of V2G-Capable PEVs: A Comprehensive Model for Battery Aging and Customer Service Quality. *IEEE Trans. Transp. Electrif.* **2020**, *6*, 1026–1034. [CrossRef]
7. Vermeer, W.; Chandra Mouli, G.R.; Bauer, P. A Comprehensive Review on the Characteristics and Modeling of Lithium-Ion Battery Aging. *IEEE Trans. Transp. Electrif.* **2022**, *8*, 2205–2232. [CrossRef]
8. Darling, R.; Newman, J. Modeling Side Reactions in Composite LiyMn2 O 4 Electrodes. *J. Electrochem. Soc.* **1998**, *145*, 990. [CrossRef]
9. Dalverny, A.-L.; Filhol, J.-S.; Doublet, M.-L. Interface electrochemistry in conversion materials for Li-ion batteries. *J. Mater. Chem.* **2011**, *21*, 10134–10142. [CrossRef]
10. Tasaki, K.; Harris, S.J. Computational Study on the Solubility of Lithium Salts Formed on Lithium Ion Battery Negative Electrode in Organic Solvents. *J. Phys. Chem. C* **2010**, *114*, 8076–8083. [CrossRef]
11. Chaoui, H.; Ibe-Ekeocha, C.C. State of Charge and State of Health Estimation for Lithium Batteries Using Recurrent Neural Networks. *IEEE Trans. Veh. Technol.* **2017**, *66*, 8773–8783. [CrossRef]
12. Degradation Cost Analysis of Li-Ion Batteries in the Capacity Market with Different Degradation Models. Available online: https://www.mdpi.com/2079-9292/9/1/90 (accessed on 18 January 2025).
13. Ecker, M.; Gerschler, J.B.; Vogel, J.; Käbitz, S.; Hust, F.; Dechent, P.; Sauer, D.U. Development of a lifetime prediction model for lithium-ion batteries based on extended accelerated aging test data. *J. Power Sources* **2012**, *215*, 248–257. [CrossRef]

batteries

MDPI

Article

Inductor-Based Active Balancing Topology with Wide Voltage Range Capability

Hourong Song *, Branislav Hredzak * and John Fletcher

School of Electrical Engineering and Telecommunications, University of New South Wales, Sydney, NSW 2052, Australia; john.fletcher@unsw.edu.au
* Correspondence: hourong.song@student.unsw.edu.au (H.S.); b.hredzak@unsw.edu.au (B.H.)

Abstract: With the increasing number of batteries integrated into the grid, the electrification of transportation, and the importance of reusing secondary batteries to preserve natural resources, active balancing techniques are becoming critical for optimizing battery performance, ensuring safety, and extending their lifespan. There is a demand for battery management solutions that can efficiently manage the balancing of battery cells across a wide range of voltage levels. This paper proposes a new inductor-based active balancing topology that achieves balancing by transferring energy from battery cells to the battery pack. One of its main advantages over existing designs is that it can operate over a wide battery cell voltage range. Moreover, multicell balancing with a balancing current independent of the imbalance level can be achieved by adjusting the width and interval of pulses. The proposed topology can be implemented using traditional low-side gate driving integrated circuits, avoiding the need for expensive isolated power modules and high-side gate drivers. Sample balancer designs for low-voltage battery cells as well as higher-voltage cells are provided. The presented experimental results verify the operation of the proposed balancer on a lithium-ion battery pack.

Keywords: battery; energy storage; active balancing; driver circuits; battery management systems

1. Introduction

Energy storage systems, such as batteries and supercapacitors, are used in various commercial and consumer applications, such as electric vehicles (EVs), uninterruptible power supplies (UPS), and distribution systems. To meet the required power demand, individual cells are interconnected in both series and parallel configurations to achieve the necessary voltage and power ratings [1].

The automotive industry experienced a significant increase in the demand for lithium-ion (Li-ion) batteries, rising from 330 GWh in 2021 to 550 GWh in 2022, marking a 65% increase. This uptick in demand can be attributed to the growing sales of electric passenger cars, which saw a remarkable 55% increase in new registrations in 2022 compared to 2021 [2]. Unfortunately, this surge in demand will result in the retirement of 100–200 gigawatt-hours worth of batteries by 2030 as they are no longer suitable for EV use. This will present a significant challenge, as the toxic content and reactive properties of the batteries are hazardous.

As the demand for batteries escalates with the rise of renewable energy systems and electric vehicles, there is a growing interest in repurposing retired battery packs. Although no longer useful for their primary applications, retired batteries still contain a significant capacity and energy storage capability, which can be utilized in secondary applications such

https://doi.org/10.3390/batteries11020077

as stationary energy storage systems, off-grid solutions, and household PV units [3,4]. As a result, there has been an increasing interest in repurposing these batteries for secondary applications to extract more value from these assets and contribute to a more sustainable and circular economy.

Energy storage systems utilize a variety of different types of cells. For example, lithium-titanate batteries and supercapacitors are ideal for high C-rate applications in heavy vehicles and locomotives. Redox flow batteries are a safe alternative for energy storage facilities. However, their different electrochemical characteristics result in varying voltage levels of the cells, which makes them incompatible with current commercial battery management integrated circuits (ICs) designed for lithium-ion cells.

The battery cells capacity inconsistency is one of the main contributing factors to the imbalance. It also significantly impacts the overall battery pack capacity, with the smallest capacity cell being the limiting factor [5]. Additionally, as the number of charging and discharging cycles increases, the capacity disparity can further exacerbate the system's performance.

Therefore, developing a practical active balancer that can operate over a wide cell voltage range and, hence, be compatible with different electrochemistry cell types is of significant interest.

Research on active balancing methods can be categorized into three major categories: capacitor-based, inductor-based, and transformer-based [6,7]. The optimal switching capacitor-based balancer in [8,9] significantly reduced the number of switches required per cell, improved the balancing speed, and achieved multicell balancing compared to traditional adjacent cells balancing topology in [10]. This type of balancer has been applied to some commercial products as it has a simple driving circuit and no control is required, which simplifies implementation. However, the capacitor-based balancer exhibits three inherent limitations. Firstly, its balancing speed is contingent on capacitance value, and uncontrolled balancing currents render it incapable of eliminating inrush currents. Secondly, the necessity of employing large capacitors with adequate voltage ratings increases production cost. Lastly, when cell voltage variations are negligible, the balancing current diminishes to very low levels, which significantly elevates the energy consumption of its driving circuit in comparison to the actual balancing power utilized.

Transformer-based balancing methods can be classified into two categories: isolated DC-DC converters and multi-winding transformer-based. Current commercial solutions offered by companies such as Texas Instruments and Analog Devices typically employ flyback converters to facilitate energy transfer from individual cells to the battery pack. These solutions often rely on complex switching matrices utilizing MOSFETs and specialized integrated circuits, or alternatively, they utilize standalone converters for each cell. However, these approaches are associated with high costs and are heavily dependent on specific integrated circuit manufacturers, limiting their scalability and adaptability to different battery management systems [11,12].

Research advancements have improved balancers utilizing multi-winding transformers, making multicell balancing feasible [13–17]. The iterative refinement of driving and control methods has simplified their implementation. However, challenges persist in terms of efficiency, transformer size, automated assembly complexity during manufacturing, and transformer customization, all of which have significant drawbacks for industrial applications.

Inductor-based balancers [18–23] and L-C balancers [23] present several challenges. Early-stage designs exhibited low efficiency due to diodes being used in the current path to a cell and being used in lower voltage applications [18,19]. Additionally, the inductor-based balancer in [20] was restricted to balancing between adjacent cells only. Other approaches required a high number of floating MOSFETs, which necessitated expensive driving circuits, including isolated power modules and isolated gate driving ICs [21–23]. A high-speed simple topology was proposed in [24], but the cost of the required driving circuit and uneven charging of other cells while balancing remained significant drawbacks. Hence, further research and development are necessary to devise new topologies that are compatible with low-cost driving circuits and offer high operating efficiency.

This paper proposes a novel inductor-based low-cost balancing topology that is compatible with various cell voltage ranges, has multicell simultaneous balancing capability, controllable balancing current, a simple control strategy, and does not require specific integrated circuits or a controller. Table 1 summarizes the comparison of the proposed balancer with the existing balancing topologies, which meet more than one out of three criteria in the left column.

Table 1. Comparison of the proposed balancer with the existing balancing topologies.

	References: [11–17,19] #, [21,23] *	References: [1,8–10,18,20,22] **	References: [24] ***	Proposed
Maintaining a consistent balancing speed regardless of imbalance levels	✔		✔	✔
Eliminating isolated power modules and gate drivers		✔		✔
Avoiding complex control systems or specific controller ICs		✔	✔	✔

* Quiescent current loss, not suitable for cell level balancing, customised magnetic components. ** Slow one-by-one balancing speed or increased power losses when balancing adjacent cells. *** Bulky power modules, quiescent current loss. # Although this topology uses only one inductor, the overall cost is high as it requires isolated power modules and isolated gate drivers.

2. Proposed Balancing Topology Description

The proposed active balancing topology for an n-cell series-connected battery pack is illustrated in Figure 1. It requires n pairs of switches, n inductors, n pairs of Schottky diodes, and n dual output low side driving ICs.

The operation of the circuit is controlled by a pulse signal, which synchronously turns on and off the switch-pairs during the balancing process. The operational principle involves two sequential steps: initially, both switches are closed, facilitating energy transfer from the cell to the inductor (as indicated by the red arrow in Figure 1a). Subsequently, with the switches open, the inductor maintains a consistent current flow direction, and the discharge path of the inductor is redirected by the pair of diodes to the terminals of the battery pack (as indicated by the green arrow in Figure 1b). Consequently, energy is transferred from an individual cell to the entire battery pack. Notably, this balancing procedure operates independently for each cell, enabling the simultaneous discharge of multiple cells. This topology operates independently of the cell voltage level. As long as a cell can charge the inductor and the cell's voltage does not fall below its minimum operating voltage, the balancing current can be maintained at a desired RMS value.

In low-voltage battery packs, the pair of switches can be implemented by N-channel MOSFETs and a low forward-voltage Schottky diode to maximize efficiency. A schematic of the implementation is shown in Figure 2. In a typical low-voltage battery pack, such as a series-connected 10-cell NiMH battery pack (12 V) or a three- to four-cell series-connected lithium battery pack (typically 12.6 V to 14.8 V), a low-side driver with a maximum output voltage range of 20 V to 25 V can activate the high-side switch. This configuration provides a gate-source voltage between 5 to 10 volts, which is sufficient for MOSFETs that operate at a 4.5 V gate-source voltage. The switches are open when the driver output voltage (gate-ground voltage) is changed to zero. Consequently, both the gate-source and drain-source voltages for the lower MOSFET will not drop below the negative pack voltage, which remains significantly lower than the maximum rating. Simultaneously, the source voltage of the upper MOSFET falls to around −0.5 V due to the discharge current path through the Schottky diodes. With a gate voltage of approximately 0.5 V, the MOSFET remains inactive.

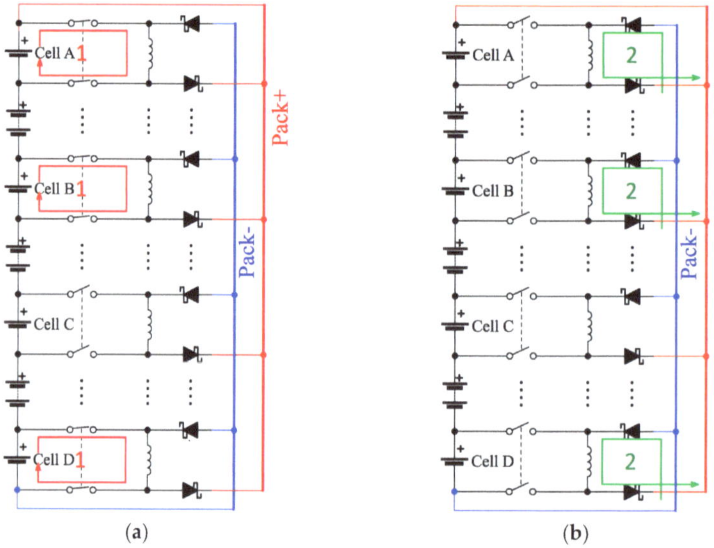

Figure 1. Proposed active balancing topology. Cells *A*, *B* and *D* are being balanced, while Cell C is within the balancing threshold and is not being balanced. *Pack*+ represents the positive terminal of the pack. *Pack*− represents the negative terminal of the pack. (**a**) *Path* 1 represents the cells charging the inductors. (**b**) *Path* 2 represents the inductors releasing energy to the battery pack.

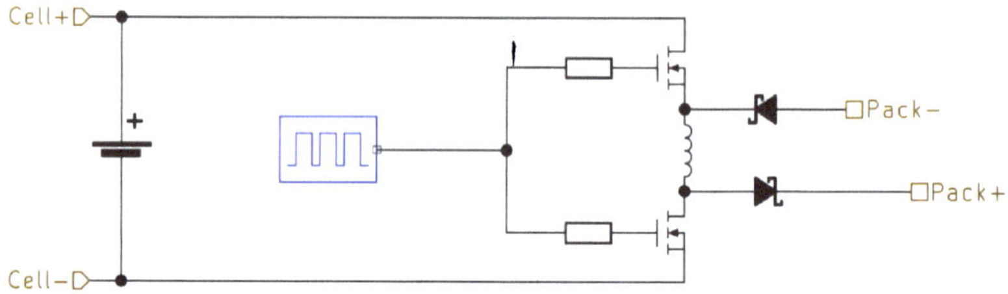

Figure 2. Implementation for low voltage cells.

Figure 3 illustrates the necessary modifications to adapt the topology for higher-voltage battery packs. Typically, commercial BMS front ends support 6–16 series-connected lithium cells with voltages ranging from 25.2 V to 67.2 V. This results in pack voltages that exceed the typical maximum operating range of low-side gate drivers, causing multiple cells to have voltage levels higher than the gate drivers' power supply.

In Figure 2, the upper switch's high side and the bottom switch's low side are connected to the positive and negative terminals of the corresponding cell, thereby providing a relatively steady voltage level reference to the negative terminal of the pack. To eliminate the floating source of the MOSFET, *P*-channel MOSFETs, and *N*-channel MOSFETs are used as replacements for the switches in Figure 3. A dual-channel low-side driver drives the complementary pair of MOSFETs with one inverting output through a bootstrap circuit.

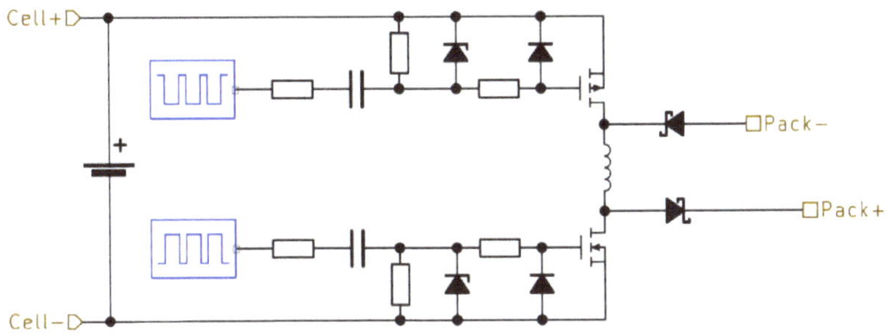

Figure 3. Implementation for higher voltage cells.

This configuration guarantees synchronous switching on and off simultaneously. Gate resistors R_g are placed in series on both sides of the bootstrap capacitor to limit peak current in the driving circuit and inrush current via the bootstrap diode during transient states when connecting the balancer to the cells. Additionally, optional Zener or TVS diodes with operating voltage between gate-driving and maximum gate-source voltage provide extra protection by capping the gate-source voltage within the maximum rating. The gate-source resistor pulls the gate-source voltage to zero when the driver IC with a push-pull output shuts down, enhancing safety and robustness in a noisy environment.

In summary, the proposed balancing topology is insensitive to the cells operating voltage range. Its performance is constrained primarily by the voltage ratings of the selected components rather than by any inherent topology limitations. In very low-voltage cells (provided the cell voltage remains above zero), the inductor can still store and transfer energy effectively. Conversely, in high-voltage cells, one can simply replace the Schottky diodes and MOSFETs with high-voltage-rated components to maintain functionality.

3. Control Signal Generation and Efficiency Analysis

This section describes the PWM control signal generation and analyses energy transfer efficiency to assist in selecting suitable operating parameters for the implementation of the proposed balancing topology. The efficiency calculation does not consider eddy current losses in the inductor and copper plane below it, MOSFET reverse recovery, diode recovery, or energy consumed by the driving circuit. All symbols are listed in Table 2.

Table 2. Nomenclature.

Symbol	Unit	Description
t_1	s	Interval when signal pulse is set to high
t_2	s	Interval when inductor is discharging
t_3	s	Deadtime
R_P	Ω	Resistance of *P*-channel MOSFET fully on
R_N	Ω	Resistance of *N*-channel MOSFET fully on
R_L	Ω	Inductor resistance
R	Ω	Sum of R_p, R_N and R_L
L	H	Inductor inductance
n	-	Number of cells
V	V	Cell voltage
V_p	V	Pack voltage
V_f	V	Diode forward voltage drop
I_L	A	Inductor current

Table 2. *Cont.*

Symbol	Unit	Description
I_{peak}	A	Inductor peak current
I_{Cavg}	A	Average cell current during balancing
I_{Davg}	A	Average diode current during balancing
E_L	J	Energy stored in inductor
$E_{R_{loss}}$	J	Energy loss due to series resistance
$E_{D_{loss}}$	J	Energy loss due to diode forward voltage

PWM Signal Generation

In Figure 4, t_1 represents the interval when the control signal pulse is high. During this time, the MOSFET pairs are turned on, allowing the cell to charge the inductor. t_2 represents the interval during which the MOSFETs are switched off, enabling the inductor to release energy to the pack via the pair of diodes. t_3 represents the dead time before the next cycle, allowing for diode reverse recovery and extra time to prevent continuous current flow through the inductor due to electromagnetic interference or propagation delay on gate drivers.

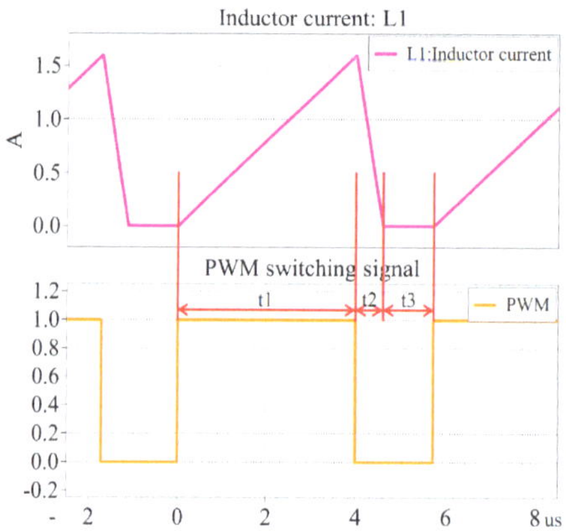

Figure 4. Simulated inductor current (A) and PWM signal vs. time (μs).

The peak current through the inductor I_{peak} can be expressed as (1), where R is the sum of the series resistance of the P-channel MOSFET R_P, the N-channel MOSFET R_N and the inductor R_L; V is the cell voltage, and L is the inductor inductance,

$$I_{peak} = \frac{V}{R} - \frac{V}{R} \times e^{-\frac{R}{L} \cdot t_1} \tag{1}$$

For the desired peak current through the inductor, the duration of t_1 is obtained from (1) as (2),

$$t_1 = -\frac{\ln\left(-\frac{I_{peak}R - V}{V}\right)L}{R} \tag{2}$$

Note: Since $t_1 \ll L/R$ in Figure 4, (1) and (2) could be simplified as $I_{peak} = \frac{V}{L}t_1$ and $t_1 = \frac{I_{peak}L}{V}$, respectively.

The energy stored in the inductor is given by (3),

$$E_L = \frac{1}{2} I_{peak}^2 L \tag{3}$$

and it is equal to the energy transfer during discharging, given by (4),

$$E_L = \int_0^{t_2} \left(V_P + V_f\right) \cdot \left(-\frac{I_{peak}}{t_2} \cdot t + I_{peak}\right) dt \tag{4}$$

Solving (3) and (4), the interval t_2 can be found as (5),

$$t_2 = \frac{I_{peak} L}{V_P + V_f} \tag{5}$$

The series resistance loss R_{loss} while the cell is charging the inductor during one cycle is given by (6),

$$E_{R_{loss}} = \int_0^{t1} \left(\frac{V}{R} - \frac{V}{R} \times e^{-\frac{R}{L} \cdot t_1}\right)^2 R\, dt \tag{6}$$

The diode loss D_{loss} while the inductor is releasing energy to pack during one cycle is given by (7),

$$E_{D_{loss}} = \int_0^{t_2} 2V_f \cdot \left(I_{Peak} - \frac{I_{peak}}{t_2} \cdot t\right) dt \tag{7}$$

The efficiency of the cell-to-pack energy transfer can be estimated by (8), and after substituting (2) and (5), the simplified expression (9) can be obtained,

$$Efficiency = \frac{E_L - E_{Dloss}}{E_{R_{loss}} + E_L} \times 100\%$$

$$= \frac{\frac{1}{2} I_{peak}^2 L - \int_0^{t_2} 2V_f \cdot \left(I_{Peak} - \frac{I_{peak}}{t_2} \cdot t\right) dt}{\int_0^{t1} \left(\frac{V}{R} - \frac{V}{R} \times e^{-\frac{R}{L} \cdot t_1}\right)^2 R\, dt + \frac{1}{2} I_{peak}^2 L} \times 100\% \tag{8}$$

$$= \frac{I_{peak}^2 \left(-V_p + V_f\right) R^2}{2\left(V_p + V_f\right) V \left(I_{peak} R + ln\left(\frac{-I_{peak} R + V}{V}\right) V\right)} \times 100\% \tag{9}$$

Assuming $V_p = n \cdot V$ and using parameters from the datasheets of the components in Table 3 as an example, (9) can be simplified as (10),

$$= \frac{0.0008405(V \cdot n - 0.4) I_{peak}^2}{\left(0.041\, I_{peak} + \ln\left(\frac{-0.041\, I_{peak} + V}{V}\right) V\right) V (V \cdot n + 0.4)} \times 100\% \tag{10}$$

Table 3. Components parameters for efficiency estimation example.

	Condition	Typical
V_f	$I_{avg} = 3A, 25\ ^\circ C$	0.4 V
R_P	$V_{GS} = -10$ V	<16.5 mΩ
R_N	$V_{GS} = +10$ V	<9.5 mΩ
R_L	-	15 mΩ
R	-	0.041 Ω

Figure 5 illustrates the efficiency of the balancing circuit. The vertical axis of the 3D graph represents the efficiency; the horizontal axes show the inductor's peak current and the cell's voltage. It displays six layers from bottom to top, illustrating the relationship between the pack voltage and balancing efficiency with the number of cells varying from 4 to 9.

The trend indicates that increasing the number of cells, and consequently the pack voltage, results in higher balancing efficiency.

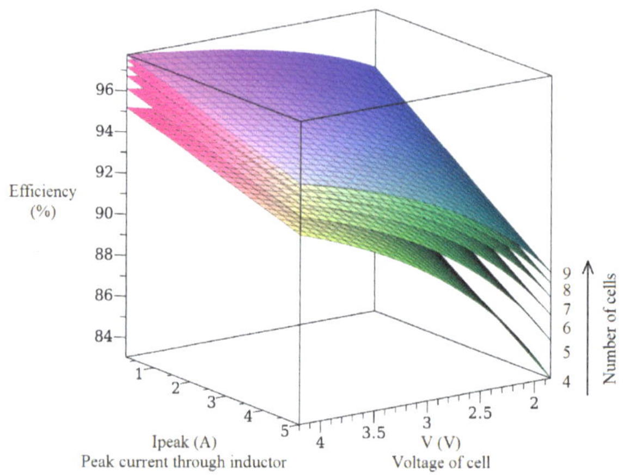

Figure 5. Efficiency analysis for various cell voltages, number of cells in a pack, and peak balancing current.

4. Experimental Setup and Test Results

4.1. Components Selection Guide and Application Note

4.1.1. MOSFET Selection

Based on the operating principle of the proposed balancer, the maximum MOSFET voltage stress occurs at the top cell's high side and the bottom cell's low side; hence, selecting MOSFETs with voltage ratings slightly higher than the maximum pack voltage suffices.

In the simplified driving circuit for low-voltage cells (Figure 2), the gate voltage rating is more critical than the pack voltage. This is because the maximum pack voltage is limited by the difference between the maximum voltage of the gate and its driver. MOSFETs typically turn on at around 4.5 V, so selecting components with voltage ratings of 12 V and 25 V establishes pack voltage limits of 7.5 V and 20.5 V, respectively.

For driving circuits designed for higher cell voltages and a wider cell voltage range (Figure 2), conventional MOSFETs with a maximum gate-source rating between ±15–±20 V are sufficient for driving ICs operating at a typical voltage of 12 V.

4.1.2. Driver IC Selection

The selection of gate drivers is critical to minimize additional energy loss outside of the balancing circuitry. A dual-channel low-side driver should be considered during the design phase. Such a driver ensures consistent internal propagation delay, hysteresis, and synchronous output while drawing less current than two standalone single-channel drivers. A pulse output current capability higher than 1 A is sufficient to rapidly turn on small MOSFETs for balancing applications.

4.1.3. Inductor Selection

The selection of inductors for the proposed topology prioritizes size, balancing speed, and efficiency. Equation (10) confirms that with a fixed value of series resistance, efficiency remains independent of inductance. However, (5) and (6) reveal that for a consistent peak current through the inductor, a lower inductance leads to a shorter cycle period, resulting in a higher average cell discharging current and improved balancing speed.

Consequently, after calculating the inductance value based on the desired balancing speed, inductors with low series resistance and fully shielded designs within physical size constraints are recommended. Furthermore, when employing non-shielded inductors, it is pertinent to note that wide copper tracks or planes beneath the surface-mounted inductor may induce eddy current loss and elevate the current through the inductor.

4.1.4. Diode Selection

Assuming $t_1 \ll L/R$, the average cell current during balancing (the average value of the inductor current) I_{Cavg} is given by (11),

$$I_{Cavg} = \frac{1}{2} \cdot I_{peak} \cdot D \quad (11)$$

The voltage rating of the Schottky should be slightly higher than the pack voltage, and the average current through the diode I_{Davg} at the PWM duty cycle D when the inductor releases energy to the pack is given by (12),

$$I_{Davg} = I_{Cavg} \cdot \frac{V}{V_p} \approx I_{Cavg} \cdot \frac{1}{n} \approx \frac{1}{2n} \cdot D \cdot I_{peak} \quad (12)$$

Since I_{Davg} is significantly lower than I_{Cavg}, the required current rating of Schottky diodes is significantly lower than I_{peak}, which is beneficial for cost reduction.

4.2. Experiment Setup

To validate the proposed balancing topology and its low-cost driving circuit, a series of experiments were conducted. The experimental setup, shown in Figure 6, consists of a five-cell series-connected battery pack, a five-cell balancer board with the gate driving circuit shown in Figure 3, and a controller board for cell voltage measurement, PWM generation, and data logging. Additionally, power supplies for the gate driving circuits and a bi-directional power supply were directly connected to the terminals of the pack for balancing tests during both charging and discharging. Table 4 lists the values of the components used for the balancer board in Figure 6.

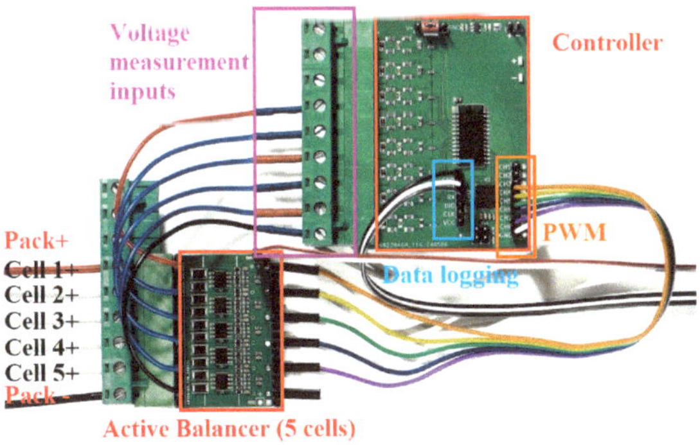

Figure 6. Balancing setup.

Table 4. Components specification and design considerations for the proposed balancer board.

	Component	Value	Number	1st Consideration	2nd Consideration
Cells	Li-ion	INR18650-25R	5	N/A	N/A
Analog front-end	Multiplexer	MUX507	1	N/A	N/A
	Op-amp	INA826	1	N/A	N/A
Controller	Microcontroller	STM32G0 series	1	ADC resolution	Cost

Table 4. *Cont.*

	Component	Value	Number	1st Consideration	2nd Consideration
Driving circuit	Dual-output Driver	UCC21525	1 *	Cost	N/A
	Capacitor (boot)	1 µF, 25 V	2 *	Cost	Voltage rating
	Resistor (gate)	2.2 Ω	2 * or 4 **	Cost	Gate current
	Resistor (gate-source)	300 kΩ	2 *	Cost	N/A
	Diode	1N4148	2 *	Cost	Recovery time
	Zener Diode (Optional)	15 V	2 *	Cost	Voltage rating
Balancing circuit	Power Inductor	10 uH	1 *	Current rating	Switching time
	MOSFET Pair	AO4614B	1 *	Voltage rating	Current rating
	Schottky Diode	PMEG4030ER	2 *	Voltage rating	Current rating

* Per cell. ** R_g connected to the driver output and C_{boot} is optional.

4.3. Balancing Circuit Verification

To verify that the topology and the driving circuit are functioning as expected, the five-cell series-connected pack was charged to 20.532 V, with the top cell at 4.2062 V and its negative terminal referenced to the *Pack−* at 16.327V. The power supply for the gate driving circuit is 11.25 V, referenced to *Pack−*. In this case, both the positive and negative terminals of the top cell have higher voltages than the supply voltage of the gate drivers.

In Figure 7, Channel 4 shows the 3.3 V, 175 kHz PWM control signal with a duty ratio of 70%, which feeds to the gate driver input and is used as the trigger of the source. Channel 1 and Channel 2 display the DC-coupled gate voltages of the *P*-channel MOSFET and *N*-channel MOSFET, respectively, for the top cell referenced to the Pack-(GND). Channel 3 shows the DC-coupled current through the inductor during balancing. (Two wires are attached to the inductor and then soldered to the PCB for inductor current measurement.)

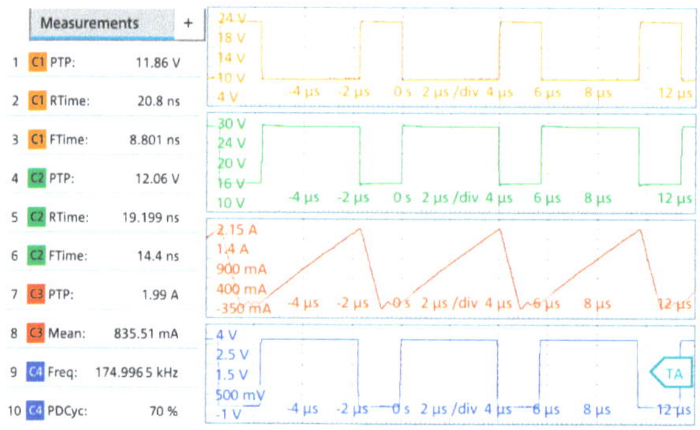

Figure 7. Experimental measurements (oscilloscope). Ch1: DC coupled gate voltage of *P*-channel MOSFET. Ch2: DC coupled gate voltage of *N*-channel MOSFET. Ch3: DC coupled inductor current Ch4: PWM control signal, trigger source.

Oscilloscope measurements show that the peak-to-peak voltage of the *P*-channel MOSFET is 11.86 V, with an 8.801 ns fall time and 20.8 ns rise time; the peak-to-peak voltage of the *N*-channel MOSFET is 12.06 V, with a 19.199 ns rise time and 14.4 ns fall time. The measured results validate that the proposed circuit can adequately drive the pair of complementary MOSFETs on the high side. The gate-source voltages of both MOSFETs are kept over 10 V for the entire on-time when using an 11.25 V (measured) non-isolated power supply. The rise and fall times of the gate voltage are short enough to avoid significant switching loss.

The current waveform in Channel 3 shows that when the control signal in Channel 4 rises to high, and the MOSFETs are turned on, the inductor is charged by the cell (between 0 to 4 µs), then releases energy back to the pack when the control signal drops too low, and the MOSFETs turn off (between 4 to 4.9 μs). The current ripple between 4.9 to 5.7 μs represents the diode reverse recovery and oscillation of parasitic components before the next cycle begins. The measured result matches the simulation result shown in Figure 4. The slight difference in peak current through the inductor is caused by the difference in cell voltage between the simulation and the actual test, as well as the tolerance of the inductor. The additional dead time (t_3) mentioned in Figure 4, inserted before the next pulse of PWM, may vary for each specific battery pack and can be finalized during this stage.

4.4. Control Algorithm

The control algorithm for the proposed active balancer is voltage-based and designed to ensure efficient and balanced energy distribution among cells in a battery pack.

As outlined in the flowchart Figure 8a, the battery cell balancing process begins with the system's initialization and calibration. Then, the system scans the voltages of all battery cells to identify any variations. If the variation does not exceed the threshold initially, the system takes no action and waits before repeating the voltage scan. If the voltage variation exceeds a pre-defined threshold, the system configures the Pulse Width Modulation (PWM) timer based on the cell voltages. This allows the balancing process to adapt to the specific conditions of the cells, whether it involves a single cell or multiple cells requiring balancing. Once configured, the balancing process starts and continues for a specified duration, during which the system waits to allow the cells with higher voltages to be discharged and transfer the energy to the pack to equalize. The balancing process could be performed cell-by-cell or on multiple cells simultaneously. After the waiting period, the balancing is stopped, and a relaxation phase follows to account for any transient effects for more accurate measurements of the next scan. This iterative process ensures that balancing is achieved efficiently across either individual or multiple cells as needed.

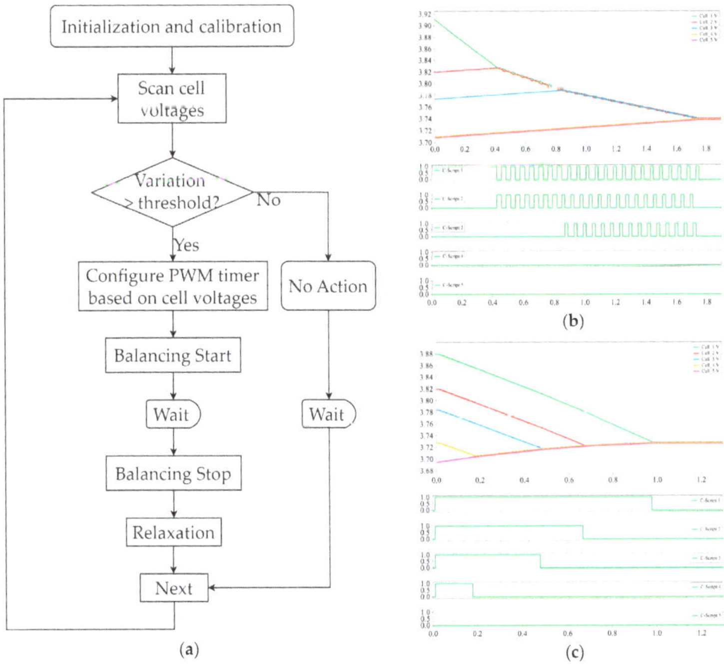

Figure 8. (**a**) Workflow of cell balancing algorithm; (**b**) cell voltages and balancing status vs. time for balancing cell-by-cell (simulation results); (**c**) cell voltages and balancing status vs. time for balancing multiple cells (simulation results).

Figure 8b illustrates the cell voltages (top plot) and the balancing status of each cell (bottom plot) during the simulation of balancing a single cell. As depicted in the bottom plot, only the cell with the highest voltage, which exceeds the balancing threshold, is discharged, releasing energy back to the battery pack. Figure 8c presents the cell voltages (top plot) and the balancing status of each cell (bottom plot) during the simulation of balancing multiple cells. As shown in the bottom plot, all cells whose voltages exceed the balancing threshold are discharged simultaneously, releasing energy to the battery pack.

4.5. Experimental Results

The experiment results presented in Figures 9–12 demonstrate the effectiveness of battery cell balancing under various operating conditions.

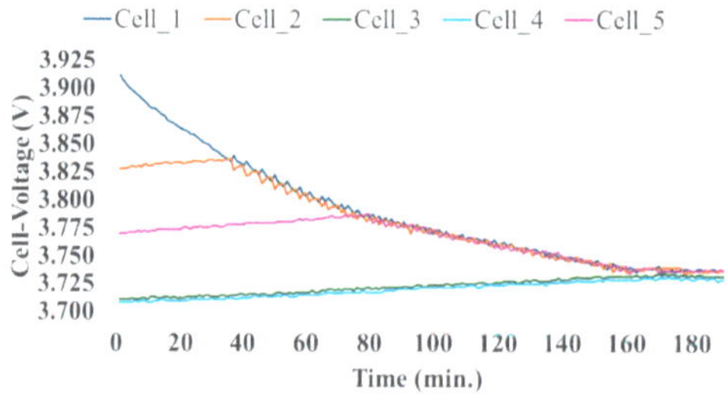

Figure 9. Experimental results illustrating cell voltages vs. time (balancing cell-by-cell, 170 min are required to reduce the cells' voltages difference to below a threshold of 10 mV).

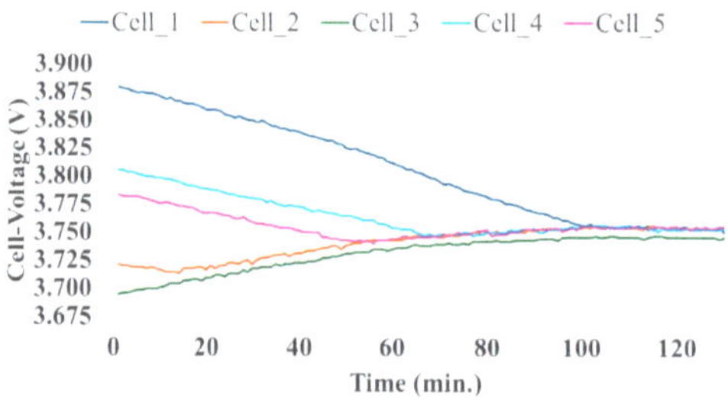

Figure 10. Experimental results illustrating cell voltages vs. time (balancing multiple cells, 100 min are required to reduce the cell voltage difference to below a threshold of 10 mV).

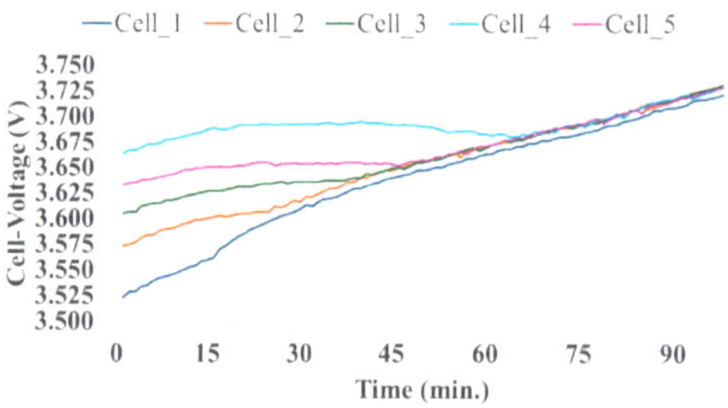

Figure 11. Experimental results illustrating cell voltage vs. time (balancing multiple cells while charging, 67 min are required to reduce the cell voltage difference to below a threshold of 10 mV).

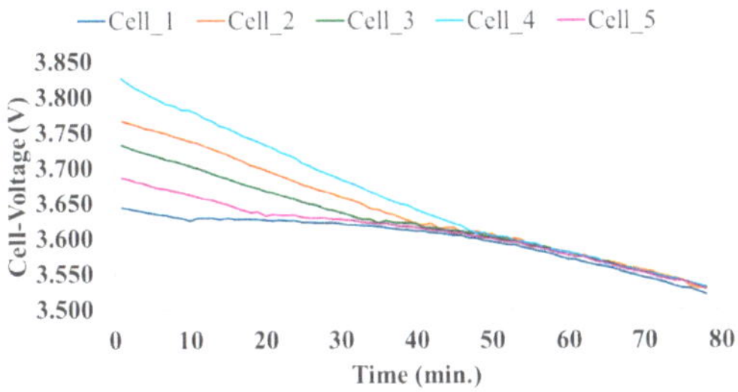

Figure 12. Experimental results illustrating cell voltage vs. time (balancing multiple cells while discharging, 47 min are required to reduce the cell voltage difference to below a threshold of 10 mV).

In Figure 9, the cell voltage variation over time during the cell-by-cell balancing is shown. In this test, only the cell with the highest cell voltage is discharged and releases the energy to the pack. Initially, cell 1 had a significantly higher voltage than the other cells. As the balancing process progresses, the voltage of cell 1 decreases steadily while the other cells are being charged until its voltage converges with the voltage of cell 2. Now, cells 1 and 2 are the cells with the highest voltage, and subsequently, they are being alternatively discharged one after the other until their voltages converge to the voltage of cell 5. By the end of the process, all cell voltages converge to the same voltage within the balancing threshold. The steady increase in cell voltages indicates that the balancing process does not discharge the cells but rather evenly charges them.

Cells 3 and 4 have the lowest voltage initially, and since their voltage variation remains within the balancing threshold, they are only being charged during the whole balancing process. This proves that the balancer does not introduce additional imbalance within the pack.

Figure 10 illustrates the result for multiple cell balancing. Except for cell 3, which has the lowest cell voltage at the beginning of the test, all other cells (1, 2, 4, and 5) are being discharged and release energy to the pack until their voltages are within the balancing threshold. By the end of the process, all cell voltages converge, indicating successful multicell balancing. This test also demonstrates the system's capability to balance multiple cells simultaneously, thereby ensuring a faster and more uniform voltage distribution throughout the battery pack.

Figure 11 shows the result of balancing multiple cells during the charging process. The imbalanced pack is charged with an external power supply during the test. Balancing is not enabled for the first 15 minutes, and significant cell voltage differences remain while charging. Once the balancing is started, the voltage increases of cells with higher voltages become slower as they are being discharged by the balancer, and the cells with lower voltages are being charged faster. In this case, the system is not only balancing the cells but also compensating for the cell voltage increases due to charging. Despite the overall rise in the cell voltage, the voltages gradually converge, demonstrating the system's ability to perform cell balancing in a dynamic charging condition.

Additionally, Figure 12 presents the balancing result while discharging. Similar to the previous experiments, the system equalizes the voltages of cells while the battery pack is connected to a discharging load. This showcases the system's robustness in maintaining balanced cell voltages under both charging and discharging scenarios.

5. Conclusions

This paper reviewed various existing solutions for active balancing and proposed a novel, inductor-based, cell-to-pack active balancing topology capable of operating across a wide range of cell voltages. Compared to conventional balancers, this topology integrates the necessary driving circuits, thus reducing overall system costs by eliminating the need for costly isolated power modules or high-side gate drivers while maintaining a simpler control structure. The proposed system efficiently transfers energy from individual cells to the battery pack, enabling multicell balancing with a balancing current that remains unaffected by the level of imbalance. Furthermore, it employs a straightforward control algorithm, simplifying implementation. The system's effectiveness is confirmed by the experimental results, which demonstrate the convergence of cell voltage during both charging and discharging operations.

Future work will focus on adapting the circuit design for series-connected pack-level balancing. Additionally, a simple and cost-effective controller will be developed to ensure continuous conduction mode (CCM) operation in higher-voltage applications, thereby increasing average balancing power. Experimental validation will be extended to larger battery modules and real-world applications, such as PV energy storage systems, to evaluate scalability and long-term reliability.

Author Contributions: Conceptualization, H.S.; methodology, H.S.; software, H.S.; validation, H.S.; formal analysis, H.S.; investigation, H.S.; resources, H.S., B.H. and J.F.; data curation, H.S.; writing—original draft preparation, H.S.; writing—review and editing, B.H.; visualization, H.S.; supervision, B.H.; project administration, B.H.; funding acquisition, B.H. and J.F. All authors have read and agreed to the published version of the manuscript.

Funding: This research received no external funding.

Data Availability Statement: Data are available on request from the corresponding author.

Conflicts of Interest: The authors declare no conflicts of interest.

References

1. Ye, Y.; Lin, J.; Li, Z.; Wang, X. Double-Tiered Cell Balancing System with Switched-Capacitor and Switched-Inductor. *IEEE Access* **2019**, *7*, 183356–183364. [CrossRef]
2. Global EV Outlook 2023—Analysis. Available online: https://www.iea.org/reports/global-ev-outlook-2023 (accessed on 9 January 2025).
3. Wang, H.; Rasheed, M.; Hassan, R.; Kamel, M.; Tong, S.; Zane, R. Life-Extended Active Battery Control for Energy Storage Using Electric Vehicle Retired Batteries. *IEEE Trans. Power Electron.* **2023**, *38*, 6801–6805. [CrossRef]
4. Zhu, J.; Mathews, I.; Ren, D.; Li, W.; Cogswell, D.; Xing, B.; Sedlatschek, T.; Kantareddy, S.N.R.; Yi, M.; Gao, T.; et al. End-of-Life or Second-Life Options for Retired Electric Vehicle Batteries. *Cell Rep. Phys. Sci.* **2021**, *2*, 100537. [CrossRef]
5. Wang, H.; Bai, H.; Fu, Y.; Tao, Z.; Xiao, H.; Si, N.; Bai, H.; Deng, S. The Capacity Effect on Consistency of Energy Storage Batteries. In Proceedings of the 2018 International Conference on Electronics Technology (ICET), Chengdu, China, 23–27 May 2018; pp. 1–5.
6. Omariba, Z.B.; Zhang, L.; Sun, D. Review of Battery Cell Balancing Methodologies for Optimizing Battery Pack Performance in Electric Vehicles. *IEEE Access* **2019**, *7*, 129335–129352. [CrossRef]
7. Izadi, Y.; Beiranvand, R. A Comprehensive Review of Battery and Supercapacitor Cells Voltage-Equalizer Circuits. *IEEE Trans. Power Electron.* **2023**, *38*, 15671–15692. [CrossRef]
8. Ye, Y.; Cheng, K.W.E.; Fong, Y.C.; Xue, X.; Lin, J. Topology, Modeling, and Design of Switched-Capacitor-Based Cell Balancing Systems and Their Balancing Exploration. *IEEE Trans. Power Electron.* **2017**, *32*, 4444–4454. [CrossRef]
9. Shang, Y.; Lu, F.; Xia, B.; Zhang, C.; Cui, N.; Mi, C. A Switched-Coupling-Capacitor Equalizer for Series-Connected Battery Strings. In Proceedings of the 2017 IEEE Applied Power Electronics Conference and Exposition (APEC), Tampa, FL, USA, 26–30 March 2017; pp. 1425–1429.
10. Ye, Y.; Cheng, K.W.E. Modeling and Analysis of Series–Parallel Switched-Capacitor Voltage Equalizer for Battery/Supercapacitor Strings. *IEEE J. Emerg. Sel. Top. Power Electron.* **2015**, *3*, 977–983. [CrossRef]
11. Lu, J.; Wang, Y.; Li, X. Isolated Bidirectional DC–DC Converter with Quasi-Resonant Zero-Voltage Switching for Battery Charge Equalization. *IEEE Trans. Power Electron.* **2019**, *34*, 4388–4406. [CrossRef]
12. Conway, T. An Isolated Active Balancing and Monitoring System for Lithium Ion Battery Stacks Utilizing a Single Transformer Per Cell. *IEEE Trans. Power Electron.* **2021**, *36*, 3727–3734. [CrossRef]

13. Li, S.; Mi, C.C.; Zhang, M. A High-Efficiency Active Battery-Balancing Circuit Using Multiwinding Transformer. *IEEE Trans. Ind. Appl.* **2013**, *49*, 198–207. [CrossRef]
14. Lee, K.-M.; Lee, S.-W.; Choi, Y.-G.; Kang, B. Active Balancing of Li-Ion Battery Cells Using Transformer as Energy Carrier. *IEEE Trans. Ind. Electron.* **2017**, *64*, 1251–1257. [CrossRef]
15. Zilio, A.; Fogagnolo, D.; Gallo, E.; Biadene, D.; Mattavelli, P. A Multiport Converter for Flexible Active Balancing in Li-Ion Batteries. *IEEE Trans. Ind. Electron.* **2024**, *71*, 7085–7094. [CrossRef]
16. Al-Smadi, M.K.; Abu Qahouq, J.A. Evaluation of Current-Mode Controller for Active Battery Cells Balancing with Peak Efficiency Operation. *IEEE Trans. Power Electron.* **2023**, *38*, 1610–1621. [CrossRef]
17. Xu, G.; Chen, E.; Liu, F.; Liu, Y.; Xiong, W.; Su, M. Multicell-to-Multicell Equalizer with Hybrid Pulsewidth Modulation and Phase-Shift Modulation. *IEEE Trans. Power Electron.* **2024**, *39*, 4400–4411. [CrossRef]
18. Kutkut, N.H. A Modular Nondissipative Current Diverter for EV Battery Charge Equalization. In Proceedings of the APEC '98 Thirteenth Annual Applied Power Electronics Conference and Exposition, Anaheim, CA, USA, 15–19 February 1998; Volume 2, pp. 686–690.
19. Park, S.-H.; Kim, T.-S.; Park, J.-S.; Moon, G.-W.; Yoon, M.-J. A New Buck-Boost Type Battery Equalizer. In Proceedings of the 2009 Twenty-Fourth Annual IEEE Applied Power Electronics Conference and Exposition, Washington, DC, USA, 15–19 February 2009; pp. 1246–1250.
20. Ding, X.; Zhang, D.; Cheng, J.; Wang, B.; Chai, Y.; Zhao, Z.; Xiong, R.; Luk, P.C.K. A Novel Active Equalization Topology for Series-Connected Lithium-Ion Battery Packs. *IEEE Trans. Ind. Appl.* **2020**, *56*, 6892–6903. [CrossRef]
21. Wang, S.; Yang, S.; Yang, W.; Wang, Y. A New Kind of Balancing Circuit with Multiple Equalization Modes for Serially Connected Battery Pack. *IEEE Trans. Ind. Electron.* **2021**, *68*, 2142–2150. [CrossRef]
22. Phung, T.H.; Collet, A.; Crebier, J.-C. An Optimized Topology for Next-to-Next Balancing of Series-Connected Lithium-Ion Cells. *IEEE Trans. Power Electron.* **2014**, *29*, 4603–4613. [CrossRef]
23. Liu, F.; Zou, R.; Liu, Y. An Any-Cell-to-Any-Cell Battery Equalizer Based on Half-Bridge LC Converter. *IEEE Trans. Power Electron.* **2023**, *38*, 4218–4223. [CrossRef]
24. He, Y.; Liu, X.; Liu, X.; Zheng, X.; Su, L. Research on the High Speed Inductor-Based Active Battery Equalization Schemes. In Proceedings of the 2016 IEEE 8th International Power Electronics and Motion Control Conference (IPEMC-ECCE Asia), Hefei, China, 22–26 May 2016; pp. 1389–1393.

batteries

MDPI

Article

Evaluating the Role of Entropy Change in Lithium-Ion Battery Electro-Thermal Modelling

Félix-Antoine LeBel [1,2], Pascal Messier [1,2], Mathieu Blanchard [2] and João Pedro F. Trovão [1,3,4,*]

1 electric-Transport, Energy Storage and Conversion (e-TESC) Lab., Université de Sherbrooke, Sherbrooke, QC J1K 2R1, Canada; felix-antoine.lebel@usherbrooke.ca (F.-A.L.); pascal.messier@usherbrooke.ca (P.M.)
2 SysNergie Inc., Sherbrooke, QC J1N 0G2, Canada; mathieu.blanchard@sysnergie.ca
3 Institute for Systems Engineering and Computers at Coimbra (INESC Coimbra), Department of Electrical and Computer Engineering, University of Coimbra, Polo II, 3030-290 Coimbra, Portugal
4 Coimbra Institute of Engineering, Polytechnic of Coimbra (IPC-ISEC), 3030-199 Coimbra, Portugal
* Correspondence: joao.trovao@usherbrooke.ca

Abstract: The accurate estimation of lithium-ion cell internal temperature is crucial for the safe operation of battery packs, especially during high discharge rates, as operating outside the safe temperature range can lead to accelerated degradation or catastrophic failures. Heat generation in lithium-ion cells arises primarily from ohmic losses and entropy change (ΔS), yet the latter remains frequently overlooked in battery modelling. However, the impact of considering or discarding ΔS from electro-thermal modelling remains subject to debate. This research highlights the critical role of ΔS in improving the accuracy of electro-thermal models for lithium-ion batteries, particularly in high-fidelity thermal simulations. It presents a systematic integration, ΔS, into electro-thermal models, leveraging the energetic macroscopic representation (EMR) approach to enhance predictive accuracy, a methodology not previously structured in this manner. This paper addresses this issue by performing a comparative analysis of an electro-thermal model (ETM) with and without ΔS. The findings provide clear insights into the role of entropy in electro-thermal modelling, demonstrating that while entropy change has a minimal impact on electrical behaviour prediction, it plays a crucial role in accurately capturing temperature dynamics, helping define the conditions under which it must be considered in simulations. While entropy can be neglected for coarse heat generation estimation, its inclusion enhances temperature prediction accuracy by up to 4 °C, making it essential for applications requiring precise thermal management. This study offers a detailed analysis of the conditions under which ΔS becomes critical to model accuracy, providing actionable guidance for battery engineers and researchers.

Keywords: lithium-ion battery; heat generation; entropic coefficient; electro-thermal model; electric equivalent circuit

1. Introduction

Lithium-ion batteries (LIBs) are widely used across various applications, from electric vehicles (EVs) to battery energy storage systems (BESS), where safety, durability, and reliability are of paramount concern. Temperature is a critical factor influencing the performance and longevity of LIBs. Temperature directly influences key parameters such as cell impedance, capacity retention, and permissible charge/discharge rates. Furthermore, temperature accelerates the degradation of electrode materials, a process known as aging. Several recent events have highlighted the dramatic consequences of the violent catastrophic failure of LIBs. Excessive heat buildup is a known failure causing thermal runaway (TRA), posing significant safety risks [1].

To ensure the safe and reliable operation of LIBs, it is essential to maintain temperature and voltage within specified safe operational windows. The accurate assessment of temperature dynamics and heat generation across diverse operating scenarios by the battery management system (BMS) is crucial for preventing accelerated degradation or TRA [2]. Accurate thermal estimation is particularly vital in the battery pack design phase, influencing both thermal management strategies and overall system design.

Unlike other types of electrical components, LIBs are non-linear electrochemical systems in which heat generation cannot be directly inferred from the nominal internal resistance value. Their behaviour and properties are simultaneously influenced by temperature, the state of charge (SOC), and aging, all of which interact in complex ways. Additionally, structural changes caused by the intercalation and de-intercalation of lithium ions in the electrodes during charge and discharge cycles, in turn causing changes in entropy and enthalpy that alter both the thermal and electrical properties of the cells. The phenomena has been studied by [3], and it has been found that the thermodynamics of lithium intercalation into graphite is governed by a complex interplay of mixing entropy, vibrational entropy, and lithium–lithium interactions.

In lithium-ion batteries, entropy (S) and entropy change (ΔS) are related concepts, but they represent different aspects of the thermodynamics within the battery system. Entropy in the context of a lithium-ion battery refers to the inherent thermodynamic property that quantifies the degree of disorder or randomness of the battery's system in a specific state. Each component and phase of a lithium-ion battery—such as the cathode, anode, and electrolyte—has a certain entropy value based on its structure and composition. In general, higher entropy signifies a greater degree of disorder.

Entropy change (ΔS) is a dynamic concept that measures the difference in entropy between two states, particularly during chemical reactions or physical processes, like charging and discharging. For LIBs, this is especially relevant because entropy change can reveal valuable insights into electrochemical processes, like lithium-ion intercalation and de-intercalation. Entropy change is often calculated for reactions occurring at the electrodes and can provide information on the energy efficiency and temperature dependence of the battery's operation. Entropy is a static property of a given state, while entropy change is dynamic, reflecting the shift between two states (e.g., before and after lithium-ion intercalation).

Entropy change impacts thermal management in lithium-ion batteries. During charging and discharging, changes in entropy contribute to heat generation within the battery. Understanding these entropy changes is crucial in designing systems to manage temperature and avoid overheating. Entropy changes can be experimentally measured through thermodynamic methods and provide insights into how temperature affects the electrochemical potential of the battery, which is useful for optimizing operating conditions.

Since the characterization of ΔS is a long and tedious process, it is often overlooked in system-level models, such as electric vehicle simulation. Omitting this fundamental behaviour can potentially lead to discrepancies between measurements and model predictions. This paper investigates whether accounting for entropy change can improve the accuracy of heat generation and temperature estimations in lightweight electro-thermal models used in the BMS and system-level performance assessement.

Estimating the various states of an LIB is a critical function of modern BMSs and a key consideration in LIB design. Contemporary modelling approaches differ in complexity, with the choice of model depending on the desired balance between accuracy, spatial resolution, and computational feasibility.

Physics-based models (PBMs) are designed to improve the accuracy of battery modelling by capturing detailed internal dynamics. The foundation of fully physics-based lithium-ion battery models is the pseudo-two-dimensional (P2D) porous electrode model, also known as the Doyle–Fuller–Newman model, named after the researchers who pioneered its development [4–6]. This model is based on porous electrode theory, concentrated solution theory, and the Butler–Volmer kinetic equations.

Variants of the P2D model include the single-particle model (SPM) [7], which simplifies calculations to reduce computational demands for embedded applications, and the multi-dimensional multi-physics model (MuDiMod) [8,9], which increases model complexity to investigate the intricate internal behaviour of lithium-ion cells. PBMs offer insights into internal battery processes such as lithium-ion diffusion, Ohmic effects, and electrochemical kinetics. These models enable the exploration of battery degradation mechanisms, the prediction of the SOC and state of health (SOH) with aging considerations, and the development of optimal charging strategies. However, PBMs are composed of numerous partial differential equations (PDEs), requiring significant computational power to solve. This inherent complexity makes the real-time integration of PBMs into low-power controllers found in BMSs impractical [10].

Equivalent-circuit models (ECMs) [11–17] represent LIBs as networks of electronic components such as resistors and capacitors. These resistor–capacitor networks are used to capture battery behaviour over various time constants associated with diffusion and charge-transfer processes [18]. ECMs are based on empirical insights and experimental data, making them the predominant model for real-time SOC estimation in electric vehicle BMSs. Their popularity stems from their ability to predict electrical behaviour quickly and with minimal computational load, as most information can be pre-computed and stored in lookup tables.

Two main methods are commonly used to determine ECM parameters: electrochemical impedance spectroscopy (EIS) [19,20] and the galvanostatic intermittent titration technique (GITT) [21], also known as pulse testing and referred to by various names in the prior literature [22]. ECMs, however, typically do not account for temperature and heat generation effects, so they are often coupled with thermal models of varying complexity to address these factors.

Entropy change (ΔS) represents the heat absorbed or released due to microstructural changes during lithium intercalation. This reversible heat, which differs from non-reversible heat generation related to impedance, is traditionally measured using potentiometric methods [23–25]. However, these measurements are labour-intensive, prompting a common practice of excluding ΔS in the practical thermal modelling of lithium-ion batteries [26].

Neglecting ΔS is a common engineering practice in the thermal modelling of lithium-ion cells. For instance, the models found in [26] all neglect entropy. Ye et al. argued that there is a reversible heat source during discharge at higher currents, thus concluding that entropy can be neglected in such cases [27].

As battery materials evolve to minimize Ohmic losses and new chemistries are developed, the consideration of ΔS could become increasingly relevant to improve model accuracy. This paper aims to offer battery engineers a nuanced perspective on when and to what extent ΔS should be incorporated into electro-thermal models, considering its impact on model accuracy under different operational scenarios.

2. Methodology

In this study, a rigorous methodology was employed to investigate the impact of entropy change. Firstly, a comprehensive model was constructed using MatLab/Simulink, providing a computational framework for analysis. Subsequently, the model's parameters, crucial for accurate simulation, were determined through various experimental methods. Parameters such as impedance, open-circuit voltage, the coefficient of convection, and entropy change were precisely quantified. To validate fidelity and predictive capability, the model was benchmarked against real-world data obtained from continuous discharge and actual driving cycles. The accuracy of the proposed model was evaluated under constant current discharge and dynamic profile scenarios, offering valuable insights into the importance of considering ΔS. Finally, leveraging the validated model, simulations were conducted to explore various scenarios of continuous currents and power discharge at different rates. The insights obtained from these simulations are intended to help guide engineers in their decision process of whether to include entropy change or not in simulations.

The experimental setup used for this work is pictured in Figure 1. Cells were held in place and connected by ZKE Battery Rack cell fixtures, which had independent probes for voltage measurement and current carrying [28]. Each cell holder was connected to one of the 48 channels of an LBT-2000 (Arbin Instruments, College Station, TX, USA) cell cycler with the four-wire Kelvin method [29]. The precision of each channel was 0.5 mV. This equipment is well suited for detailed battery characterization and Coulombic efficiency studies due to its very high level of precision. The sampling rate of the channels was set to 10 Hz. Throughout all tests, constant ambient temperature in the samples was ensured by placing the samples inside a CSZ MicroClimate (Cincinnati Sub-Zero, Cincinnati, OH, USA) test chamber [30]. The temperature of each cell was measured with Type-T thermocouples, placed midway along the length of the cells, held in place with a thin polyamide film tape.

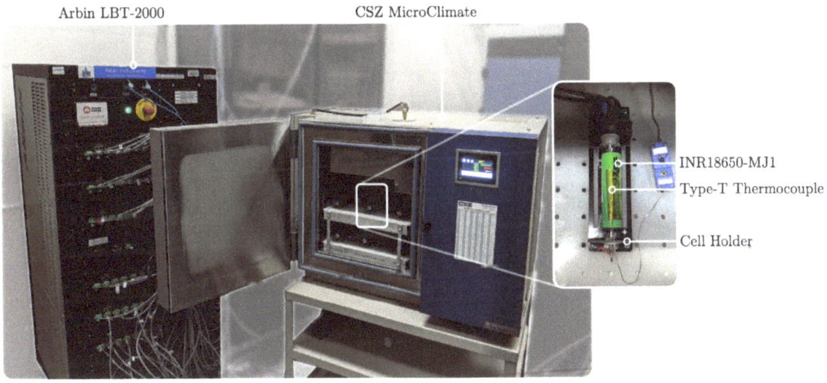

Figure 1. Experimental setup.

For this work, 3.5 Ah INR18650-MJ1 (LG Chem, Seoul, South Korea) cells composed of a nickel–manganese–cobalt cathode and silicon–graphite anode were used. The choice of the INR18650-MJ1 was based on the fact that its behaviour and construction has already been studied in the prior literature [31–38], so it can be tied to several other publications.

The model developed in this work was composed of a lumped thermal model with convection and a first-order RC ECM electrical model. Figure 2 shows the basic representation of the thermal and electrical models. The sign convention used in this work was such that positive current was considered flowing out of the cell to provide power to the load it was connected to.

It was assumed that properties and behaviour were homogeneous throughout the cell body and electrodes. The electrical boundary of the system was the load, and the thermal boundary was natural convection with the surrounding air. The coupling of the two models forming the electro-thermal model (ETM) could be visualized conceptually using energetic macroscopic representation (EMR). Figure 3 illustrates the interactions between the various physical behaviours within a lithium-ion cell. Similar representations have been depicted in prior work [39].

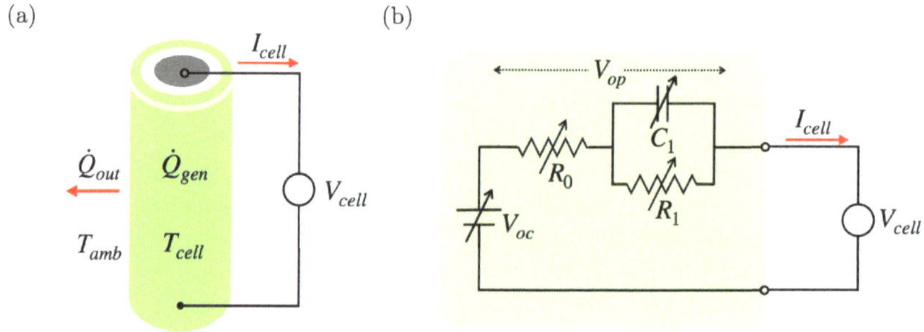

Figure 2. Schematic representation of lithium-ion cell. (**a**) Thermal. (**b**) Electrical.

EMR is a systemic representation for modelling multi-physical systems. It is a well-suited representation for structuring models in a graphical programming environment such as MatLab/Simulink. The model was implemented in MatLab/Simulink R2021b.

In this representation, the behaviour of the cell is broken down in three main components: the electrochemical energy source (designated as Electrodes), the losses (depicted as a multi-physics coupling block), and the thermal inertia. A load model and a thermal boundary conditions model (designated as Load/Supply and Air) complete the ETM. In the model, the electrochemical energy source defines the open-circuit potential of the cell V_0. It is assumed that V_0 is the theoretical voltage of the cell without any current flowing through the cell. The value of V_0 was measured in prior work [22] at temperatures, T, ranging from −20 °C to 60 °C, at increments of 1% SOC, as shown in Figure 4. Its value was estimated by bilinear interpolation within the empirical response surface.

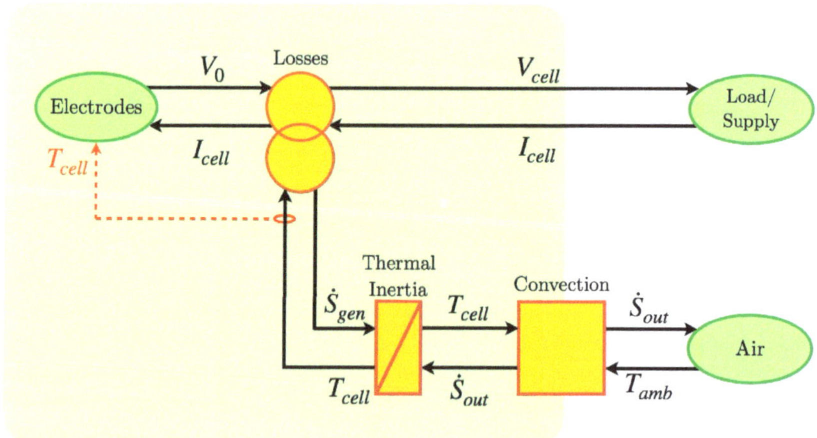

Figure 3. Equivalent energetic macroscopic representation (EMR) of lithium-ion cell.

The electrochemical potential V_0 of a cell is a function of the amount of lithium ions present in the electrodes, or SOC. As ions are intercalated in the anode and cathode, the resulting potential varies accordingly. While the evolution of potential with regard to the degree of lithiation is somewhat linear, phase changes in the microstructure of the electrodes may cause fluctuations to the slope at certain specific SOCs [40]. Note that V_{cell} and V_{op} are functions of the cells SOC θ, temperature T_{cell}, current I_{cell}, and time t, while V_0 is not influenced by the current and time. SOC, θ, is defined as the ratio of the remaining quantity of electrical charges within the electrodes to the theoretical total amount of useful charges of said electrodes C_0 at the reference temperature (25 °C). The remaining quantity of charges can be computed by the integration of I_{cell} over t. For (1) to be computed correctly, C_0 is expressed in Coulombs (C).

$$\theta = \frac{\int I_{cell} dt}{C_0} \tag{1}$$

A first-order ECM is used to represent overpotentials in this work. The overpotential voltage of a resistor–capacitor (RC) ECM is given in (2) in the general form where the electrical network is composed of the entirely resistive component R_0 and n RC couples, designated by R_k and C_k, where k is the couple instance subscript.

$$V_{op}(\theta, T, t) = R_0(\theta, T)I(t) + \sum_{k=1}^{n} R_k(\theta, T)I(t)e^{-t/R_k(\theta,T)}C_k(\theta, T) \tag{2}$$

This translates to the following transfer function in the Laplace domain (3).

$$V_{op}(\theta, T) = IR_0(\theta, T) + I\sum_{k=1}^{n} \frac{R_k(\theta, T)}{R_k(\theta, T)C_k(\theta, T)s + 1} \tag{3}$$

In the model, V_{cell} is the difference between the potential and overpotential V_{op} caused by the inefficiencies of the various processes (4).

$$V_{\text{cell}}(\theta, T, I, t) = V_{oc}(\theta, T) - V_{\text{op}}(\theta, T, I, t) \tag{4}$$

The effective electrical power of the cell, P_{cell}, is the product of the electrical current I_{cell} and cell voltage V_{cell} (5).

$$P_{\text{cell}} = V_{\text{cell}} I_{\text{cell}} \tag{5}$$

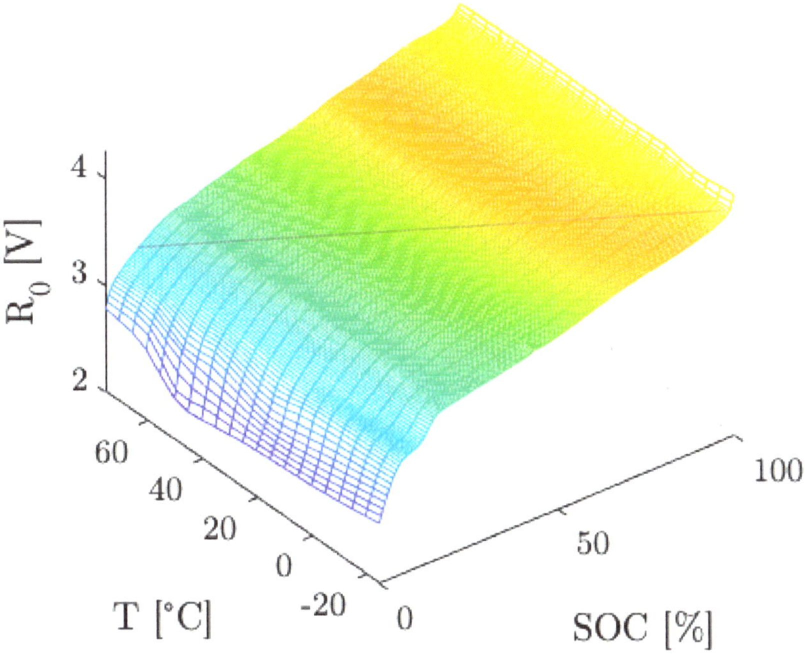

Figure 4. Open-circuit voltage interpolation table.

The complexity of the non-linear property changes with regard to θ and T can be difficult to model. Therefore, empirical mapping as described in [22] remains an accurate and straight-forward approach. Figure 5 shows the response surface for the different impedance parameters of the model. Impedance characteristics are unique to each lithium-ion cell model and manufacturer. They are influenced by factors such as the chemical composition of the electrode's active materials, as well as numerous design- and production-induced parameters, including electrode thickness and particle size [41].

Irreversible losses generated by overpotential (Q_{op}) can be expressed by (6)

$$\dot{Q}_{op} = V_{op} I_{cell} \tag{6}$$

However, the total heat generation inside the cell ($\dot{Q}_{gen}$) is equal to the sum of irreversible losses, $\dot{Q}_{op}$, and reversible losses from entropy change, $\dot{Q}_{ent}$. While ΔS can be seen as a variation in temperature caused by a change in potential, the opposite is also true. Equation (7) describes this relationship, where F is the Faraday constant. Therefore, using the direct correlation between temperature change and voltage change, one can obtain the entropic coefficient of a cell by applying a temperature change to a cell, while measuring the resulting voltage change. This method is known as the potentiostatic method [23–25].

$$\Delta S = F \frac{dV}{dT} \tag{7}$$

In a controlled temperature chamber, T_{cell} was changed from 20 °C to 40 °C with an acclimatization period of approximately 30 min after each temperature change to allow the full entropic phase transformation. Using the experimental setup described earlier, two samples were tested along the entire capacity by discharging the cell at a rate of C/10 and resting at equally spaced intervals of 5% SOCs to perform the test. A similar methodology was proposed by Geifes et al. [42].

This property is dependent of the degree of lithiation of the electrodes or θ at the cell level and has been shown to be independent of the temperatures at which it is measured [25,43]. The entropy change coefficient ΔS is shown in Figure 6.

Assuming linear interpolation between the measurement points, reversible heat flow due to ΔS was computed with Equation (8).

$$\dot{Q}_{\text{ent}} = \frac{-I T_{\text{cell}} \Delta S}{F} \tag{8}$$

The thermal output of the multi-physics losses model was as is stated in (9).

$$\dot{Q}_{\text{gen}} = \dot{Q}_{\text{op}} + \dot{Q}_{\text{ent}} \tag{9}$$

However, as discussed in German et al. [39], losses must be expressed in the flow of entropy as defined by (10) to respect the principles of EMR.

$$\dot{S} = \frac{\dot{Q}}{T} \tag{10}$$

Therefore, (9) becomes (11).

$$\dot{S}_{gen} = \frac{\dot{Q}_{gen}}{T_{cell}} \tag{11}$$

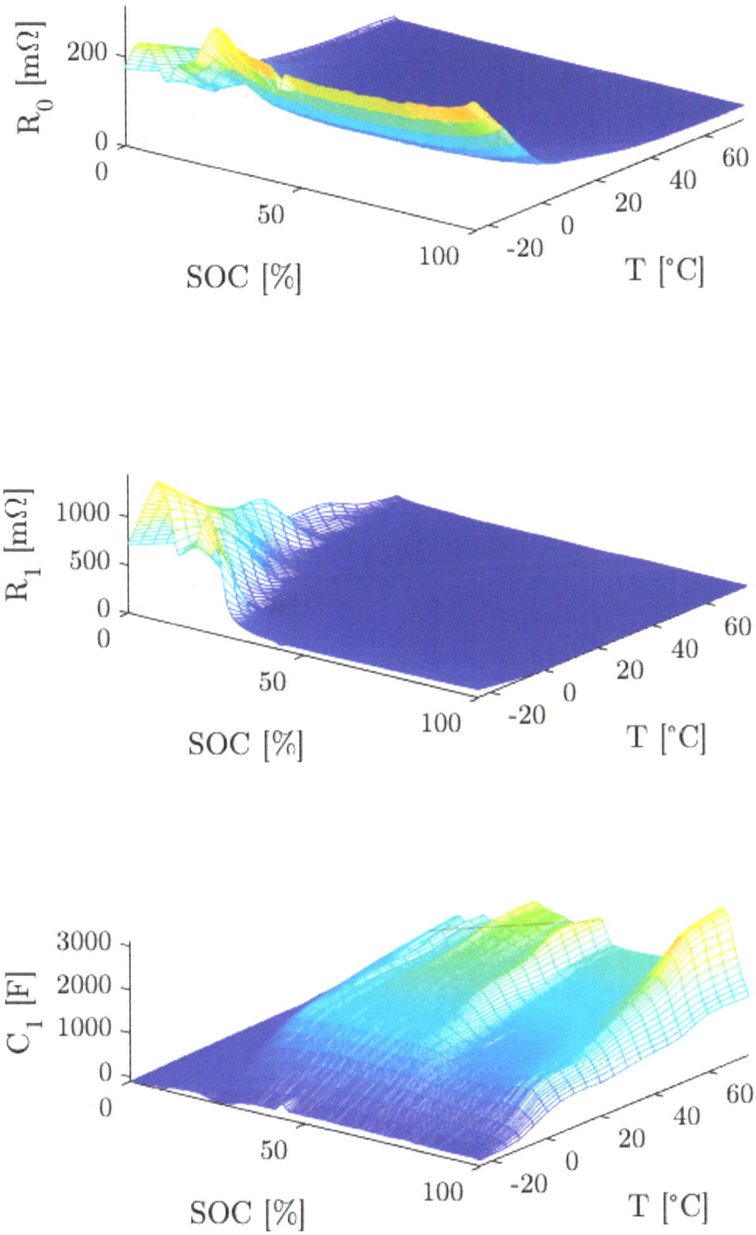

Figure 5. Map of parameters of electrical model.

This result was then fed as an input of the thermal inertia model (Thermal Inertia block in Figure 3). Unless the cell was placed in perfectly adiabatic boundary conditions, heat would leak out of the cell to its surroundings through radiative, convective, and conductive heat transfer. Radiation losses ($\dot{Q}_{rad}$) were considered negligible under normal operation. Conduction heat losses ($\dot{Q}_{cond}$) were also neglected from this work. Only convection heat transfer ($\dot{Q}_{conv}$) to the surrounding air was considered. To respect EMR formalism, convection heat transfer was expressed as entropy flow, S_{conv}. It was computed by (12).

$$\dot{S}_{\text{conv}} = \frac{hA(T_{\text{cell}} - T_{\text{amb}})}{T_{\text{cell}}} \tag{12}$$

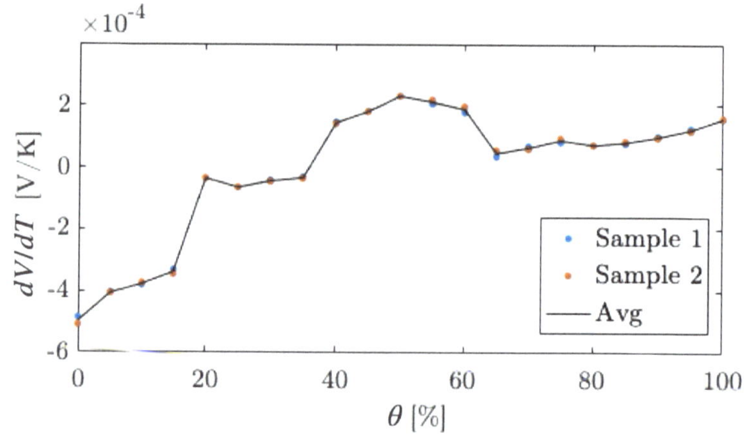

Figure 6. Entropic coefficient of cell at different states of charge.

While the surface area of a cell (A) can be estimated from geometry, the convection coefficient (h) is typically determined experimentally. Its value depends on various parameters such as fluid viscosity, surface finish, object shape, and orientation. The total entropy flow to the environment is given by (13).

$$\dot{S}_{\text{out}} = \dot{S}_{\text{rad}} + \dot{S}_{\text{cond}} + \dot{S}_{\text{conv}} \tag{13}$$

Cell temperature for the next iteration step T^*_{cell} was computed with (14),

$$T^*_{cell} = T_{cell} + \frac{\int(\dot{S}_{gen} - \dot{S}_{out})dt}{m\,Cp\,T_{cell}} \tag{14}$$

Thermal inertia is the product of the cell mass (m_{cell}) and a specific heat (Cp_{cell}). According to a review by Steinhardt et al., Cp_{cell} of cylindrical cell is between 884 J kg^{-1} K^{-1} and 1172 J kg^{-1} K^{-1}, with a lower median of 912 J kg^{-1} K^{-1} [44]. The model assumed a constant mass, m, in kg and specific heat, Cp, in J kg^{-1} K^{-1} in the cell. Assuming a constant convection coefficient, the value of h and Cp_{cell} could be computed by the least-square curve fitting of Equation (15).

$$T_{\text{cell}}(t) = T_{\text{amb}} e^{-\left(\frac{hA}{mcp} t\right)} \tag{15}$$

The values of h and Cp_{cell} were determined experimentally by heating test samples to 50 °C and then placing them at 23 °C, while recording T_{cell} throughout the cooldown at 10 Hz until the temperature stabilized to an ambient temperature T_{amb}. Figure 7 shows the temperature of the samples during the cooldown test. A value of 31.2 W m^{-2} K^{-1} was found for h and a value of 912 J kg^{-1} K^{-1} was determined for Cp_{cell}.

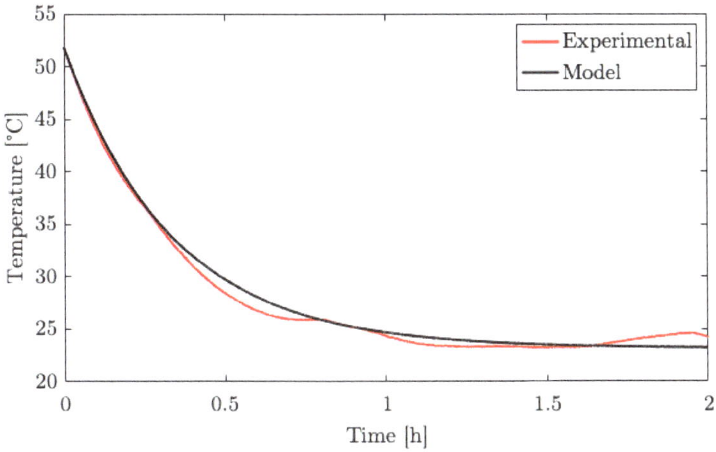

Figure 7. The cooldown of the samples from 50 °C to 23 °C compared with the model.

Other characteristics of the lithium-ion cell used in this work are gathered at Table 1.

Table 1. Cell specifications.

Name	Symbol	Value	Unit
Manufacturer	-	LG Chem	-
Model	-	INR18650MJ1	-
Chemistry	-	NMC-G	-
Nominal voltage	V_{nom}	3.65	V
Capacity	C_0	3.5	Ah
Energy	E_{nom}	12.7	Wh
Weight	m	48	g
Specific heat	Cp	912	$\mathrm{J\,kg^{-1}\,K^{-1}}$
Convection	h	31.2	$\mathrm{W\,m^{-2}\,K^{-1}}$

3. Results

The different parameters of the ETM were measured by a series of independent characterization procedures. The experimental validation of the model therefore had to be performed to ensure the proper implementation of the model. The validation of the model was performed using the same setup as shown in Figure 1 with a constant current discharge (1 C) and dynamic current conditions. Figure 8 shows the results of the experiment, benchmarked against the model.

The temperature in the climate chamber for this test was 23 °C. The slight fluctuations in the experimental measurement can be explained by the normal variations in temperature in the climate chamber. The results show that entropic heating mostly influenced temperature estimation. Regardless of entropic heating, the voltage error between the model and experiment was inferior to 10 mV throughout most of the discharge except below $\theta = 5\%$. Without consideration of entropic heating, the error between the model and experiment was superior to 2 °C for most of the discharge. Meanwhile, the measured error between the experimental and simulated temperatures was within 1 °C, while the overall evolution was preserved much more accurately. Thus, it can be concluded that the effect of entropy had a negligible effect on the accuracy of the electrical model but had a measurable effect on the thermal model.

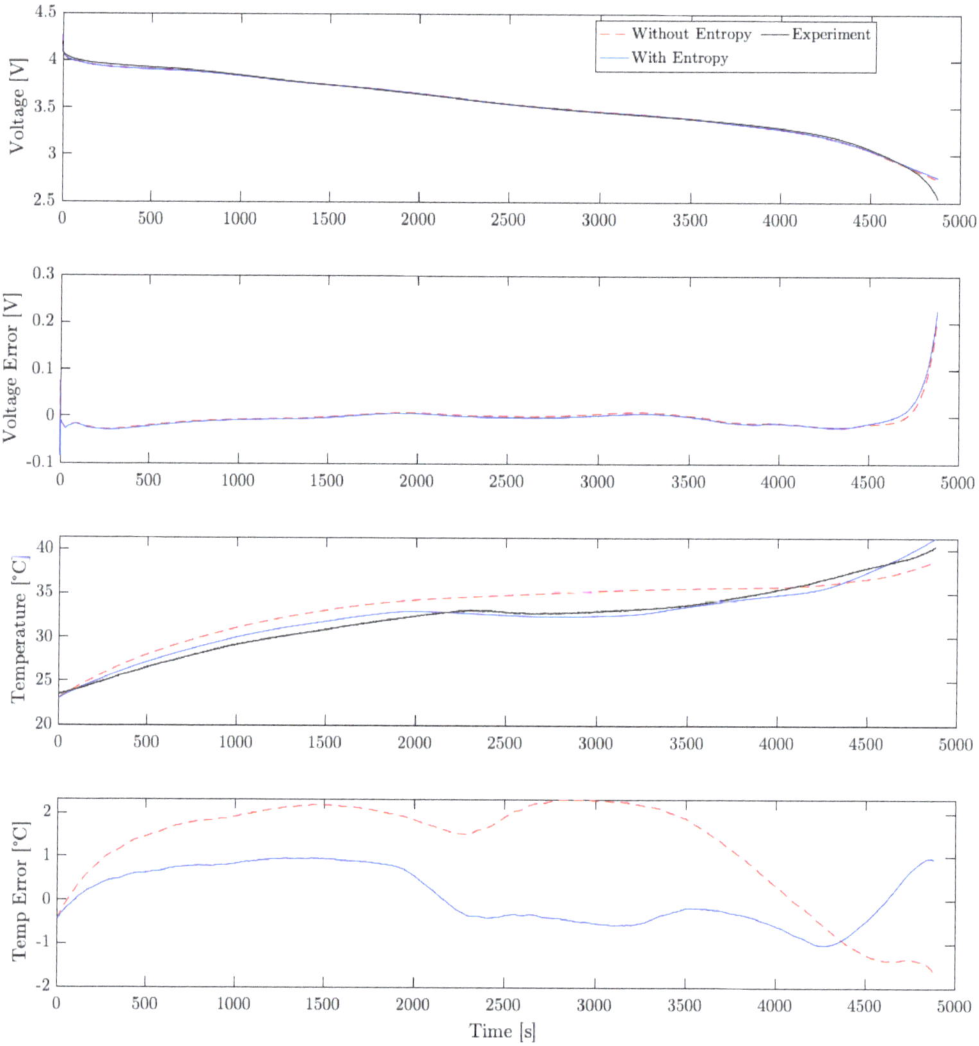

Figure 8. Constant current validation with and without entropy.

Further validation of the model was performed on a cell using a dynamic cycle from Messier et al. [45], representative of light electric vehicle drive cycle. The same experimental setup was used for this test. The vehicle current was scaled down to a cell-level equivalent current for the experiment. Figure 9 shows the applied current and the resulting voltage and temperature for the simulation and experiment.

The results show that the voltage error of the model was inferior to 70 mV throughout most of the discharge, except at θ, where it was inferior to 15%. The temperature error was inferior to 1 °C, with the highest error occurring when cell θ was between 50% and 70%, which corresponded to a rapid change in the entropic coefficient. It is likely that increasing the resolution of the entropic coefficient interpolation table could improve the thermal accuracy of the model.

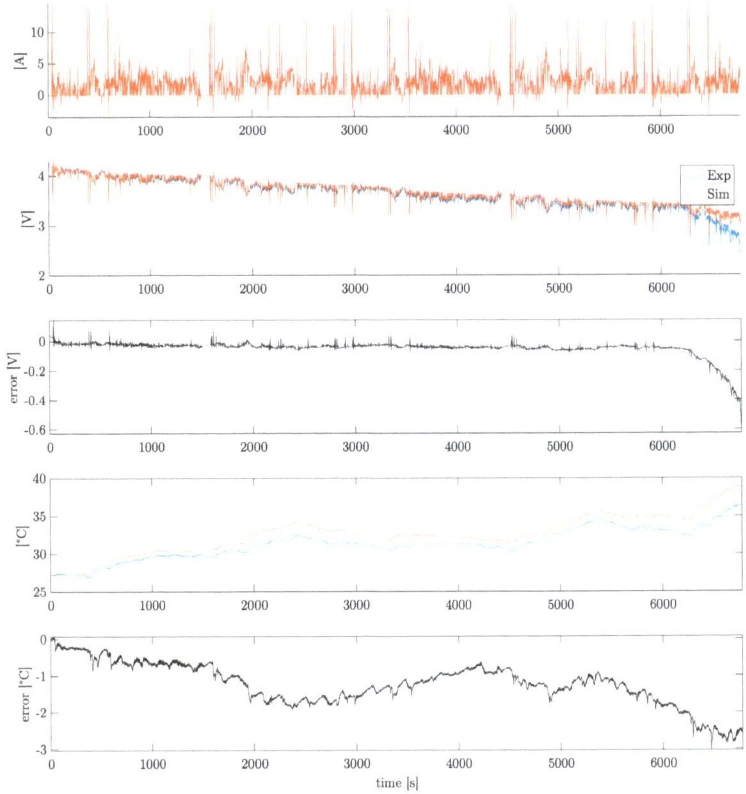

Figure 9. Electro-thermal model dynamic simulation.

4. Simulations

Having validated the model, we then explored the impact of the entropic effect in comparison with Joule losses. Though it is often stated that the entropic effect can be neglected from models since it is directly proportional to the current, it can be observed in Figures 8 and 9 that the accuracy of the thermal model was greatly improved when entropic heating was considered. To test this assumption, a constant current and constant power discharge were simulated with and without considering entropy.

4.1. Constant Current Discharge

To further our analysis, simulation results for the constant current discharge for C-rates ranging from 0.5 C to 3 C are shown in Figure 10. The results indicate that the entropic effect had a negligible effect at low discharge rates and caused a maximum error of 4 °C at higher rates. Figure 11 shows that despite the large error in heat generation seen mid-discharge, the average losses were nearly identical with and without entropy. When considering Joule losses, only the heat generation rate was much more stable, as indicated by Figure 11. Although the heat generation rate did increase at low SOCs from irreversible losses, as one would expect from the sharp increase in impedance, this rate was increased by 20% when considering entropic contribution under a constant C-rate. It is important to note here that simulation allowed us to visualize results, which would be impossible under real conditions due to the temperature limits imposed on a real lithium-ion cells because of the risks of TRA. Under real conditions, only a discharge inferior to 2 C would be acceptable.

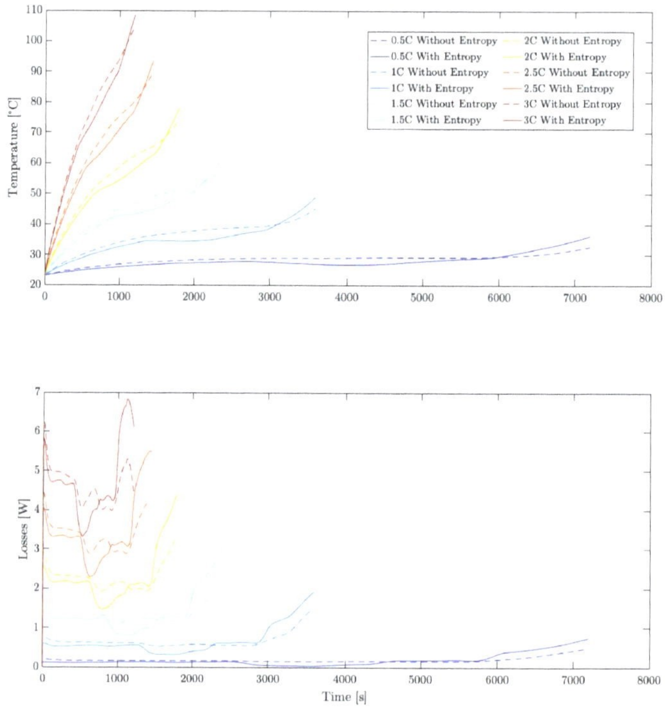

Figure 10. Constant current sensitivity analysis.

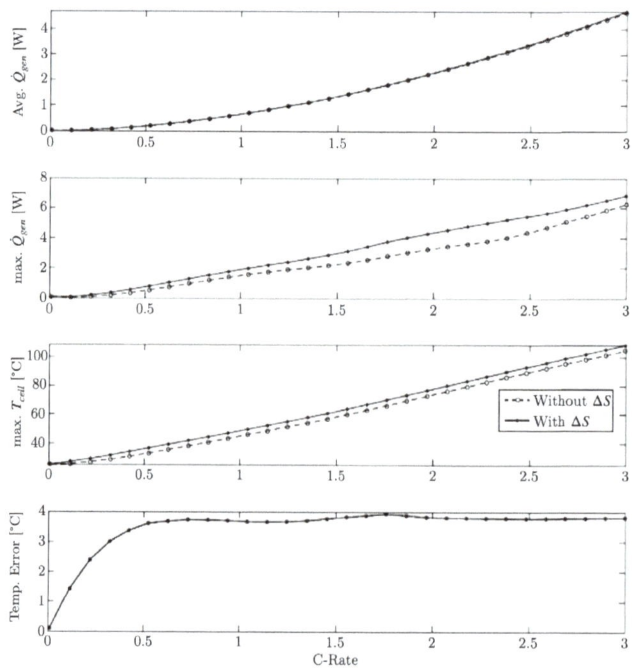

Figure 11. Sensitivity analysis of impact of C-Rate on thermal model results with and without consideration of entropic losses.

4.2. Constant Power Discharge

While galvanostatic discharge is most usual in the field of electrochemistry and the like, a constant power discharge is more representative of most engineering use cases, such as EVs. This power can be expressed in the E-rate, or multiples of the nominal energy of a cell. The nominal energy in a cell, E_{nom}, or energy storage system is equivalent to the piecewise integral of the constant current discharge curve, as stated by (16). Nominal capacity C_{nom} is defined as the total amount of charges, given in Ah, that can be stored in a cell for the electrochemical reaction to be stable.

$$E_{nom} = \int_0^{C_{nom}} V\,dC \tag{16}$$

A more practical method to estimate E_{nom} is from the product of V_{nom} and C_{nom}, typically given in product datasheets. Since V_{oc} decreases as the cell becomes depleted, the current will increase to maintain constant power. Results for discharge with constant E-rates ranging from 0.5 E to 3 E are shown in Figure 12. The boundary condition and properties were kept identical. By comparison with results of Figure 10 for a constant current discharge, the differences caused by the consideration of entropic heat generation were much less important to the proper estimation of the temperature and total heat generation. From the simulation results, it can be observed that the maximum power that could be sustained in these poor cooling conditions was 1.5 E or 19.8 W. At this E-rate, the losses could reach up to 6.5 W with entropy, while they only reached 5.3 W without entropy. This difference in heat generation was even greater at higher E-rates. As can be seen in Figure 12, Q_{gen} could reach up to 12.6 W when considering entropy. However, this specific case could not be sustained in a real application since the temperature would be significantly above the safe operational window.

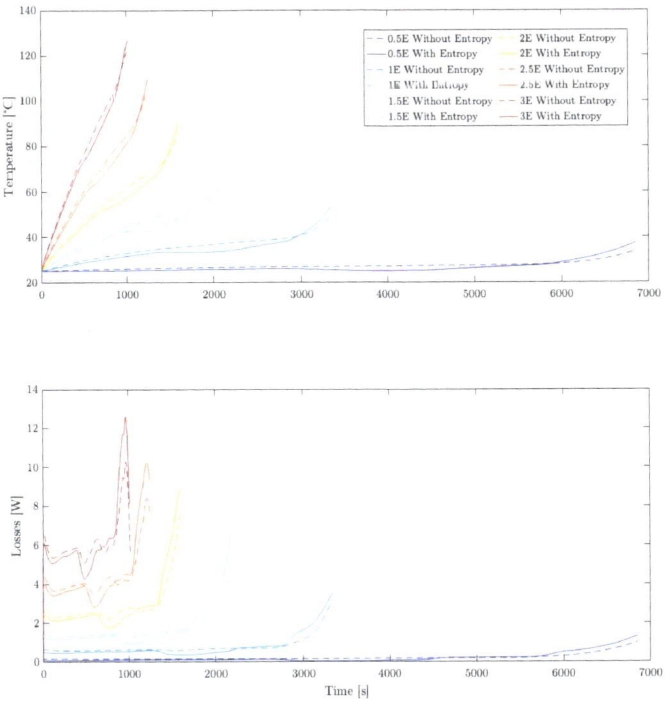

Figure 12. Constant power sensitivity analysis.

4.3. Discussion About Entropy

The estimation of heat generation is one of the primary inputs for the proper sizing of thermal management systems. Only a coarse approximation of heat generation can be made from the nominal internal resistance value. Considering the temperature and SOC dependence of internal resistance can greatly improve the accuracy of models. Further gains can be achieved by considering transient impedance instead of constant resistance. While the link between overpotentials and losses can quite easily be measured, entropy change is often overlooked due to a lack of understanding of its mechanisms. Since entropic heating is convoluted with Joule heating, it is less intuitive to measure than impedance. Simulation allowed us to deconvolute the Joule losses from entropic heating. Due to its reversibility, entropic heating cannot be considered a loss in energy per se. The exothermic phase transformation in electrodes during discharge is equally endothermic during recharge. Thus, only overpotential heating can be classified as a loss from the strict definition of losses being irreversibly lost to the environment.

The decision to consider entropy or not from thermal models can be motivated by plenty of factors. The time required to calibrate the lookup tables is one. The measurement of changes in entropy requires time, patience, well-controlled conditions, and high-precision measurement equipment. ΔS is not dependent on temperature, but it varies with SOC. Therefore, it is important to characterize this property along the entire depth of the discharge of a cell, for as many points as time allows. For instance, each data point of entropy change may take up to 1 day to obtain, simply to allow the full thermodynamic stabilization of the samples. It took 10 days to obtain the entirety of the entropic coefficient points presented in Figure 6. However, this procedure could be automated by linking the climate chamber and cell-cycler. Electro-thermal impedance spectroscopy as introduced by Schmidt et al. [46] will be considered a path to improve the speed and accuracy of the characterization of ΔS in future work.

5. Conclusions

With the increasing use of lithium-ion batteries in power-intensive applications and environments, the estimation of the thermal behaviour of lithium-ion cells through electro-thermal models becomes even more necessary to design battery packs that are safer, longer lasting, and more reliable.

This research fills a critical gap in understanding ΔS in electro-thermal modelling, offering a systematic approach to incorporating entropy change in predictive frameworks. In this work, the importance of considering ΔS in the estimation of heat generation in electro-thermal models was addressed. The electro-thermal model used for this work was structured using the energetic macroscopic representation method. The cell impedance used in the model was obtained by GITT at temperatures ranging from -20 °C to 70 °C along the entire depth of the discharge of the cells, with a resolution of a 1% SOC. The entropic coefficient was measured at increments of a 5% SOC for the entire range. A lumped thermal model with convection boundary conditions was used to estimate cell temperature.

The voltage error of the model was found to be inferior to 50 mV, while the temperature error was inferior to 1 °C when considering entropic heating, even with a dynamic drive cycle. From this validated model, sensitivity analysis was performed at various C-rates and E-rates to determine the impact of entropic heating on the temperature and total heat generation. Based on our observations, discarding entropy from the simulations could yield an error in temperature of up to 4 °C. However, the thermal boundary conditions of the present work were sub-optimal for high-power applications.

However, entropy does not seem to affect total heat generation over a full discharge. Entropy change predominantly affects the precise tracking of temperature dynamics, offering critical accuracy enhancements for high-fidelity thermal simulations. It was found that it plays no significant role on the accuracy of electrical behaviour prediction. While entropy can be neglected for coarse heat generation estimation, it must be considered when temperature must be estimated with a greater level of accuracy, such as in high-fidelity thermal simulations. Future work could focus on automating entropy measurement processes and applying the findings to new chemistries or advanced thermal management systems. Incorporating entropy change into a model hinges on the desired accuracy level, available laboratory resources, and specific simulation objectives.

Author Contributions: F.-A.L.: Investigation, Methodology, Software, Validation, Writing—original draft. P.M.: Conceptualization, Validation, Writing—review and editing. M.B.: Validation, Writing—review and editing. J.P.F.T.: Resource, Conceptualization, Validation, Funding acquisition, Supervision, Writing—review and editing. All authors have read and agreed to the published version of the manuscript.

Funding: This work is supported by the Canada Research Chairs Program (950-230672) and by the Mitacs Accelerate Program under project IT30373. Authors would like to express their gratitude to SysNergie Inc. for supporting this research.

Data Availability Statement: The original contributions presented in the study are included in the article, further inquiries can be directed to the corresponding author.

Conflicts of Interest: Authors Félix-Antoine LeBel, Pascal Messier and Mathieu Blanchard were employed by the company SysNergie Inc. The remaining author declare that the research was conducted in the absence of any commercial or financial relationships that could be construed as a potential conflict of interest.

References

1. Hendricks, C.; Williard, N.; Mathew, S.; Pecht, M. A failure modes, mechanisms, and effects analysis (FMMEA) of lithium-ion batteries. *J. Power Sources* **2015**, *297*, 113–120. [CrossRef]
2. Raijmakers, L.; Danilov, D.; Eichel, R.-A.; Notten, P. A review on various temperature-indication methods for Li-ion batteries. *Appl. Energy* **2019**, *240*, 918–945. [CrossRef]
3. ouj Reynier, Y.; Yazami, R.; Fultz, B. The entropy and enthalpy of lithium intercalation into graphite. *J. Power Sources* **2003**, *119*, 850–855. [CrossRef]
4. Doyle, M.; Fuller, T.F.; Newman, J. Modeling of Galvanostatic Charge and Discharge of the Lithium/Polymer/Insertion Cell. *J. Electrochem. Soc.* **1993**, *140*, 1526. [CrossRef]
5. Fuller, T.F. Simulation and Optimization of the Dual Lithium Ion Insertion Cell. *J. Electrochem. Soc.* **1994**, *141*, 1. [CrossRef]
6. Doyle, M.; Newman, J. The use of mathematical modeling in the design of lithium/polymer battery systems. *Electrochim. Acta* **1995**, *40*, 2191–2196. [CrossRef]
7. Moura, S.J.; Argomedo, F.B.; Klein, R.; Mirtabatabaei, A.; Krstic, M. Battery State Estimation for a Single Particle Model with Electrolyte Dynamics. *IEEE Trans. Control Syst. Technol.* **2017**, *25*, 453–468. [CrossRef]
8. Erhard, S.V.; Osswald, P.J.; Wilhelm, J.; Rheinfeld, A.; Kosch, S.; Jossen, A. Simulation and Measurement of Local Potentials of Modified Commercial Cylindrical Cells. *J. Electrochem. Soc.* **2015**, *162*, A2707–A2719. [CrossRef]
9. Erhard, S.V.; Osswald, P.J.; Keil, P.; Höffer, E.; Haug, M.; Noel, A.; Wilhelm, J.; Rieger, B.; Schmidt, K.; Kosch, S.; et al. Simulation and Measurement of the Current Density Distribution in Lithium-Ion Batteries by a Multi-Tab Cell Approach. *J. Electrochem. Soc.* **2017**, *164*, A6324–A6333. [CrossRef]
10. Jokar, A.; Rajabloo, B.; Désilets, M.; Lacroix, M. Review of simplified Pseudo-two-Dimensional models of lithium-ion batteries. *J. Power Sources* **2016**, *327*, 44–55. [CrossRef]
11. Couto, L.D.; Schorsch, J.; Nicotra, M.M.; Kinnaert, M. SOC and SOH estimation for Li-ion batteries based on an equivalent hydraulic model. Part I: SOC and surface concentration estimation. In Proceedings of the 2016 American Control Conference (ACC), Boston, MA, USA, 6–8 July 2016; pp. 4022–4028. [CrossRef]
12. Couto, L.D.; Schorsch, J.; Job, N.; Léonard, A.; Kinnaert, M. State of health estimation for lithium ion batteries based on an equivalent-hydraulic model: An iron phosphate application. *J. Energy Storage* **2019**, *21*, 259–271. [CrossRef]
13. Hu, X.; Li, S.; Peng, H. A comparative study of equivalent circuit models for Li-ion batteries. *J. Power Sources* **2012**, *198*, 359–367. [CrossRef]

14. Seaman, A.; Dao, T.-S.; McPhee, J. A survey of mathematics-based equivalent-circuit and electrochemical battery models for hybrid and electric vehicle simulation. *J. Power Sources* **2014**, *256*, 410−423. [CrossRef]
15. Salda na, G.; San Martín, J.I.; Zamora, I.; Asensio, F.J.; O nederra, O. Analysis of the Current Electric Battery Models for Electric Vehicle Simulation. *Energies* **2019**, *12*, 2750. [CrossRef]
16. Wu, W.; Wang, S.; Wu, W.; Chen, K.; Hong, S.; Lai, Y. A critical review of battery thermal performance and liquid based battery thermal management. *Energy Convers. Manag.* **2019**, *182*, 262–281. [CrossRef]
17. Mousavi, G.S.M.; Nikdel, M. Various battery models for various simulation studies and applications. *Renew. Sustain. Energy Rev.* **2014**, *32*, 477–485. [CrossRef]
18. Nejad, S.; Gladwin, D.T.; Stone, D.A. A systematic review of lumped-parameter equivalent circuit models for real-time estimation of lithium-ion battery states. *J. Power Sources* **2016**, *316*, 183–196. . [CrossRef]
19. Agudelo, B.; Zamboni, W.; Monmasson, E.; Spagnuolo, G. *Identification of Battery Circuit Model from EIS Data*; HAL: Bengaluru, India, 2019.
20. Vyroubal, P.; Kazda, T. Equivalent circuit model parameters extraction for lithium ion batteries using electrochemical impedance spectroscopy. *J. Energy Storage* **2018**, *15*, 23–31. [CrossRef]
21. Nickol, A.; Schied, T.; Heubner, C.; Schneider, M.; Michaelis, A.; Bobeth, M.; Cuniberti, G. GITT Analysis of Lithium Insertion Cathodes for Determining the Lithium Diffusion Coefficient at Low Temperature: Challenges and Pitfalls. *J. Electrochem. Soc.* **2020**, *167*, 090546. [CrossRef]
22. Lebel, F.A.; Messier, P.; Sari, A.; Trovao, J.P.F. Lithium-Ion cell Equivalent Circuit Model Identification by Galvanostatic Intermittent Titration Technique. *J. Energy Storage* **2022**, *54*, 105303. [CrossRef]
23. Al Hallaj, S.; Venkatachalapathy, R.; Prakash, J.; Selman, J.R. Entropy Changes Due to Structural Transformation in the Graphite Anode and Phase Change of the $LiCoO_2$ Cathode. *J. Electrochem. Soc.* **2000**, *147*, 2432. [CrossRef]
24. Viswanathan, V.V.; Choi, D.; Wang, D.; Xu, W.; Towne, S.; Williford, R.E.; Zhang, J.G.; Liu, J.; Yang, Z. Effect of entropy change of lithium intercalation in cathodes and anodes on Li-ion battery thermal management. *J. Power Sources* **2010**, *195*, 3720–3729. [CrossRef]
25. Bazinski, S.J.; Wang, X. The Influence of Cell Temperature on the Entropic Coefficient of a Lithium Iron Phosphate (LFP) Pouch Cell. *J. Electrochem. Soc.* **2014**, *161*, A168–A175. [CrossRef]
26. MathWorks. Generic Battery Model—Simulink. 2018. Available online: https://www.mathworks.com/help/physmod/sps/powersys/ref/battery.html (accessed on 13 February 2025).
27. Ye, Y.; Shi, Y.; Cai, N.; Lee, J.; He, X. Electro-thermal modeling and experimental validation for lithium ion battery. *J. Power Sources* **2012**, *199*, 227–238. [CrossRef]
28. Quick Review: ZKETECH ZKE Battery Rack (4-Wire Kelvin AAA/AA/18650 Test Fixture). Available online: https://bit.ly/418qDso (accessed on 13 February 2025).
29. Wilson, P. Passive Components. In *The Circuit Designer's Companion*; Elsevier: Amsterdam, The Netherlands, 2017; pp. 93–157. [CrossRef]
30. CSZ MicroClimate Benchtop Test Chambers. Available online: https://www.cszindustrial.com/Products/Temperature-Humidity-Chambers/Micro-Climate-Benchtop.aspx (accessed on 13 February 2025).
31. Zilberman, I.; Ludwig, S.; Jossen, A. Cell-to-cell variation of calendar aging and reversible self-discharge in 18650 nickel-rich, silicon–graphite lithium-ion cells. *J. Energy Storage* **2019**, *26*, 100900. [CrossRef]
32. Zilberman, I.; Sturm, J.; Jossen, A. Reversible self-discharge and calendar aging of 18650 nickel-rich, silicon-graphite lithium-ion cells. *J. Power Sources* **2019**, *425*, 217–226. [CrossRef]
33. Zilberman, I.; Schmitt, J.; Ludwig, S.; Naumann, M.; Jossen, A. Simulation of voltage imbalance in large lithium-ion battery packs influenced by cell-to-cell variations and balancing systems. *J. Energy Storage* **2020**, *32*, 101828. [CrossRef]
34. Zilberman, I.; Ludwig, S.; Schiller, M.; Jossen, A. Online aging determination in lithium-ion battery module with forced temperature gradient. *J. Energy Storage* **2020**, *28*, 101170. [CrossRef]
35. Sturm, J.; Ludwig, S.; Zwirner, J.; Ramirez-Garcia, C.; Heinrich, B.; Horsche, M.F.; Jossen, A. Suitability of physicochemical models for embedded systems regarding a nickel-rich, silicon-graphite lithium-ion battery. *J. Power Sources* **2019**, *436*, 226834. [CrossRef]
36. Berg, P.; Spielbauer, M.; Tillinger, M.; Merkel, M.; Schoenfuss, M.; Bohlen, O.; Jossen, A. Durability of lithium-ion 18650 cells under random vibration load with respect to the inner cell design. *J. Energy Storage* **2020**, *31*, 101499. [CrossRef]
37. Lammer, M.; Königseder, A.; Hacker, V. Holistic methodology for characterisation of the thermally induced failure of commercially available 18650 lithium ion cells. *RSC Adv.* **2017**, *7*, 24425–24429. [CrossRef]
38. Heenan, T.M.; Jnawali, A.; Kok, M.; Tranter, T.G.; Tan, C.; Dimitrijevic, A.; Jervis, R.; Brett, D.J.; Shearing, P.R. Data for an Advanced Microstructural and Electrochemical Datasheet on 18650 Li-ion Batteries with Nickel-Rich NMC811 Cathodes and Graphite-Silicon Anodes. *Data Brief* **2020**, *32*, 106033. [CrossRef]

39. German, R.; Solis, N.; Reyes, L.; Silva, L.; LeBel, F.-A.; Trovao, J.P.; Bouscayrol, A. Coupled electric and thermal batteries models using energetic macroscopic representation (EMR) for range estimation in electric vehicles. In Proceedings of the 2017 19th European Conference on Power Electronics and Applications (EPE'17 ECCE Europe), Warsaw, Poland, 11–14 September 2017; Volume 2017, pp. 1–8. [CrossRef]
40. Barai, A.; Uddin, K.; Dubarry, M.; Somerville, L.; McGordon, A.; Jennings, P.; Bloom, I. A comparison of methodologies for the non-invasive characterisation of commercial Li-ion cells. *Prog. Energy Combust. Sci.* **2019**, *72*, 1–31. [CrossRef]
41. Beck, D.; Dechent, P.; Junker, M.; Sauer, D.U.; Dubarry, M. Inhomogeneities and Cell-to-Cell Variations in Lithium-Ion Batteries, a Review. *Energies* **2021**, *14*, 3276. [CrossRef]
42. Geifes, F.; Bolsinger, C.; Mielcarek, P.; Birke, K.P. Determination of the entropic heat coefficient in a simple electro-thermal lithium-ion cell model with pulse relaxation measurements and least squares algorithm. *J. Power Sources* **2019**, *419*, 148–154. [CrossRef]
43. Onda, K.; Kameyama, H.; Hanamoto, T.; Ito, K. Experimental Study on Heat Generation Behavior of Small Lithium-Ion Secondary Batteries. *J. Electrochem. Soc.* **2003**, *150*, A285. [CrossRef]
44. Steinhardt, M.; Barreras, J.V.; Ruan, H.; Wu, B.; Offer, G.J.; Jossen, A. Meta-analysis of experimental results for heat capacity and thermal conductivity in lithium-ion batteries: A critical review. *J. Power Sources* **2022**, *522*, 230829. [CrossRef]
45. Messier, P.; Nguyen, B.H.; LeBel, F.A. and Trovão, J.P.F. Disturbance observer-based state-of-charge estimation for Li-ion battery used in light electric vehicles. *J. Energy Storage* **2020**, *27*, 101144. [CrossRef]
46. Schmidt, J.P.; Weber, A.; Ivers-Tiffée, E. A novel and precise measuring method for the entropy of lithium-ion cells: ΔS via electrothermal impedance spectroscopy. *Electrochim. Acta* **2014**, *137*, 311–319. [CrossRef]

batteries

MDPI

Article

Advanced Load Cycle Generation for Electrical Energy Storage Systems Using Gradient Random Pulse Method and Information Maximising-Recurrent Conditional Generative Adversarial Networks †

Steven Neupert *,‡, Jiaqi Yao ‡ and Julia Kowal *

Department of Electrical Energy Storage Technology, Technische Universität Berlin, Einsteinufer 11, 10587 Berlin, Germany; jiaqi.yao@tu-berlin.de

* Correspondence: s.neupert@tu-berlin.de (S.N.); julia.kowal@tu-berlin.de (J.K.)

† This paper is an extended version of our paper published in Neupert, S.; Yao, J.; Kowal, J. Load cycle design and analysis for energy storage technologies utilising micro-trip methods and machine learning approaches. In Proceedings of the Energy Storage Conference 2023 (ESC 2023), Glasgow, UK, 15–16 November 2023; pp. 63–68.

‡ These authors contributed equally to this work.

Abstract: The paper presents two approaches to generating load cycles for electrical energy storage systems. A load cycle is described as the operation of an energy storage system. The cycles can include different metrics depending on the storage application. Load cycle analysis using the rainflow counting method is employed to understand and validate the metrics of the load cycles generated. Current load cycle generation can involve clustering methods, random microtrip methods, and machine learning techniques. The study includes a random microtrip method that utilises the Random Pulse Method (RPM) and enhances it to develop an improved version called the Gradient Random Pulse Method (gradRPM), which includes the control of stress factors such as the gradient of the state of charge (SOC). This method is relatively simple but, in many cases, it fulfills its purpose. Another more sophisticated method to control stress factors has been proposed, namely the Information Maximising-Recurrent Conditional Generative Adversarial Network (Info-RCGAN). It uses a deep learning algorithm to follow a machine learning-based, data-driven load cycle generation approach. Both approaches use the measurement dataset of a BMW i3 over multiple years to generate new synthetic load cycles. After generating the load cycles using both approaches, they are applied in a laboratory environment to evaluate the stress factors and validate how similar the synthetic data are to a real measurement. The results provide insights into generating simulation or testing data for electrical energy storage applications.

Keywords: loadcycle analysis; load cycle design; simulation; testing

1. Introduction

The number of electrical energy storage applications has increased significantly in recent years. The push for electric vehicles and stationary systems is enormous. No matter which application the electrical energy storage system is used for, significant effort is put into improving the system's control and enhancing the understanding of the application and its requirements. For this purpose, the systems are either tested in laboratories, applying different load cycles, or simulated. Either approach requires adequate knowledge of the system's requirements and behaviours during the application. Furthermore, the appropriate input data are required to represent the given application's stress factors.

Recent research by Ref. [1] showed that a varying dynamic discharge profile led to an increase of up to 38% in equivalent full cycles at the end of life. In Ref. [2], differences in the degradation were recorded, especially for low temperatures. A review of battery data in Ref. [3] revealed that dynamic ageing has often been overlooked in degradation analysis, primarily due to the complexity of assessing ageing mechanisms under dynamic stress factors and the limited availability of real-world application data for dynamic cycles. Additionally, degradation studies are frequently accelerated to reduce time and effort. This underscores the usefulness and necessity of load cycle generation approaches. This work shows an approach to extracting the required information from measurements and using them to generate new (simplified) load cycles through load cycle analysis. Two different methods for load cycle generation are introduced and compared. This work is developed based on our preliminary work [4], upon which the content is significantly expanded. The expansion includes the tuning of the two methods introduced and makes it possible to compare the approaches based on their influences on actual laboratory measurements instead of relying on comparing synthetically generated load cycles. In particular, the influence of gradRPM on secondary stress factors, such as temperature and voltage, can be evaluated and validated through the Info-RCGAN approach. The preliminary work does not include any validation in a laboratory environment.

Load cycle analysis is necessary to evaluate the data, identify the relevant stress factors to design a degradation schedul, and validate the load cycles generated. Different approaches can be found in the literature. Ref. [5] summarised papers utilising various approaches to load spectrum analysis. The methods included instantaneous value counting, rainflow counting, range pair mean counting, total charge throughput, and half-cycle counting. The techniques can be used for different stress factors. Instantaneous value counting is a method where the appearance of each value belongs to a bin, and the specific bin counter is incremented. Every value is counted. Charge throughput calculates the charge that is charged or discharged instead of occurrences. The occurrences do not include varying sample times. Rainflow counting is more complex and counts stress cycles and their mean values and ranges. Rainflow counting can be applied to different stress factors such as SOC, temperature, voltage, and current. The results can be displayed as histograms or spectra. In Ref. [6], the rainflow counting was extended using a fuzzy logic approach to address the quantisation error implicit in the histogram generation that could have introduced incorrect representations of the stress signal. In addition, it has been used as the basis for battery degradation models to account for temperature and c-rate dynamics, such as in Ref. [7]. Ref. [8] highlighted the drawbacks of applying rainflow counting to real-time data and introduced a faster method to approximate the full cycles that a battery endured during a frequency response application. In Ref. [9], the problem of applying the rainflow counting approach to real-time data, as well as in an optimisation problem, was addressed by counting the half cycles generated by each concavity change. In Ref. [10], machine learning approaches were applied to analyse the stress factors and rank them. The least absolute shrinkage and selection operator was used for the analysis, and a random forest approach was employed to perform the ranking. In general, variations of the rainflow counting method have been widely used as approaches to stress analysis. This method breaks down complex load cycles into stress cycles, capturing the range and mean of each cycle. It can easily be applied to multiple stress factors that are critical for battery degradation processes, such as state of charge, temperature, voltage, and current. Its representation through histograms or spectra is explicitly useful for the easy comparison and interpretation of measurement data and the load cycles generated. Due to its versatility, rainflow counting is used in this work for the analysis of the profiles generated and the investigation of the stress factors of the measurement data.

Load cycle generation is often described using driving cycles. In the case of driving cycles, vehicle kinematics are used to design cycles, which are then run with a simulation. Afterwards, a battery profile can be calculated. Load cycles generally pertain to the load profile applied to batteries and battery systems. The literature concerning load cycles is sparse, whereas there is more literature using different approaches for driving cycles. Nevertheless, approaches to driving cycle construction can be used for load cycle generation. In general, cycle generation can be divided into three different categories. Simple methods based on employing known profiles to generate new variations using, for example, battery models or vehicle models. In Ref. [11], a battery model was used to generate the synthetic data of an end-of-line test, which was then used to train a machine learning algorithm. Clustering methods can be used to help identify features and categorise measurement data to reduce the data information to a single drive cycle like a standard drive cycle, i.e., the New European Driving Cycle (NEDC). In Ref. [12], principal component analysis, clustering, and an optimisation algorithm were used to obtain a driving cycle that was typical for the data acquired during the case study, whereas in Ref. [13], a relatively raw clustering approach used principal component analysis to reduce the dimensionality, as well as clustering to categorise the microtrips. Afterwards, the microtrips were randomly picked to construct a new cycle. Random microtrip methods include restructuring the measurement data randomly to new cycles, like in Ref. [14]. In Ref. [15], the generation of a load cycle for electrical vehicle application was based on the data from a real electrical vehicle, where different methods were used to classify the uses of the vehicle, including, for example, K-Means clustering; the dimensionality of the features was reduced using principal component analysis. For the characterisation of driving cycles in Ref. [16], the energy consumption per unit of time was used as an indicator to classify the cycles into semi-urban or semi-highway cycles. The categories were characterised, and probability functions were designed to represent the categories. Afterwards, the functions were sampled with a random seed to generate new profiles. In Ref. [17], cycles were generated for a stationary application. The data were divided into intervals, and twelve different metrics were calculated to describe the intervals. A dispatch interval matrix was designed for further analysis. Principal component analysis was used to reduce the dimensions before unsupervised K-Means clustering was applied for the categorisation. For the cycle generation, a number of characteristic duty cycles were extracted. Other approaches include machine learning methods. Sometimes, they are used to improve the existing clustering methods. In Ref. [18], an autoencoder was added to the clustering method to improve the generation of a new cycle by half a percent, whereas in Ref. [19], a Time Generative Adversarial Network was applied to generate more synthetic data on different driving conditions and operational parameters like varying temperatures for the training of a state of charge prediction model. So, there are multiple approaches to generating usable load cycles, from simple to advanced. To reach the goal of creating realistic controllable load cycles for application in a laboratory environment, two approaches are implemented during this study. The first approach belongs to the group of approaches that reuses segments of the original measurement data. The simple algorithm is based on the RPM from Ref. [14] and was chosen as one of the approaches due to its ease of implementation, low requirements for computational resources, and the amount of data needed to generate adequate cycles. These advantages make the RPM a good candidate for straightforward load cycle generation. Expanding the RPM to include the possibility of controlling different stress factors leads to the development of the gradRPM. However, not all of the stress factors can be controlled by the gradRPM; to do so, a parameterised battery model would be needed to model the voltage and temperature responses to the generated load cycle. That would increase the complexity and information needed to generate load cycles, which is not part of this work for the gradRPM. The second approach is aimed at using the Info-RCGAN. The Info-RCGAN is used to generate load cycles that resemble the measurement data as much as possible, with advanced stress factor control for secondary stress factors such as temperature and voltage. Using a large dataset for training increases adaptability and flexibility. Furthermore, this approach is able to capture more complex dependencies between the stress factors than a simple approach. These advantages make it a perfect cross-comparison for the gradRPM method. Both approaches need a real-world measurement dataset, where the amount of data may vary. The gradRPM can work even using just a small amount of data, where the Info-RCGAN needs a larger amount to effectively train the whole model. By combining a simpler, computationally efficient method (gradRPM) with an advanced, high-fidelity approach (Info-RCGAN), we aim to provide a more comprehensive solution for load cycle generation. These methods are tested using real-world data from a BMW i3 and validated in a laboratory environment.

The paper is structured as follows: First, we describe the dataset used as a reference and the basis of our study, see Section 2. Next, we present the methodology, which is divided into two approaches, the gradRPM in Section 3.1 and the Info-RCGAN in Section 3.3. This is followed by the results of the two approaches and a discussion of the findings in Section 4. Finally, we conclude the paper with key insights and implications in Sections 5 and 6.

2. Dataset

The dataset used as the basis for this data-driven pulse generation comprises logged data from a BMW i3 (BMW AG, Munich, Germany). It includes data recorded over 4 years, with shorter trips in most cases. The BMW i3 belongs to the Fraunhofer Institute for Transportation and Infrastructure Systems (Dresden, Germany) and is regularly used as a passenger car. Consequently, the gathered data closely resemble the driving patterns of a private BMW i3 owner. Figure 1 displays a summary of the dataset for the range of SOC per trip, its mean SOC, and the mean temperature. The colours, starting from yellow to violet, in the scatter plot indicate the time of acquisition, where yellow is the oldest measurement, i.e., the cell had a higher SOH at this point. In general, the trips are shorter, with an SOC range of up to 20%, and they mostly cycled around an SOC of 75%, which is rather high. Over the ageing of the battery, the mean SOC decreases. The temperature typically remains between 20 °C and 30 °C, with brief deviations occurring only at cold or hot ambient temperatures immediately after vehicle startup. Therefore, the temperature is not displayed here. Each measurement includes voltage, current, SOC, time and temperature. The Fraunhofer Institute for Transportation and Infrastructure Systems provided the data for the joint research project "Field Data-Based Battery Diagnosis and Lifetime Prediction (FeBaL)". The project goal was the utilisation of field data to improve battery assessment by learning the ageing behaviour, including stress factors and interactions, examining the ageing without a capacity test, and understanding the usage patterns for application-specific lifetime predictions.

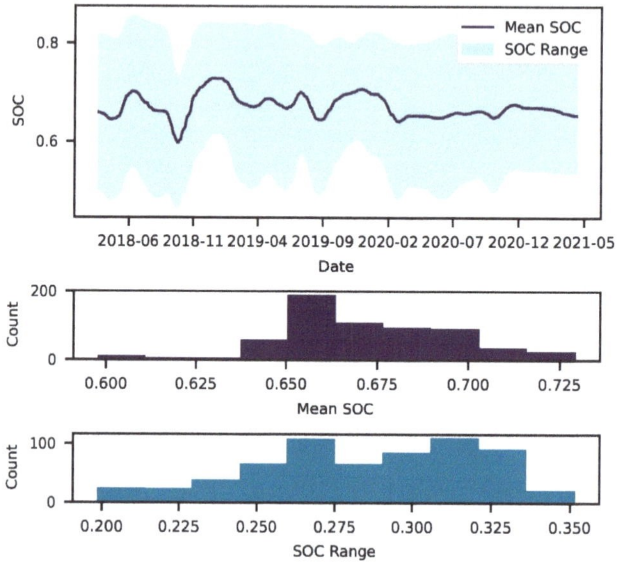

Figure 1. Diagram of the dataset visualising the SOC range and the mean SOC of every trip.

3. Methodology

The methodology is subdivided into three different parts. Load cycle analysis is the first part, the improved random pulse method is the second, and the third is the machine learning-based approach to cycle generation. To compare and validate the generation of the profiles with the measurements, there needs to be a way of analysing the profiles generated. Some approaches have been mentioned in Section 1. In this work, the rainflow counting method is used to analyse all the generated and measurement data concerning the stress factors, SOC, temperature, voltage, and current or C-rate. Regarding the generation, the only analysable factor is the C-rate because, for example, the RPM (Section 3.1) only generates a C-rate profile. On the other hand, the machine learning approach (Section 3.3) can generate profiles for the different stress factors, but in the end, only the C-rate profile is applied.

3.1. Gradient Random Pulse Method (gradRPM)

The Random Pulse Method (RPM) is highly dependent on the measurements and the number of measurements within the dataset used to extract information on the application. In Ref. [14], the method was introduced to construct driving schedules for high-performance batteries used in racing applications on specified racing tracks. Ref. [14] described segmenting the dataset into smaller parts of the dataset. A segment's start and end were defined by the current direction change, either from the discharging to the charging direction or vice versa. Therefore, the data were analysed for zero crossings and split into segments. The segments were then saved to a database, and the charging and discharging databases were subdivided. Furthermore, the databases could be arranged into racetrack-specific segments. Then, the databases were analysed considering predefined target parameters, where applicable. The target parameters described different characteristics of the data. They included the mean power discharge and charge, as well as the mean absolute power and net discharge power. In addition, they considered parameters describing the fraction of charging or discharging of the pulse. Different duration metrics were included in the target parameters and the overall duration. Some parameters were not calculated for the segments since they did not make sense, like the fraction of charging or discharging. Because of the segmentation, a segment was either a charging or discharging pulse. The target parameters were the criteria for the load cycles constructed to decide whether they were helpful. The databases of the charging and discharging segments were sorted by the duration of the segments. The generation of the profile used by Ref. [14] started with generating a random number within the range of the number of segments in the discharging database. The corresponding discharging and charging segments with similar durations were added to the cycle. The reason for adding both a discharging and charging segment with similar durations was to have real profiles and to avoid the occurrence of a short discharge pulse followed by a long charging pulse that may lead to overcharging. When the target duration for the duty cycle was reached, the cycle was evaluated in relation to the target parameters. The error of the target parameters had to be below 10% in the case of Ref. [14]. About 200 iterations were needed to generate 10 candidates using the 10% margin [14]. The cycle was stored as a candidate if the target parameters were within the intended range. The procedure was repeated until ten candidates were stored. The overall error for each candidate was calculated, and the candidate with the lowest error was accepted. The general approach of the RPM is displayed in Figure 2a.

Based on the explained method utilised by Ref. [14], the RPM was improved to reduce the number of iterations and enhance its controllability. This approach is called gradRPM, is based on the RPM, and is tested in this paper. Controllability is especially important to generate load cycles that are used for dynamic ageing in laboratories. Because in ageing tests, the degradation is planned for specific stress factors. The stress factors that are investigated include the SOC range, the mean SOC or mean voltage, and the current rates for cyclic degradation. Target parameters and control parameters are used to control the stress factors. A minimum and maximum C-rate is included to maintain at least the boundaries of the current rates. Furthermore, an SOC gradient and range are part of the parameters used to control the mean SOC and range. The target parameters include the SOC range, that describes the maximum change of the SOC over the load cycle and the maximum duration of the load cycle.

The parameters are distinguished into control and target parameters because the process of the RPM has been altered. The new process of the SOC gradient-oriented RPM starts with filtering the database for pulses that are within the boundaries given by the parameters $CRate_{\text{max}}$, $CRate_{\text{min}}$ and within ±5% of SOC_{gradient}. The information concerning the control parameters was already calculated when the segmentation of the actual data was performed, and they were saved to the pulse database as additional information, which is helpful for the generation of the load cycle. The database filtering reduces the pulses to usable ones to generate the load cycle for the given parameters. Then, a random number is generated in the range of the filtered database. A single segment is chosen. The segment could either be in the charge or in the discharge direction. The SOC gradient is calculated for the current load cycle and extracted from the database's chosen segment. If the current gradient of the load cycle differs from the control parameter, only a segment that improves the gradient in the direction of the control parameter is accepted.Therefore, the resulting cycle has a gradient around the set SOC gradient by construction. The generation is finished if the candidate load cycle exceeds the requested SOC range or duration. The process of the basic RPM utilised by Ref. [14] and the adapted gradient-based approach are depicted in Figure 2.

The described approach of the gradRPM can generate more complex gradients as long as the database provides the required data. So, complex load cycles containing multiple segments with different SOC gradients can be generated by providing a sequence of parameters. An example is displayed in Figure 3. As expected, the higher the gradient, the higher the number of pulses with higher currents that reach the desired gradient for the load cycle.

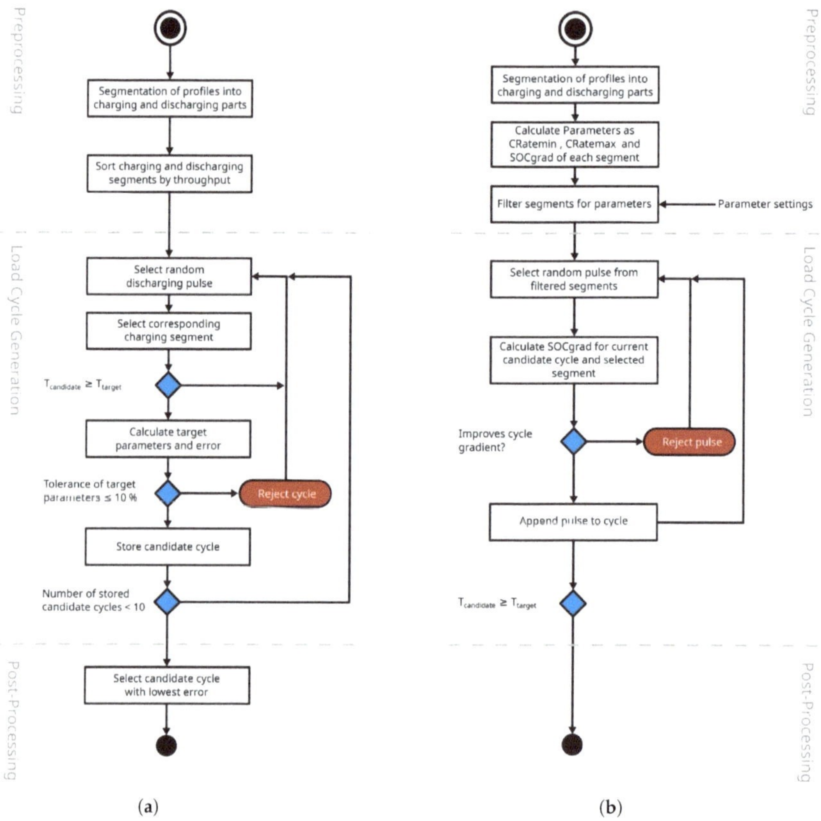

Figure 2. The figure displays the flowchart of the general RPM and the introduced gradRPM. (**a**) RPM scheme by Ref. [14]. (**b**) gradRPM.

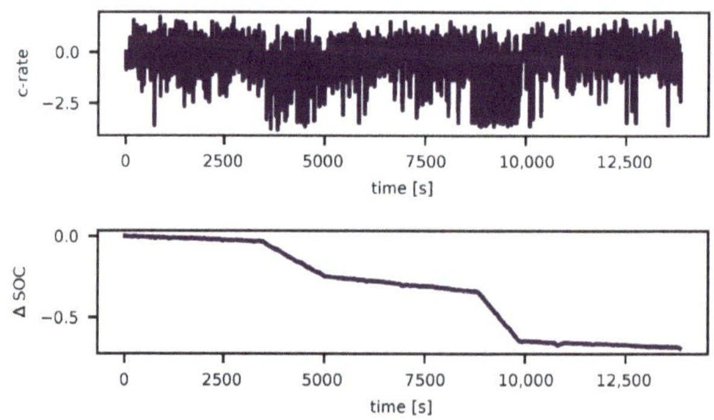

Figure 3. Generated profile using the gradRPM and the corresponding change in the SOC within a sequence of control parameters.

3.2. Process of Analysing Dynamic Load Cycles

The analyses of dynamic data and actual driving data are carried out through the following steps:

- Preprocessing,
- Downsampling,
- And the analysis.

Preprocessing manages the analysis' noise, outliers, and unimportant segments. Unimportant segments can be, in the case of a driving schedule, a prolonged time of data logging after the car is parked. In this work, another approach is used, combining preprocessing and downsampling. A unique sampling approach, a so-called importance sampling method, is used. In this specific case, it is the Largest-Triangle-Three-Bucket approach (LTTB). The reason for the use of an importance sampling approach is, on the one hand, to reduce the amount of data and thus shorten the calculation time of the analysis and, on the other hand, not lose essential segments of the data, like the peaks and valleys. These peaks are significant in identifying the stress factors of the C-rate. The LTTB sampling refers to the use of a shifting window over the data that is meant for downsampling. It divides the window into three buckets containing a predefined number of samples. A sample is chosen for each centre bucket. For the right bucket, a temporary point is calculated as the mean of the samples in that bucket. The left bucket is the old centre bucket with a previously chosen sample. These three samples form a triangle, and the LTTB aims to maximise the area of this triangle by selecting the sample for the centre bucket accordingly. Afterwards, the window shifts for one bucket size. The schematic in Figure 4a visualises the steps of the LTTB approach [20]. By applying the LTTB to the BMW data, the data are compressed by a factor of 100.

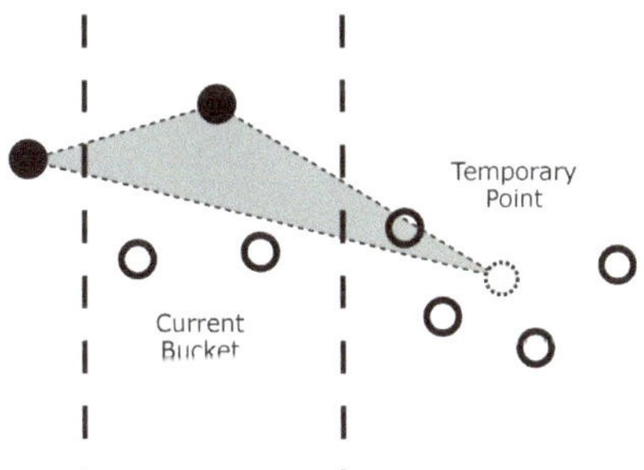

(**a**) LTTB scheme

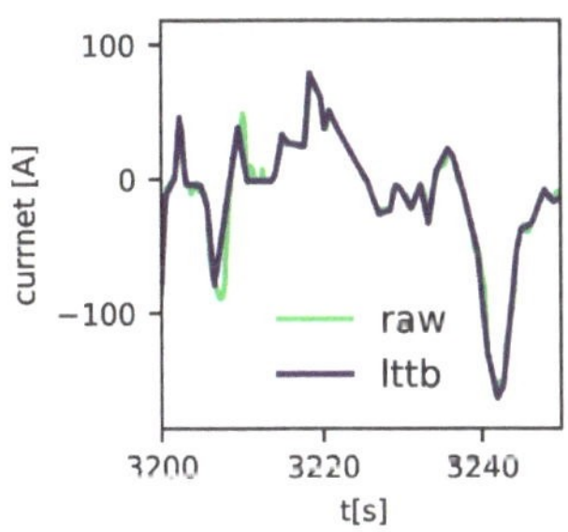

(**b**) Example after LTTB

Figure 4. Schematic of LTTB (**a**) and an example of output after downsampling (**b**).

After preprocessing using the LTTB, the data concerning the stress factors can be analysed. There are different approaches for evaluating the data. They are divided into the following two categories: one-parametric and multi-parametric analysis. One-parametric analysis investigates the signal for only one parameter, like the mean or peaks. Multi-parametric analysis investigates the signal, for example, for both the mean and the range, visualising more signal information. For example, there is instantaneous value counting, where the signal is segmented into specified bin sizes, and every value of the signal is assigned a bin so that it is counted, resulting in a histogram. If the signal is sampled with one specified frequency, each bin count represents the time spent in this stress factor. Multi-parametric approaches, in many cases, consider two parameters because of the ability to be visualised. An example is rainflow counting, which analyses the reversals of signals to identify the stress cycles. It counts the cycles and categorises them based on their mean and range, considering two parameters [5]. Sometimes, another similar approach is mentioned, half-cycle counting, which is, in fact, part of a typical implementation of rainflow counting because during rainflow counting, the signal is analysed for peaks and valleys, where a half cycle is from one valley to one peak or vice versa. This half cycle is also counted as a half cycle in typical rainflow counting. During this work, instantaneous value counting and rainflow counting are applied to analyse the load cycles generated and the measurement data.

3.3. *Data-Driven Approach (Info-RCGAN)*

A machine learning approach for the generation of battery load profiles considering multiple stress factors is proposed in this work. Not only are the generated load profiles of the model expected to have the same fundamental properties as the real ones, but certain features of the generated profiles can also be regulated through the model's conditional input.

3.3.1. Data Preprocessing

The sampling period of the original data is 0.01 s; however, the samples are usually not even, which means there could be no samples for a period that is longer than the sampling time. Sometimes, the device is also shut down for a longer time, during which the batteries are charged. In order to obtain independent load profiles, the load profiles that include long stand-by periods are segmented into separate profiles; otherwise, the increase in voltage, state of charge, and temperature will be considered as a part of the individual load profile. The segmented profiles are then interpolated. Downsampling is applied to improve the efficiency of the training, with a sampling rate of 1 Hz. Since different features have different scales, the input features are normalised into the same range. Research has also shown that data normalisation is able to improve the performance of various networks [21]. The normalisation approach applied is min–max normalisation, with the expected data range as $[-1, 1]$. The equation for the transformation is given in Ref. [22].

3.3.2. Sequence Processing

Recurrent Neural Networks (RNNs) [23] are commonly used for the processing of time series [24,25]. However, a vanilla RNN suffers from the problem of a vanishing gradient, which can be solved by the refined RNN model, Long Short-term Memory (LSTM) [26]. LSTM introduces a gate mechanism to control which information the LSTM units should remember and which information they should forget to learn the sequences' long-term dependencies. Because of its strong capability to capture temporal features, it has become one of the most used architectures in time series analysis [27]. Therefore, LSTM is utilised as a basic unit for sequence processing in this work.

3.3.3. Generative Model

As one of the basic types of deep learning models, generative models aim to generate synthetic data by learning the data distribution throughout the space. In this work, a Generative Adversarial Network (GAN) [28] is utilised as the overall framework for load profile generation.

As shown in Figure 5, GAN consists of two components—the generator (G) and the discriminator (D). The generator takes random noise z as input and produces synthetic data x_{fake}. Given that the prior probability distribution of the input noise vector z is $p_z(z)$, the generator aims to learn the distribution $p_g(z)$ such that its output resembles the ground truth data x. The transformation from noise space to data space is defined by the function $G(z;\theta_g)$, in which G represents the generator's neural network with the parameter θ_g [28]. The discriminator, on the other hand, receives either actual data x_{real} or fake data x_{fake} and outputs a probability indicating the likelihood that the input belongs to the ground truth.

This mapping is represented by the function $D(x; \theta_d)$, where D denotes the discriminator's neural network with the parameter θ_d [28].

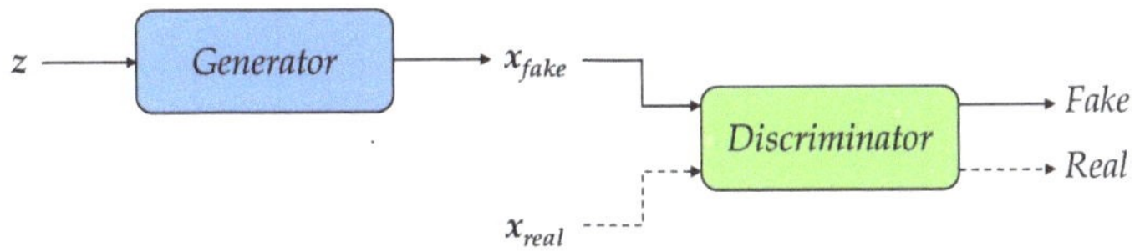

Figure 5. Generative Adversarial Network.

The generator and discriminator engage in a two-player minimax game and are jointly trained. The discriminator aims to maximise its accuracy in distinguishing between the real and generated data. In contrast, the generator aims to minimise the discriminator's ability to identify the generated data as fake [28]. The value function $V(G, D)$ of the game is as follows:

$$\min_G \max_D V(G, D) = \mathbb{E}_{x \sim p_{\text{data}}(x)}[logD(x)] + \mathbb{E}_{z \sim p_z(z)}[log(1 - D(G(z)))] \quad (1)$$

To incorporate conditional constraints into the load profile generation, the extended framework known as the Conditional Generative Adversarial Network (cGAN), introduced by Mehdi Mirza et al. [29], is utilised. In this framework, the naive GAN is modified to include conditional information y, which is provided as additional input to both the generator and discriminator. This conditioning information alters the value function of the two-player minimax game as follows:

$$\min_G \max_D V(G, D) = \mathbb{E}_{x \sim p_{\text{data}}(x)}[logD(x|y)] + \mathbb{E}_{z \sim p_z(z)}[log(1 - D(G(z|y)))] \quad (2)$$

There are different approaches to embedding conditioning information into inputs, such as conditioning by concatenation [29], conditioning using an auxiliary classifier [30], and conditioning with projection [31]. After conducting several experiments, the conditioning approach by concatenation is chosen for this work. The inputs into the LSTM unit at each moment, both in the generator and the discriminator, are concatenated with the respective additional information.

The methods used in previous research, such as C-RNN-GAN [32] and RCGAN [33], have focused on generating continuous sequences using GANs by implementing the generator in a synchronous sequence-to-sequence manner [34]. In these models, the length of the generated sequence can be controlled by specifying the sequence length of the input noise. Building on this concept, Figure 6 illustrates the schematic diagrams of the generator and the discriminator. At each time step, the inputs into the LSTM unit in both the generator and the discriminator are concatenated with the corresponding additional information. To highlight the difference, the generator's conditional input is denoted as c, while the actual data's condition labels are denoted as y.

Most conditional GANs (cGANs) predominantly focus on categorical conditions [35]. In contrast, this work deals exclusively with continuous conditions, considerably complicating the learning process. Additional strategies are required to enhance learning outcomes. InfoGAN [36], initially designed for the unsupervised learning of disentangled representations, inspired the approach presented in this work. Specifically, the idea of strengthening the learning of additional information by integrating an auxiliary network and imposing extra penalties on condition reconstruction is adopted. By shifting the auxiliary model's learning process to a supervised manner, controlling which specific information the auxiliary network extracts becomes possible. In this work, the auxiliary model is referred to as the conditioner. Its schematic diagram is depicted in Figure 7.

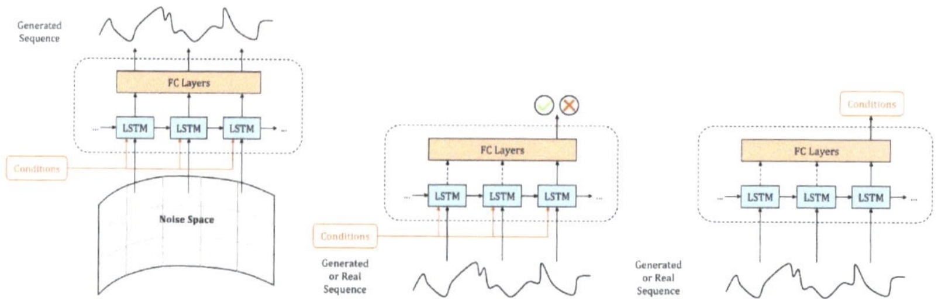

(**a**) Generator scheme (**b**) Discriminator scheme (**c**) Conditioner scheme

Figure 6. The schematic diagrams of the three submodels: (**a**) generator scheme with synchronous sequence-to-sequence implementation; (**b**) discriminator scheme with sequence-to-class implementation; (**c**) conditioner scheme with sequence-to-class implementation.

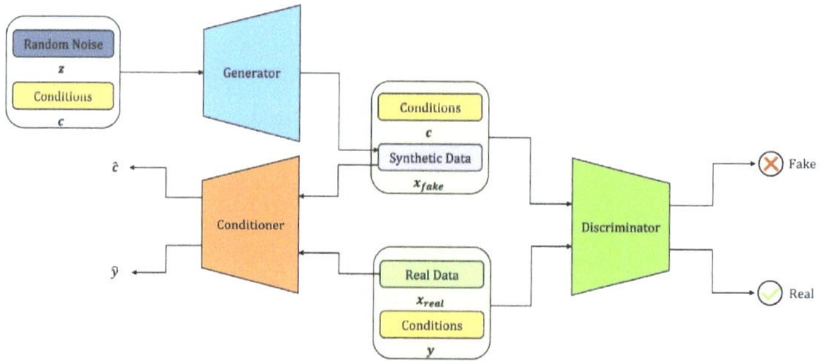

Figure 7. Overall architecture of Info-RCGAN.

Figure 7 presents the overall architecture of the deep learning model utilised, Info-RCGAN. This work uses the following desired features as the conditional input ***c***:

- ΔSOC: the change in SOC after applying the load profile;
- ΔT: the change in temperature of the battery pack;
- *min*(C-rate): the minimum C-rate within the load profile;
- *max*(C-rate): the maximum C-rate within the load profile;
- *min*(ΔU): the minimum voltage change throughout the load profile;
- *max*(ΔU): the maximum voltage change throughout the load profile.

Since validating desired features unrelated to the current on the dataset with a single generated C-rate output is challenging, the generated sequence is expanded to include three features: [C-rate, ΔU, ΔT]. This approach allows better insights into how temperature and voltage vary with the load profile under different conditional inputs.

3.3.4. Loss Function Design

Since the architecture of the Info-RCGAN has multiple submodels, the proper design of loss functions also plays an essential role in this work. The loss functions that are used include Binary Cross Entropy (BCE) and Mean Squared Error (MSE).

The discriminator is expected to not only distinguish the generated fake data from the actual ground truth data but also to learn the condition information for distinguishing the different data types. To achieve this goal, the loss functions of the discriminator are as follows:

- Fake loss $BCE(D(\boldsymbol{x}_{\text{fake}}, \boldsymbol{c}), \boldsymbol{0})$: to distinguish fake data with the respective conditional input as fake;
- Real loss $BCE(D(\boldsymbol{x}_{\text{real}}, \boldsymbol{y}_{\text{real}}), \boldsymbol{1})$: to distinguish real data with the respective real labels as real;
- Unmatched loss $BCE(D(\boldsymbol{x}_{\text{real}}, \boldsymbol{y}_{\text{fake}}), \boldsymbol{0})$: to distinguish real data with the unmatched labels as fake.

The following applies to the above-mentioned functions:

$$\boldsymbol{x}_{\text{fake}} = G(\boldsymbol{z}, \boldsymbol{c}) \tag{3}$$

The conditioning input of the generator $\boldsymbol{c}$ is chosen to be the ground truth labels $\boldsymbol{y}_{real}$ from the same batch, and $\boldsymbol{y}_{fake}$ are some randomly generated labels.

The conditioner aims to reconstruct the regression labels of the input sequence data. It is expected to learn specific desired features, as mentioned before, so the conditioner is trained on the ground truth data and its respective label in a supervised manner.

- Regression loss $MSE(C(\boldsymbol{x}_{\text{real}}), \boldsymbol{y}_{\text{real}})$: to reconstruct the ground truth labels from the ground truth sequence data.

The generator is designed to achieve the following two goals: generate fake data that can fool the discriminator's output into a "real" outcome, as well as generate data in which the condition input $\boldsymbol{c}$ can be reconstructed by the conditioner. The loss functions used to train the generator are as follows:

- Generation loss $BCE(D(\boldsymbol{x}_{\text{fake}}, \boldsymbol{c}), \boldsymbol{1})$: to generate data that can fool the discriminator;
- Condition loss $MSE(C(\boldsymbol{x}_{\text{fake}}), \boldsymbol{c})$: to generate data from which the input conditions can be reproduced by the conditioner.

The training of the generator is indirect performed through both the discriminator and the conditioner.

3.3.5. Training of the Model

GANs are notoriously difficult to train. We applied the hyperparameter optimisation framework Optuna [37] to search for the best learning rates for the generator, the discriminator, and the conditioner, respectively. Figure 8a shows the total training loss curves of the generator, the discriminator, and the conditioner. The training loss of the discriminator decreases rapidly in the beginning of the training process and then starts to decrease more gradually. The loss of the generator stays relatively steady throughout the whole training process. As for the conditioner, there is no obvious improvement in performance in the first half of the training; in contrast, the loss value drops significantly and stays stable at a low level in the second half of training.

Figure 8b shows the loss values directly related to the competition between the generator and the discriminator, namely the generation loss of the generator, as well as the sum of the fake loss and the real loss of the discriminator. The two curves are strongly related to each other in an adverse way but, in general, both stay relatively steady throughout the whole training, which is a direct indication of the competition between the two submodels that both parties are learning to improve their own abilities but neither can prevail. A detailed demonstration of the real and fake losses of the discriminator is shown in Figure 8c. Real loss displays a decreasing tendency overall, while fake loss displays an increasing tendency, which is another indication that both the discriminator and the generator are improving during the training process; the generator gets better at fooling the discriminator and, in spite of this, the discriminator still manages to get better at distinguishing the real data.

The learning of conditions can be evaluated by Figure 8d and Figure 8e. The unmatched loss of the discriminator, shown in Figure 8d, quickly decreases in the early stages of training and converges throughout the training process. It indicates an improvement in the discriminator's learning of the extra conditioning information. Figure 8e shows the condition loss of the generator and the loss of the conditioner. Although, in general, the condition loss of the generator only decreases slightly with high oscillation, the significant decrease in the conditioner's loss indirectly indicates an improvement in the generator's ability to generate data using the extra conditioning information.

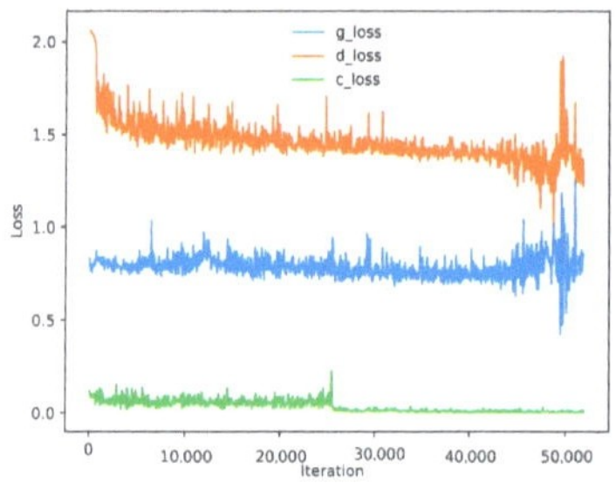

(**a**) Toal losses of the model

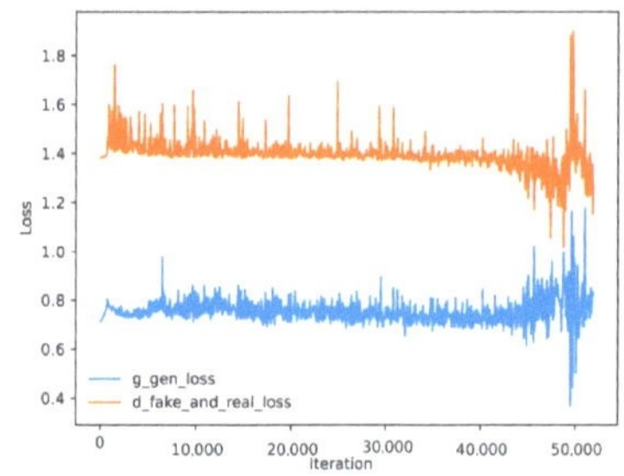

(**b**) Loss of competition

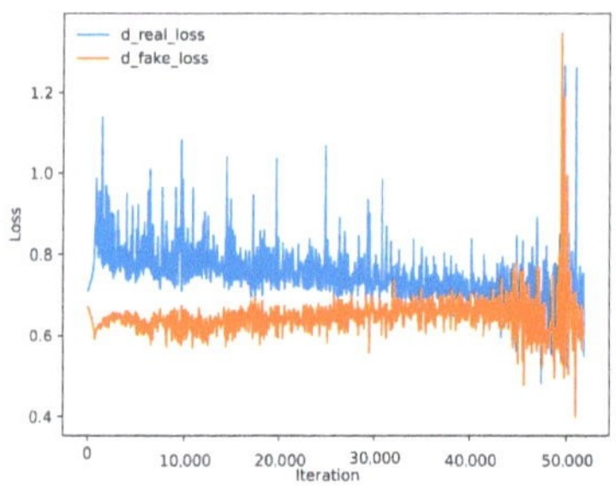

(**c**) Real loss and fake loss of discriminator

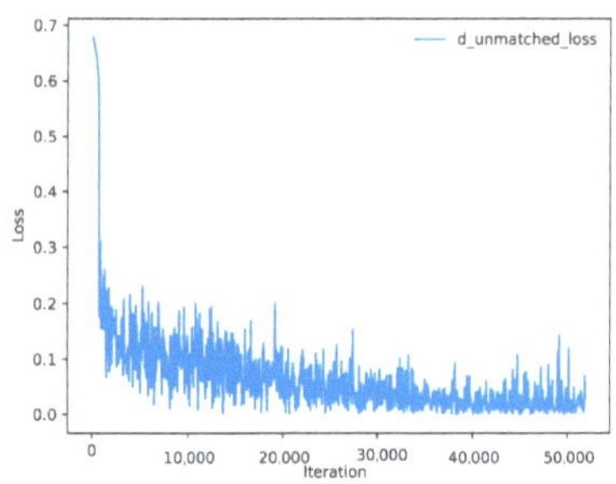

(**d**) Unmatched loss of discriminator

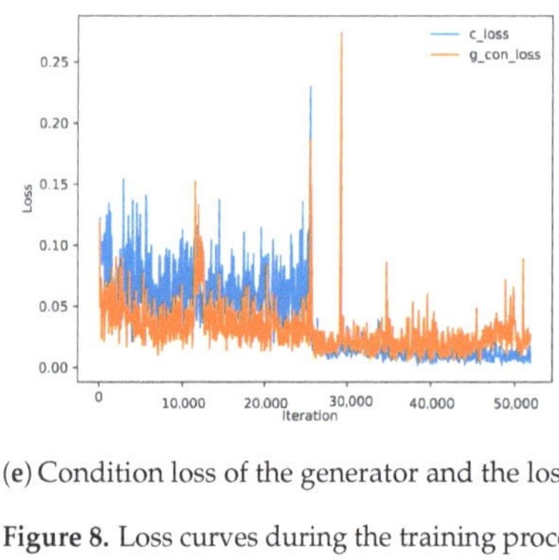

(**e**) Condition loss of the generator and the loss of the conditioner

Figure 8. Loss curves during the training processes: (**a**) total losses of the generator, the discriminator, and the conditioner; (**b**) generation loss of the generator and the sum of fake loss and real loss of the discriminator; (**c**) real loss and fake loss of the discriminator; (**d**) unmatched loss of the discriminator; (**e**) condition loss of the generator and the loss of the conditioner.

After the joint training of the three submodels, the generator is utilised for load profile generation on demand, since it has learned to generate data with a similar distribution as the ground truth dataset, with the imposed extra conditioning information considered.

4. Results

The testing and validation of the approaches are performed in two steps. First, the generated load cycles of both methods are compared with two standard stress tests, the Federal Urban Driving Schedule (FUDS) and the Dynamic Stress Test (DST). Afterwards, load cycles are designed to represent the stress from driving the BMW i3 (BMW AG, Munich, Germany). These cycles are used as inputs for an Arbin battery tester (Arbin Instruments, College Station, TX, USA), with 5 V channels and up to ±10 A during the simulation part of the testing schedule. The lithium-ion battery cell used is a cylindrical Panasonic NCR18650GA (Panasonic, Kadoma, Japan); its characteristics are stated in Table 1.

By designing the representative cycles of the BMW i3 dataset shown in Figure 1, the areas that were very common in the dataset become easy to identify. After investigating the range and mean of the SOC, it was decided that the profiles should consist of three dynamic parts. Two parts described the behaviour up to 90% SOC but with different discharge depths down to 70% SOC, while the other part was in the range of 85% to 75% SOC. Based on the plot of the BMW data, it can be seen that, while driving, the battery was mainly in the upper range of the SOC. Furthermore, the general schedule was designed to achieve about three full cycle equivalents per day to have an increased number of cycles because if the data were used for battery cell degradation in a laboratory, it would need to be designed to lead to faster degradation. Otherwise, the measurements would take too long. In addition to the number of cycles per day and the three dynamic parts, the parts were assigned weights. These weights described how large this part's share was in the overall scenario. The weights were assigned in decreasing order, starting from the shallowest to the deepest parts, regarding the depth of discharge. Given a C-rate of 1 C for charging the cell, the expected duration, the number of repetitions, and the SOC gradient can be estimated. The calculations are just estimations and do not consider the coulombic efficiency or if the cut-off voltage is reached during charging, for example.

The same ranges of the SOC are used for the FUDS- and DST-based cycles. Instead of repeating the standard schedules from full charge to the discharge cut-off voltage, they are stopped upon reaching the specified SOC range. Figure 9 displays the generated cycles and the cycles based on the DST and FUDS.

The generated load cycles show some differences. Overall, no load cycle exceeded the boundaries for the C-rates in the charging or the discharging directions. The counts of the different C-rates were normalised to be comparable. Furthermore, the DST- and FUDS-based load cycles had repetitive sequences in the cycle, whereas the two introduced approaches did not. Despite the repetitions in the FUDS, the gradRPM-based load cycle and the FUDS were similar concerning the general structure of the pulses, which were often short at about 1 s, and the height. The DST mainly involved the constant currents at specific C-rate heights that were lower than the others and repeated the identical pulses. Figure 9 displays the histogram of the different types of cycles concerning the C-rates. Specifically, this plot demonstrates the repeated constant current of the DST load cycle with high peaks in the histogram. The histogram shows that the Info-RCGAN-generated profile consists primarily of discharging pulses and only small charging currents. The discharging C-rates are very similar to those of the gradRPM. Overall, the different load cycles have different SOC gradients. The gradRPM method allowed for the desired gradient of −0.28 SOC/h. The other load cycles have a gradient of −0.5098 SOC/h, −0.5136 SOC/h, and −0.1823 SOC/h for FUDS, DST, and the Info-RCGAN, respectively. Therefore, the FUDS- and DST-based load cycles would lead to faster discharging of the battery cell and shorter dynamic parts, as well as to more full cycle equivalents per day. On the other hand, the load cycle by the Info-RCGAN is slower, although it does not include that many charging pulses, as can be seen in the histogram. Concerning the Info-RCGAN, various experiments with different parameter settings were conducted. The results indicate that the conditional inputs are highly entangled, and the model's learning of the conditional information is focused on the common scenarios of the ground truth dataset, which is why the model cannot deal with the circumstances of abnormal input conditions well.

The generated load cycles are assembled to form a schedule to be tested on the NCRGA lithium-ion battery cell. There is one schedule each for the gradRPM and Info-RCGAN approaches, where each cycle is repeated as stated in Table 2. If necessary, each repetition is interspersed by a constant current and constant voltage charging that charges the lithium-ion battery cell with the same amount of charge discharged during the dynamic load cycle. The discharged charge is logged during the load cycle and saved to a variable that is accessed during the charging step.

The schedule consists of approximately seven cycles, each equivalent to three full cycles.

Figure 10 compares the BMW measurement data with the results of the laboratory measurements of the gradRPM and the Info-RCGAN approaches. The first row contains the analysis results based on the voltage, the second row includes the analysis of the C-rates, and the third comprises the temperature analysis. The graph displays the load cycle spectra extracted using the rainflow counting approach. It is two-parametric; it extracts the detected cycle's range and mean and counts all the appearances within the dataset. The darker the colour, the higher the count. The voltage's load cycle spectra display the typical battery behaviour connected to the maximum voltage. The maximum voltage leads to the line reaching higher mean values because the higher the voltage, the lower overpotentials are allowed, and the lower the currents, especially in the charging direction, lead to shallow voltage ranges. Overall, the BMW data show a homogenous area in the load cycle spectra because the amount of data is high, and the noise is higher than in the laboratory. The voltage spectra of the two approaches are similar to the spectra of the BMW concerning the covered area but not as homogenous. That is based on the SOC range and repetitions of the three load cycles designed. In the voltage spectra of the two approaches, there is one outlier. This outlier is connected to a full charge and discharge during the cycling to estimate the change in the capacity over cycling. Because the BMW data provided do not contain single-cell voltages, the pack voltage is rescaled to a single cell for comparison, assuming that the series connections are equal. The C-rate spectra show the most significant difference from the BMW data. Firstly, the BMW spectra are homogenous, likely due to the noise and the amount of data. In addition, the spectra display straight lines, which may result from regulating the currents, leading to more consistent means and ranges. One of the lines is also visible in the spectra of both approaches. However, neither the gradRPM spectra nor the Info-RCGAN spectra include as many charging currents as the actual measurement data. Furthermore, the Info-RCGAN focuses entirely on the line with only minor deviations. A reason for the spectra obtained with only the few charging currents is that the load cycles are, in general, designed to show accelerated ageing, so the gradient of the SOC is steeper than in the actual data to reach the specified equivalent of three full cycles per day, leading to more discharging segments in the load cycle designed. In addition, the C-rate spectra of the load cycles designed, especially for the gradRPM, differ from the laboratory measurements. The charging segments are missing. That is not the case for the Info-RCGAN, which suggests that the charging current is regulated for the gradRPM because the battery cell reaches its voltage limit. Meanwhile, the Info-RCGAN does not include high-charging segments during the dynamic load cycles because, during training, the model learned to focus mainly on the most predominant feature of the training set. The temperature spectra are expected to be different from the BMW data because the car was used outside and experienced low temperatures, as we began recording the data in winter. In contrast, the laboratory measurements were taken out in a controlled environment, within a thermal chamber with a constant temperature of 25 °C. Therefore, the BMW spectra are broader and focus on the area between 18 °C and 25 °C; for both approaches, the temperature is always near to 25 °C ± 1 °C.

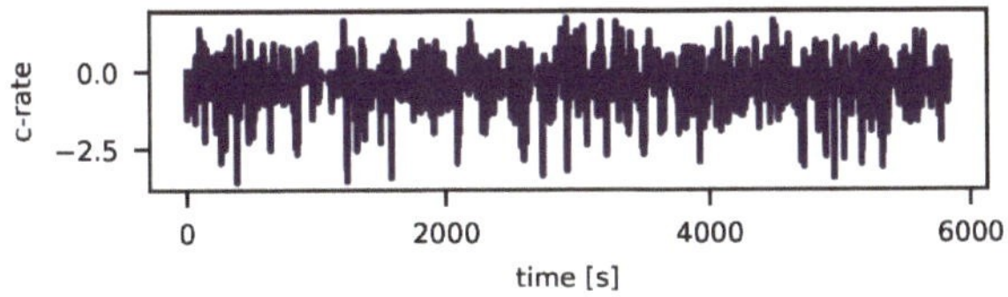

(**a**) gradRPM DLC3

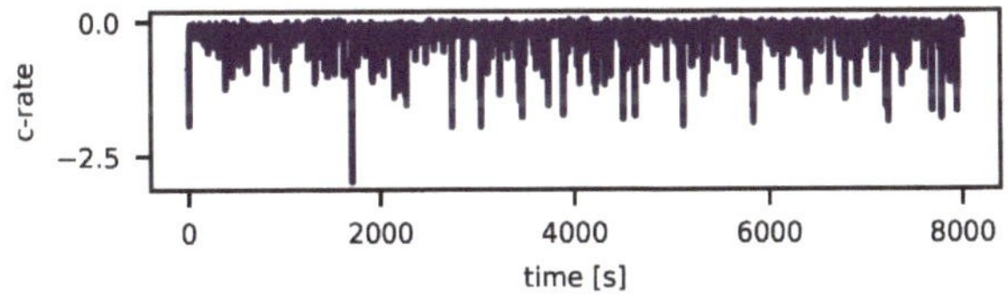

(**b**) Info-RCGAN DLC3

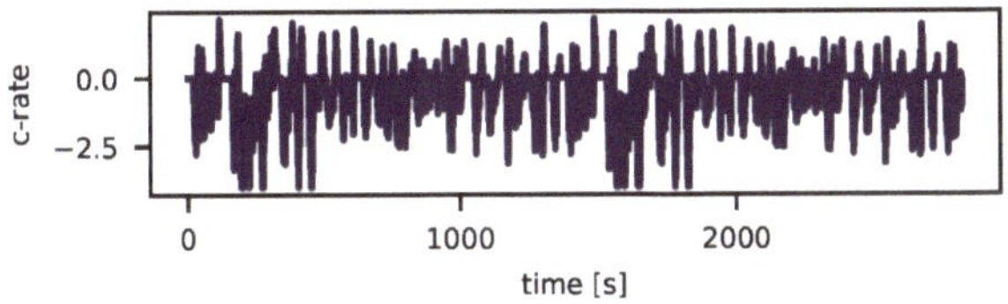

(**c**) FUDS DLC3

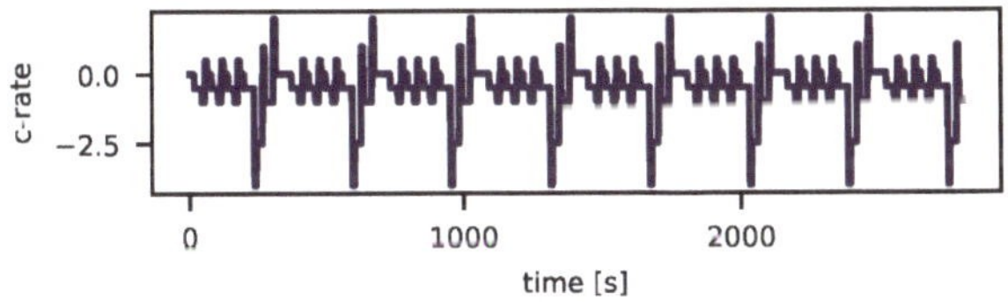

(**d**) DST DLC3

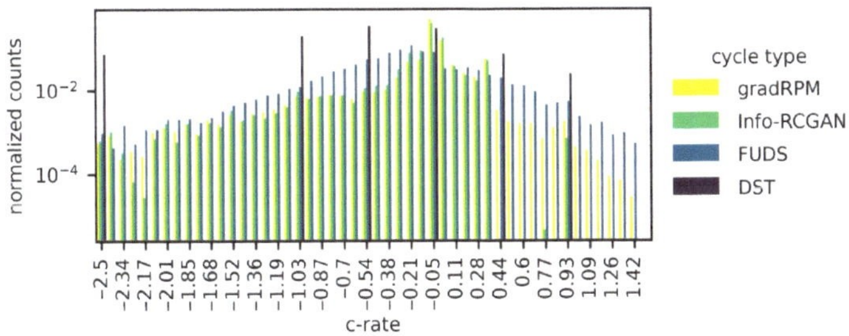

(**e**) Histogram of C-rates of the different cycle types

Figure 9. The figure displays the cycles generated based on (**a**) the gradRPM method, (**b**) the Info-RCGAN, (**c**) the FUDS and (**d**) the DST. Subfigure (**e**) displays the histogram of C-rates of the different cycles with normalised counts.

Table 1. Panasonic NCR18650GA.

Characteristic	Value
Rated capacity	3.3 Ah
Typical capacity	3.45 Ah
Nominal Voltage	3.6 V
Charging cut off voltage	4.2 V
Charging current	1.475 A
Discharging current	10 A

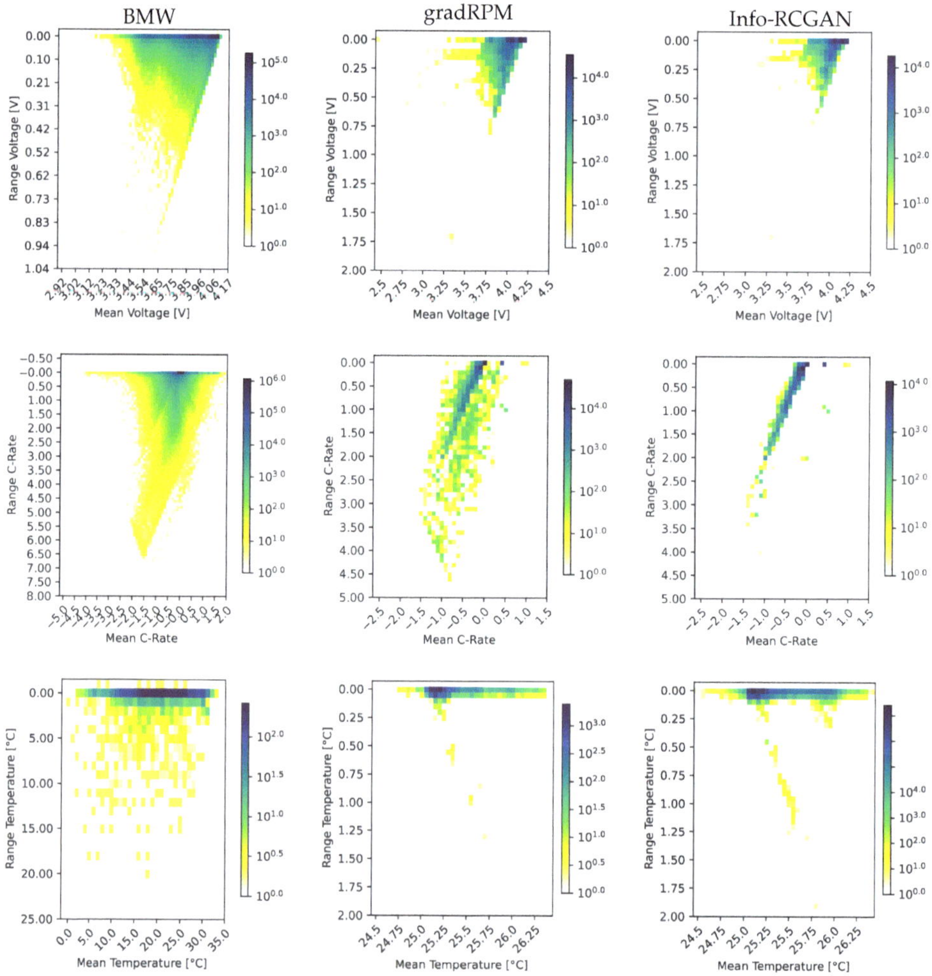

Figure 10. The figure displays the different spectra of each dataset (in the columns from left to right) of the BMW i3, the gradRPM, and the Info-RCGAN. In the rows, the analysed parameters (from top to bottom) are the voltage, the C-rate, and the temperature. Note the difference in the scaling of the BMW spectra and the gradRPM and the Info-RCGAN spectra.

Table 2. Design parameters extracted and calculated for the generation of dynamic load cycles (DLCs) and the overall scenario based on the equivalent of three full cycles per day and a C-rate for the constant current charging of 1 C.

Dynamic Load Cycle ID	SOC Start	SOC End	Percent of the Scenario	Repetitions	Load Cycle Duration	SOC per h
DLC1	0.9	0.7	0.4	12	2520 s	0.286 SOC/h
DLC2	0.85	0.75	0.5	30	1260 s	0.286 SOC/h
DLC3	0.9	0.5	0.1	2	5040 s	0.286 SOC/h

5. Discussion

The experiments show that both the gradRPM and the Info-RCGAN approaches successfully generate very dynamic load cycles capable of replicating the stress factors identified in a given dataset. In addition, the experiments show that a more significant spectrum of stress factors can be covered when dynamic load cycles are used. However, it is always a challenge to balance between realism and usefulness.

A highly realistic load cycle can be generated if the controlling parameters are set, respectively. Still, a very realistic profile might be helpful for short-term testing like state of charge algorithm validation, where ageing effects are not the primary concern. However, it can prove disadvantageous when used for long-term investigations like degradation measurements. When used for long-term investigations, there is a tendency to focus on the usefulness of the load cycle designed to reach a specified goal of the degradation of the battery cell in a given time while maintaining relevance to the real-world conditions.

If the goal is to create the most realistic cycles, using tools like the KIT DRIVING CYCLE TOOL might be more suitable for designing real driving profiles. Furthermore, the load cycle must include many resting phases when targeting automotive applications to be realistic with regard to stress factors during degradation. The BMW i3 was driven for about 1 h per day, with the vehicle remaining parked for the rest of the time. As the aim was to generate load cycles representing similar stress factors that were still controllable to fulfill a degradation schedule, for example, the gradRPM and the Info-RCGAN proved to be valuable tools. At this point, although the Info-RCGAN shows great potential in conditional battery load profile generation with versatile regulations over the target features, the gradRPM approach has proven to be more straightforward to use with good results because the Info-RCGAN has some critical problems, like the entanglement of conditional inputs and its excessive focus on certain predominant features of the training set.

6. Conclusions

Based on the experience with the algorithms, there are different steps to take to improve the performance. For the gradRPM, changing the segmentation from zero crossing to other methods, like clustering the data to specific behaviours, especially for automotive applications, might be helpful. However, it might not be that controllable anymore because of the length of the segments. As for the Info-RCGAN, future work regarding the entanglement of the conditional inputs, the tracking of the predefined conditions, and the validation of the features related to the voltage and temperature would be beneficial to improve the quality of the load profiles generated and the control over specific features for studying stress factors. It would be helpful to further investigate how the model learns the data distribution by designing innovative evaluation metrics for the training process.

In general, it is worth investigating how dynamic a load cycle has to be to show degradation behaviour similar to that of a dynamic load cycle, as a dynamic load cycle takes some effort to run without issues. In addition, they lead to a high amount of data because the load changes every second; to measure the response of the battery, the sampling rate must be at least 500 ms.

Author Contributions: Conceptualisation, S.N.; methodology, S.N. and J.Y.; software, S.N. and J.Y.; validation, S.N. and J.Y.; formal analysis, S.N. and J.Y.; investigation, S.N. and J.Y.; resources, S.N. and J.Y.; data curation, S.N. and J.Y.; writing—original draft preparation, S.N. and J.Y.; writing—review and editing, S.N., J.Y. and J.K.; visualisation, S.N. and J.Y.; supervision, J.K.; project administration, J.K.; funding acquisition, S.N. and J.K. All authors have read and agreed to the published version of the manuscript.

Funding: This paper is based on the project FeBaL, which is supported by the German Federal Education and Research under grant number 03XP0308B. The responsibility for the content of this publication lies with the authors.

Data Availability Statement: The original contributions presented in the study are included in the article, further inquiries can be directed to the corresponding authors.

Acknowledgments: We would like to express our sincere gratitude to the Fraunhofer Institute for Transportation and Infrastructure Systems (IVI) for providing the essential measurement data of the BMW i3 electric vehicle. We acknowledge support by the German Research Foundation and the Open Access Publication Fund of TU Berlin.

Conflicts of Interest: The authors declare no conflicts of interest.

Abbreviations

The following abbreviations are used in this manuscript:

RPM	Random Pulse Method
gradRPM	Gradient-Based Random Pulse Method
GAN	Generative Adversarial Network
RCGAN	Recurrent Conditional GAN
Info-RCGAN	Information Maximizing-RCGAN
SOC	State of Charge
RNN	Recurrent Neural Network
LSTM	Long Short-Term Memory
LTTB	Largest-Triangle-Three-Bucket
BCE	Binary Cross Entropy
MSE	Mean Squared Error

References

1. Geslin, A.; Xu, L.; Ganapathi, D.; Mory, K.; Chueh, W.; Onori, S. Dynamic cycling enhances battery lifetime. *Nat. Energy* **2025**, *10*, 172–180. [CrossRef]
2. Kirkaldy, N.; Samieian, M.; Offer, G.; Marinescu, M.; Patel, Y. Lithium-ion battery degradation: Comprehensive cycle ageing data and analysis for commercial 21,700 cells. *J. Power Sources* **2024**, *603*, 234185. [CrossRef]
3. dos Reis, G.; Strange, C.; Yadav, M.; Li, S. Lithium-ion battery data and where to find it. *J. Energy AI* **2021**, *5*, 100081. [CrossRef]
4. Neupert, S.; Yao, J.; Kowal, J. Load Cycle Design and Analysis for Energy Storage Technologies Utilising Micro-Trip Methods and Machine Learning Approaches. In Proceedings of the Energy Storage Conference 2023 (ESC 2023), Glasgow, UK, 15–16 November 2023; pp. 63–68. [CrossRef]
5. Gewald, T.; Reiter, C.; Lin, X.; Baumann, M.; Krahl, T.; Hahn, A.; Lienkamp, M. (Eds.) *Characterization and Concept Validation of Lithium-Ion Batteries in Automotive Applications by Load Spectrum Analysis*; Society of Automotive Engineers of Japan, Inc.: Chiyoda, Japan, 2018.
6. Barragán-Moreno, A.; Gomez, P.I.; Dragičević, T. Enhancement of Stress Cycle-counting Algorithms for Li-Ion Batteries by means of Fuzzy Logic. In Proceedings of the 2022 IEEE Transportation Electrification Conference & Expo (ITEC), Anaheim, CA, USA, 15–17 June 2022; IEEE: Piscataway, NJ, USA, 2022. [CrossRef]

7. Fioriti, D.; Scarpelli, C.; Pellegrino, L.; Lutzemberger, G.; Micolano, E.; Salamone, S. Battery lifetime of electric vehicles by novel rainflow-counting algorithm with temperature and C-rate dynamics: Effects of fast charging, user habits, vehicle-to-grid and climate zones. *J. Energy Storage* **2023**, *59*, 106458. [CrossRef]
8. Gundogdu, B.; Gladwin, D.T. A fast battery cycle counting method for grid-tied battery energy storage system subjected to microcycles. In Proceedings of the 2018 International Electrical Engineering Congress (iEECON), Krabi, Thailand, 7–9 March 2018; IEEE: Piscataway, NJ, USA, 2018. [CrossRef]
9. Nebuloni, R.; Meraldi, L.; Moretti, L.; Ilea, V.; Bovo, C.; Berizzi, A.; Raboni, P. A Real-Time Cycle Counting Method for Battery Degradation Calculation in MILP Models. In Proceedings of the 2023 IEEE International Conference on Environment and Electrical Engineering and 2023 IEEE Industrial and Commercial Power Systems Europe (EEEIC/I&CPS Europ), Madrid, Spain, 6–9 June 2023; IEEE: Piscataway, NJ, USA, 2023.
10. Saxena, S.; Roman, D.; Robu, V.; Flynn, D.; Pecht, M. Battery Stress Factor Ranking for Accelerated Degradation Test Planning Using Machine Learning. *Energies* **2021**, *14*, 723. [CrossRef]
11. Krause, T.; Nusko, D.; Rittmann, J.; Bauermann, L.; Kroll, M.; Holly, C. Synthetic Data Generation for AI-Informed End-of-Line Testing for Lithium-Ion Battery Production. *World Electr. Veh. J.* **2025**, *16*, 75. [CrossRef]
12. Qiu, H.; Cui, S.; Wang, S.; Wang, Y.; Feng, M. A Clustering-Based Optimization Method for the Driving Cycle Construction: A Case Study in Fuzhou and Putian, China. *IEEE Trans. Intell. Transp. Syst.* **2022**, *23*, 18681–18694. [CrossRef]
13. Wang, T.; Jing, Z.; Zhang, S.; Qiu, C. Utilizing Principal Component Analysis and Hierarchical Clustering to Develop Driving Cycles: A Case Study in Zhenjiang. *Sustainability* **2023**, *15*, 4845. [CrossRef]
14. Kellner, Q.; Hosseinzadeh, E.; Chouchelamane, G.; Widanage, W.D.; Marco, J. Battery cycle life test development for high-performance electric vehicle applications. *J. Energy Storage* **2018**, *15*, 228–244. [CrossRef]
15. Ben-Marzouk, M.; Pelissier, S.; Clerc, G.; Sari, A.; Venet, P. Generation of a Real-Life Battery Usage Pattern for Electrical Vehicle Application and Aging Comparison With the WLTC Profile. *IEEE Trans. Veh. Technol.* **2021**, *70*, 5618–5627. [CrossRef]
16. Etxand-Santolaya, M.; Canals Casals, L.; Corchero, C. Estimation of electric vehicle battery capacity requirements based on synthetic cycles. *Transp. Res. Part D* **2023**, *114*, 103545. [CrossRef]
17. Moy, K.; Beom Lee, S.; Harris, S.; Onori, S. Design and validation of synthetic duty cycles for grid energy storage dispatch using lithium-ion batteries. *Adv. Appl. Energy* **2021**, *4*, 100065. [CrossRef]
18. Lin, J.; Liu, B.; Zhang, L. Autoencoder-based optimization method for driving cycle construction: A case study in Fuzhou, China. *J. Ambient. Intell. Humaniz. Comput.* **2022**, *14*, 12635–12650. [CrossRef]
19. Mohanty, P.; Jena, P.; Padhy, N. imeGAN-based Diversified Synthetic Data Generation Following BERT-based Model for EV Battery SOC Prediction: A State-of-the-Art Approach. *IEEE Trans. Ind. Appl.* 2025, *early acces*. [CrossRef]
20. Steinarsson, S. Downsampling Time Series for Visual Representation. Master's Thesis, University of Iceland, Faculty of Industrial Engineering, Mechanical Engineering and Computer Science, Reykjavik, Iceland, 2013.
21. Sola, J.; Sevilla, J. Importance of input data normalization for the application of neural networks to complex industrial problems. *IEEE Trans. Nucl. Sci.* **1997**, *44*, 1464–1468. [CrossRef]
22. Pedregosa, F.; Varoquaux, G.; Gramfort, A.; Michel, V.; Thirion, B.; Grisel, O.; Blondel, M.; Müller, A.; Nothman, J.; Louppe, G.; et al. Scikit-Learn: Machine Learning in Python. *arXiv* **2018**, arXiv:1201.0490. [CrossRef]
23. Rumelhart, D.E.; Hinton, G.E.; Williams, R.J. Learning internal representations by error propagation. In *Parallel Distributed Processing: Explorations in the Microstructure of Cognition, Vol. 1: Foundations*; MIT Press: Cambridge, MA, USA, 1986; pp. 318–362.
24. Du, X.; Cai, Y.; Wang, S.; Zhang, L. Overview of deep learning. In Proceedings of the 2016 31st Youth Academic Annual Conference of Chinese Association of Automation (YAC), Wuhan, China, 11–13 November 2016; IEEE: Piscataway, NJ, USA, 2016; pp. 159–164. [CrossRef]
25. Yao, J.; Kowal, J. A Multi-Scale Data-Driven Framework for Online State of Charge Estimation of Lithium-Ion Batteries with a Novel Public Drive Cycle Dataset. *J. Energy Storage* **2025**, *107*, 114888. [CrossRef]
26. Hochreiter, S.; Schmidhuber, J. Long Short-Term Memory. *Neural Comput.* **1997**, *9*, 1735–1780. [CrossRef] [PubMed]
27. Yu, Y.; Si, X.; Hu, C.; Zhang, J. A Review of Recurrent Neural Networks: LSTM Cells and Network Architectures. *Neural Comput.* **2019**, *31*, 1235–1270. [CrossRef]
28. Goodfellow, I.J.; Pouget-Abadie, J.; Mirza, M.; Xu, B.; Warde-Farley, D.; Ozair, S.; Courville, A.; Bengio, Y. Generative Adversarial Networks. *arXiv* **2014**, arXiv:1406.2661. [CrossRef]
29. Mirza, M.; Osindero, S. Conditional Generative Adversarial Nets. *arXiv* **2014**. [CrossRef]
30. Odena, A.; Olah, C.; Shlens, J. Conditional Image Synthesis With Auxiliary Classifier GANs. *arXiv* **2017**, arXiv:1610.09585. [CrossRef]
31. Miyato, T.; Koyama, M. cGANs with Projection Discriminator. *arXiv* **2018**. [CrossRef]
32. Mogren, O. C-RNN-GAN: Continuous recurrent neural networks with adversarial training. *arXiv* **2016**. [CrossRef]
33. Esteban, C.; Hyland, S.L.; Rätsch, G. Real-valued (Medical) Time Series Generation with Recurrent Conditional GANs. *arXiv* **2017**. [CrossRef]

34. Yao, J.; Neupert, S.; Kowal, J. Cross-Stitch Networks for Joint State of Charge and State of Health Online Estimation of Lithium-Ion Batteries. *Batteries* **2024**, *10*, 171. [CrossRef]
35. Ding, X.; Wang, Y.; Xu, Z.; Welch, W.J.; Wang, J. *CcGAN: Continuous Conditional Generative Adversarial Networks for Image Generation*; The University of British Columbia: Vancouver, BC, Canada, 2021.
36. Chen, X.; Duan, Y.; Houthooft, R.; Schulman, J.; Sutskever, I.; Abbeel, P. InfoGAN: Interpretable Representation Learning by Information Maximizing Generative Adversarial Nets. *arXiv* **2016**. [CrossRef]
37. Akiba, T.; Sano, S.; Yanase, T.; Ohta, T.; Koyama, M. Optuna: A Next-generation Hyperparameter Optimization Framework. *arXiv* **2019**, arXiv:1907.10902. [CrossRef]

batteries

MDPI

Article

SOH Estimation of Lithium-Ion Batteries Using Distribution of Relaxation Times Parameters and Long Short-Term Memory Model †

Abdul Shakoor Akram, Muhammad Sohaib and Woojin Choi *

School of Electrical Engineering, Soongsil University, Seoul 06978, Republic of Korea
* Correspondence: cwj777@ssu.ac.kr

† This paper is an extended version of our paper published in Akram, A.S.; Sohaib, M.; Choi, W. SOH Estimation of the Lithium-ion Batteries using the Parameters of Distribution of Relaxation Times. In Proceedings of the KIPE Autumn Conference, Seoul, Republic of Korea, 22 November 2024; pp. 52–54.

Abstract: Lithium-ion batteries are extensively utilized in modern applications due to their high energy density, long cycle life, and efficiency. With the increasing demand for sustainable energy storage solutions, accurately estimating the State of Health (SOH) is essential to address challenges related to battery degradation and secondary life management. Electrochemical Impedance Spectroscopy (EIS) is a widely used diagnostic tool for evaluating battery performance due to its simplicity and cost-effectiveness. However, EIS often struggles to decouple overlapping electrochemical processes. The Distribution of Relaxation Times (DRT) method has emerged as a powerful alternative, enabling the isolation of key processes, such as ohmic resistance, SEI resistance, charge transfer resistance, and diffusion, thereby providing deeper insights into battery aging mechanisms. This paper presents a novel approach for estimating the State of Health (SOH) of batteries by leveraging DRT parameters across multiple State of Charge (SOC) levels. This study incorporates data from three lithium-ion batteries, each with distinct initial capacities, introducing variability that reflects the natural differences observed in real-world battery performance. By employing a Long Short-Term Memory (LSTM)-based machine learning model, the proposed framework demonstrates a superior accuracy in SOH prediction compared to traditional EIS-based methods. The results highlight the sensitivity of DRT parameters to SOH degradation and validate their effectiveness as reliable indicators for battery health. This research underscores the potential of combining a DRT analysis with AI-driven models to advance scalable, precise, and interpretable battery diagnostics.

Keywords: electrochemical impedance spectroscopy; distribution of relaxation times parameters; long short-term memory; state of health estimation

1. Introduction

Lithium-ion batteries (LIBs) have become indispensable in modern technological advancements, powering applications ranging from portable electronics to electric vehicles and large-scale renewable energy storage systems [1–3]. Their popularity stems from their high energy density, extended lifespan, and operational efficiency, which collectively address the growing demand for sustainable energy solutions [4,5]. However, with the increasing deployment of LIBs, accurately estimating their State of Health (SOH) has become crucial for ensuring safety, optimizing performance, and extending their service life [6]. Beyond first-life applications, the SOH estimation is also critical for assessing a battery's suitability for second-life uses, such as stationary energy storage or backup power systems [7]. An effective SOH estimation promotes the sustainable utilization of resources, reduces waste, and enhances the overall lifecycle value of lithium-ion batteries.

Electrochemical Impedance Spectroscopy (EIS) is widely regarded as a fundamental technique for evaluating battery health by analyzing the impedance response of a cell across a range of frequencies. This diagnostic approach captures key electrochemical properties, including internal resistance, charge transfer behavior, and capacitive dynamics, which are directly linked to the aging mechanisms of the battery. Despite its utility, EIS has notable limitations when analyzing overlapping electrochemical processes. EIS encounters limitations when overlapping processes share similar characteristic frequencies leading to merged semicircles in the Nyquist plot, the ambiguous mapping of circuit elements in equivalent circuit modeling, single-frequency-based representation rather than a more revealing time-domain breakdown, susceptibility to measurement noise and constrained frequency resolution, and ultimately an ill-posed parameter estimation process make it difficult to accurately isolate and interpretate the electrochemical mechanism of the battery systems [8,9].

The Distribution of Relaxation Times (DRT) has emerged as an advanced technique to refine the impedance analysis and overcome its limitations [10]. By deconvoluting impedance spectra, the DRT provides a higher resolution of distinct electrochemical processes and their timescales, enabling a deeper understanding of the internal dynamics that influence battery performance. Integrating EIS with the DRT allows for a more detailed analysis of the internal chemical behavior of the battery and contributes to improving the Equivalent Circuit Models (ECMs) traditionally developed using EIS methods, as highlighted in the recent literature [11], thereby creating a robust framework for assessing battery health. Also, ref. [12] apply a full-spectrum DRT-CNN with Grad-CAM and achieve better accuracy, and ref. [13] employs static regression on multiple DRT features from both cycling and calendar-aging datasets, while our approach focuses solely on the most aging-related information peak parameters across the SOC intervals and feeding them into a LSTM, demonstrating that accurate, dynamic SOH estimations can be achieved with far less and simpler data and a more generalized model. This study leverages both methods to improve the accuracy of SOH estimations in lithium-ion batteries, particularly focusing on applications for second-life usages [3,10].

This paper utilizes the DRT method to analyze EIS data, aiming to extract key parameters that reflect detailed internal processes within lithium-ion batteries. The charge transfer peak, a prominent feature in the DRT spectrum, is utilized to train a Long Short-Term Memory (LSTM) model for estimating the SOH of Galaxy S9+ battery cells. By integrating DRT-derived features into an AI-based model, this approach establishes an advanced framework to enhance the accuracy of SOH estimations, supporting data-driven strategies for monitoring the long-term performance of lithium-ion batteries. For comparison, our model was quantitatively validated against an ECM-based neuro-fuzzy model, and the notably lower Mean Absolute Error (MAE) and Root Mean Squared Error (RMSE) highlight the superior accuracy and robustness of the proposed DRT-based LSTM approach in capturing complex, overlapping electrochemical processes in battery degradation.

The overall framework for the proposed methodology is summarized in Figure 1, which outlines the key steps involved in the process. First, lithium-ion batteries undergo cycling tests and EIS measurements to evaluate their degradation behavior. The EIS data are then processed using the DRT method, which deconvolutes the impedance spectra to isolate critical processes, particularly the charge transfer peak. Key DRT parameters are extracted as quantitative features reflecting the charge transfer resistance and associated aging mechanisms [14]. These features are analyzed and subsequently used as inputs to a

LSTM model, which is trained to predict the SOH of the batteries. Finally, the performance of the LSTM model is evaluated using prediction results and error metrics, confirming the accuracy and robustness of the approach. This integrated framework demonstrates the effective combination of an advanced electrochemical analysis with AI-based modeling for accurate SOH estimations.

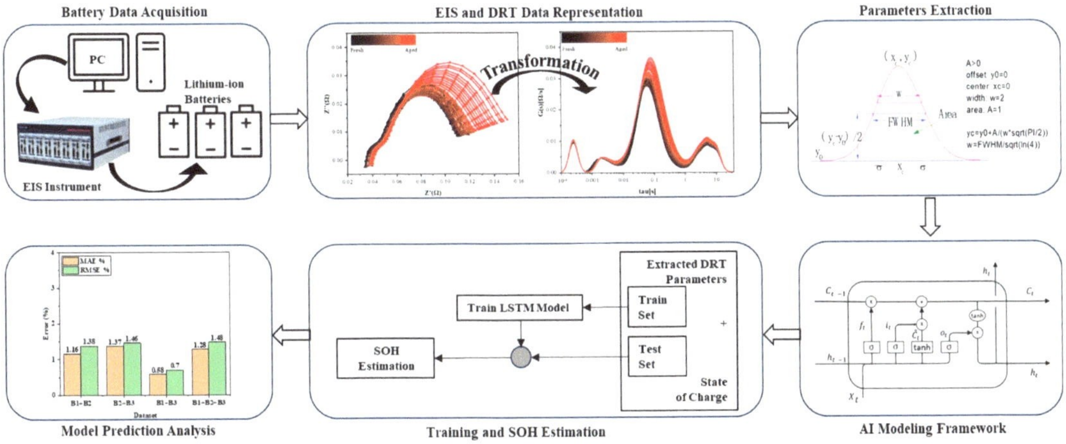

Figure 1. Overall framework of SOH estimation by LSTM.

2. Data Acquisition Method

The battery aging test was conducted on three lithium-ion batteries extracted from the Galaxy S9+ smartphone to evaluate their performance degradation over 1000 charge–discharge cycles, and specifications are shown in Table 1. Capacity and impedance spectra measurements were performed at intervals of every 20 cycles using EIS. The aging tests were conducted in a controlled environment using a HYSCLAB chamber to maintain a constant temperature of 25 °C, ensuring consistency throughout the experimental duration. WonATech WBCS3000 M2 equipment was used to perform the charge–discharge cycling, while ZIVE MP2A was employed for EIS measurements. The charge and discharge profile, as illustrated in Figure 1, included distinct phases designed to simulate real operating conditions. During the charging phase, a constant current (CC) of 2 A (0.57C) was applied until the battery reached its termination voltage of 4.4 V, at which point the process transitioned into constant voltage (CV) charging [3]. In the CV phase, charging was completed when the current decreased to 0.02C (70 mA). Discharging was performed at an accelerated rate of 1.35C (4.71 A), and the process terminated when the voltage dropped to 2.8 V. To stabilize the battery before initiating the next cycle, a 1 h rest period was introduced after each discharge phase.

Table 1. Specification of Samsung Galaxy S9+ lithium battery.

Property	Value
Chemistry	Lithium-ion polymer (NMC cathode/graphite anode)
Nominal Capacity	3500 mAh
Nominal Voltage	3.85 V
Maximum Voltage	4.4 V

Impedance spectra measurements were taken every 20 cycles, as shown in Figure 2, at six State of Charge (SOC) intervals (0%, 20%, 40%, 60%, 80%, and 100%) to capture the evolution of the battery impedance under varying states. In galvanostatic mode, a small perturbation voltage of 100 mA was applied across a frequency range from 0.1 Hz to 4 kHz to ensure accurate and reliable impedance measurements [8]. After completing EIS measurements at a 100% SOC, the battery was fully discharged at a current of 0.2C until the voltage reached 2.8 V, and the next cycle of charge–discharge testing commenced. The decision to perform EIS measurements at 20-cycle intervals strikes a balance between obtaining sufficient impedance data for analysis and minimizing the additional stress imposed on the battery during testing, which could independently accelerate battery degradation. This strategy ensured that the aging behavior of the batteries remained representative of real-world usage conditions without excessively extending the testing duration.

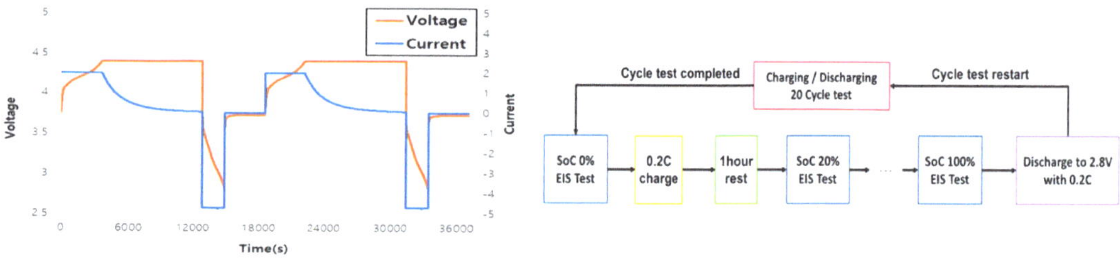

(a) (b)

Figure 2. Data acquisition network. (**a**) CCCV charge discharge profile, (**b**) EIS tests framework.

The results of the 1000-cycle aging test revealed distinct trends in capacity retention across the three battery cells. Coulombic counting was used to calculate the initial and remaining capacities relative to the nominal capacity of 3500 mAh. As presented in Figure 3, the capacities of all three batteries declined gradually over the test duration, with variations between the cells. After completing 1000 cycles, the remaining capacities for Battery 1, Battery 2, and Battery 3 were recorded as 90.76%, 93.76%, and 90.34%, respectively, as shown in Figure 3. These results demonstrate a consistent aging pattern while highlighting minor variations in capacity fading, likely due to intrinsic differences between the cells despite their identical specifications.

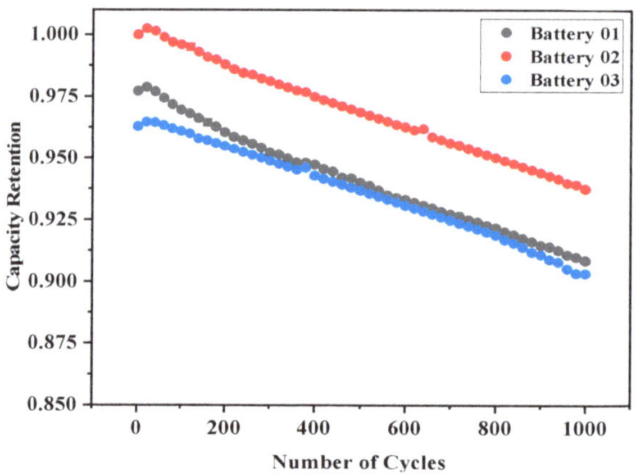

Figure 3. Capacity retention of three lithium-ion cells with cycling.

3. Analysis of Electrochemical Impedance Spectroscopy and Distribution of Relaxation Times

EIS is a fundamental technique for analyzing the internal dynamics of the battery by measuring impedance and observing electrochemical processes over a range of frequencies. In this study, EIS measurements were performed every 20 charge–discharge cycles at six State of Charge (SOC) levels (0%, 20%, 40%, 60%, 80%, and 100%) to monitor and evaluate changes in the internal processes of the three lithium-ion cells as they aged over 1000 cycles. The resulting impedance data are visualized using Nyquist plots, which are crucial for understanding the progression of the battery degradation by identifying and analyzing distinct electrochemical phenomena, such as ohmic resistance, Solid Electrolyte Interphase (SEI) layer resistance, charge transfer resistance, and diffusion processes [14].

Figure 4 displays the Nyquist plots of the impedance spectra at 100% SOC for all three cells across the cycling duration, from 0 cycles to 1000 cycles. These plots illustrate the progressive changes in impedance, which reflect the internal degradation mechanisms. The high-frequency intercept on the real axis, representing ohmic resistance, remains relatively stable during the early stages of cycling but gradually increases as cycling progresses. An increase in ohmic resistance, or contact resistance, indicates reduced electrical conductivity, often stemming from electrolyte consumption and decomposition. Similarly, the semi-circular arc in the high-to-mid-frequency range, associated with the SEI layer resistance, demonstrates a gradual increase in the diameter with cycling. This trend highlights the continuous formation and thickening of the SEI layer, which increases the internal resistance and impacts performance.

The second semi-circular arc observed in the mid-frequency region, linked to charge transfer resistance, shows a more pronounced evolution compared to other components. As cycling progresses, the charge transfer resistance increases significantly, indicating slower electrochemical reactions at the electrode–electrolyte interface [15]. This is a critical marker of aging, as it reflects the deterioration of active materials and the growing difficulty in facilitating ion transfer during charge and discharge processes. The diameter of this arc expands over time for all three cells, suggesting that charge transfer resistance plays a dominant role in the observed capacity fade and performance decline. Despite the detailed insights provided by the Nyquist plots, the low-frequency tail of the spectra, indicative of diffusion processes or Warburg impedance, is not fully captured due to the limited frequency range of the measurements. Additionally, overlapping semi-circular arcs with similar reaction time constants obscure the clear separation of electrochemical processes [16], complicating the interpretation of battery aging mechanisms solely from EIS data.

To overcome these limitations, the DRT analysis was employed for a more precise deconvolution of the impedance spectra. In our study, we adopt a Tikhonov Regularization (TR) approach with Gaussian basis functions for the DRT inversion of the Electrochemical Impedance Spectroscopy (EIS) data. Unlike Nyquist plots, the DRT analysis breaks down the impedance response into distinct peaks, each corresponding to a specific internal process, such as ohmic resistance, SEI layer resistance, charge transfer resistance, and diffusion [17]. This enables a deeper and more detailed insight into the individual contributions of various electrochemical processes over time.

As shown in Figure 4 (right), in the DRT spectrum, the peaks moving left to right correspond to the ohmic resistance, Solid Electrolyte Interphase (SEI) layer resistance, charge transfer resistance (the primary focus of this work), and diffusion (Warburg impedance) at low frequencies. The charge transfer peak, which is the most dynamic feature of the spectrum, exhibits several clear trends with the cycling across all three cells. Firstly, the peak shifts to longer relaxation times (τ), indicating slower charge transfer processes as the battery ages. This shift reflects the growing resistance at the electrode–electrolyte interface and the declining efficiency of the ion transport. Secondly, the peak height increases noticeably, confirming a steady rise in resistive behavior due to material degradation and structural changes in the electrodes. Additionally, the FWHM (Full Width at Half Maximum) of the charge transfer peak widens over the cycling duration, suggesting increased heterogeneity in the electrode reactions and reduced uniformity in the material properties. The overall area under the charge transfer peak also increases significantly, underscoring the critical role of charge transfer resistance in the aging process and its direct correlation with battery degradation.

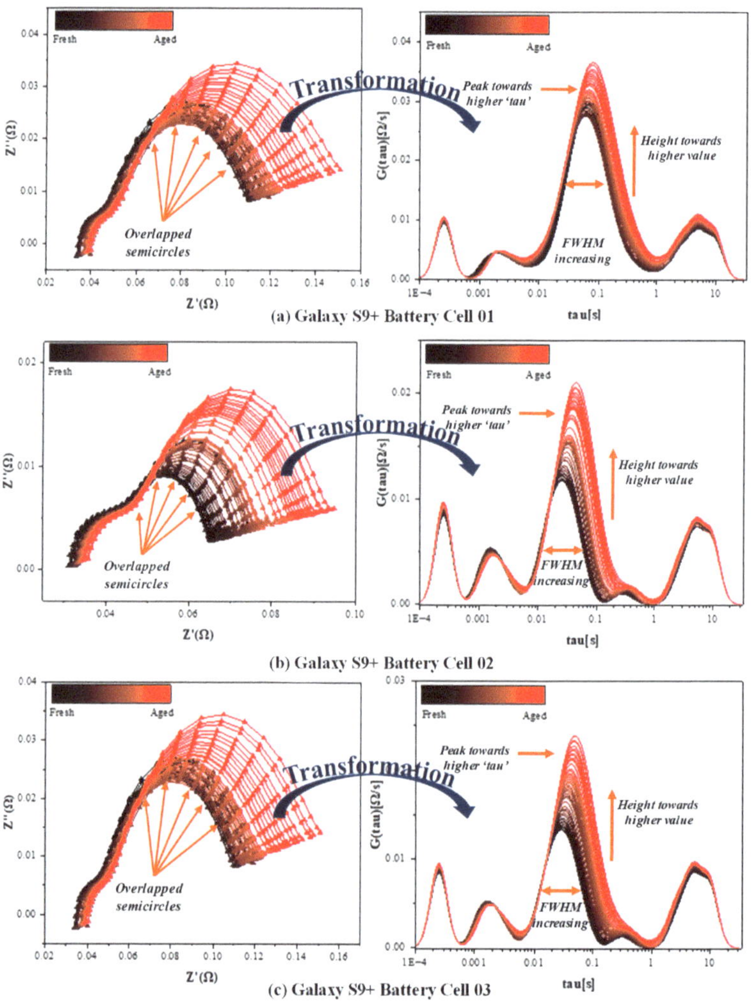

Figure 4. Nyquist (**left**) and DRT (**right**) plots of Galaxy S9+ Battery at 100% SOC.

Other peaks observed in the DRT analysis, such as those corresponding to ohmic resistance and SEI layer resistance, show comparatively minor changes during the 1000-cycle test. The ohmic resistance peak remains stable for the initial cycles but increases slightly as the aging process progresses, reflecting a growing internal resistance. The SEI layer resistance peak also demonstrates a modest increase in magnitude, which is consistent with the gradual thickening of the SEI layer [18]. The diffusion-related processes, which manifest at longer relaxation times, are observed but remain less pronounced due to the limited frequency range. However, their presence highlights the cumulative effects of aging on ion transport and mass diffusion, even if these mechanisms are less dominant compared to charge transfer resistance.

Overall, the DRT analysis reveals that charge transfer resistance is the most significant indicator of battery aging across all three cells. The progressive changes in the charge transfer peak provide critical insights into the mechanisms underlying the capacity fade and performance degradation. By enabling a clearer separation and quantification of electrochemical processes, the DRT analysis serves as a valuable tool for accurately assessing battery health and improving SOH estimation models. In summary, the enhanced clarity offered by the DRT is particularly useful for identifying key aging mechanisms and evaluating their impact on the battery performance over time.

4. Correlation of Aging and DRT Parameters Evolution

Figure 5 shows the evolution of key DRT-derived parameters, including the peak area, FWHM, center relaxation time, and peak height for the charge transfer peak in the DRT plots of Cell 01 over 1000 cycles at six different SOC levels. These parameters were extracted through the Gaussian fitting of the charge transfer peaks, ensuring a precise quantification. The results reveal consistent trends influenced by both the cycle number and SOC, reflecting the progression of battery aging mechanisms. As the battery ages, the charge transfer resistance, which dominates the mid-frequency region of the impedance spectrum, increases significantly.

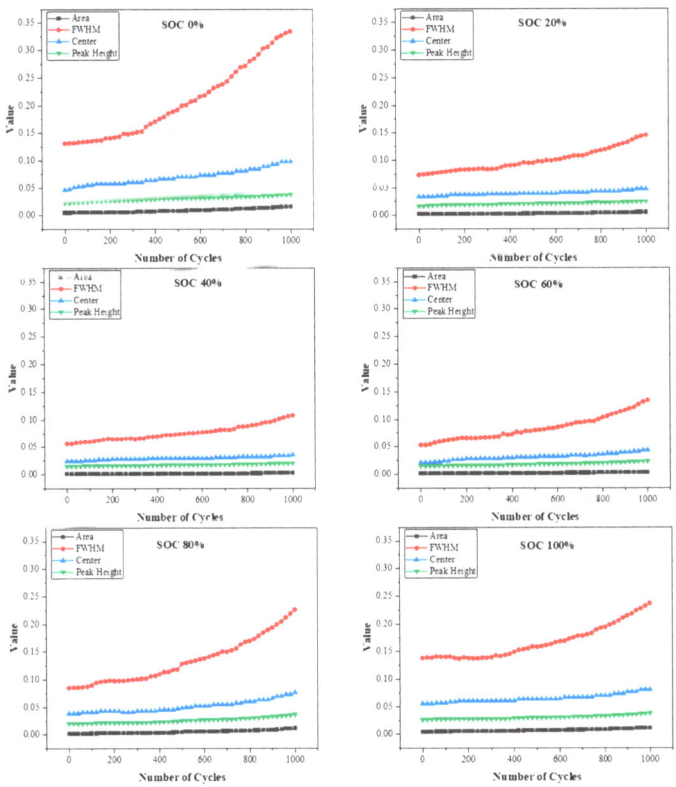

Figure 5. Variation in DRT parameters at different SOCs over 1000 cycles for Cell 01.

This resistance is indicated by the widening of the FWHM, the gradual increase in peak area and height, and the shift in the center relaxation time to longer values. The changes become more pronounced at both ends of SOC levels, where aging effects tend to accelerate, signaling slower reaction kinetics and reduced transport efficiency. These trends suggest that the charge transfer process becomes increasingly complex and heterogeneous over time, making it a key factor in understanding battery degradation [19].

Furthermore, the observed consistency across all three cells reinforces the robustness and reliability of the analysis. By monitoring DRT parameters at multiple SOC levels, a comprehensive understanding of internal battery processes and their evolution over time is achieved, offering critical insights for developing accurate battery state estimation models.

In order to clarify the links between these variables, a heatmap was created using the correlation matrix in Figure 6 using the Spearman correlation matrix. The Spearman correlation coefficient (ρ) is a non-parametric measure that evaluates the strength and direction of these relationships [20] and is calculated using below equation.

$$\rho = 1 - \frac{6\sum d_i^2}{n(n-1)} \tag{1}$$

where d_i is the rank difference between the corresponding values of two variables, and n is the total number of observations. The analysis reveals a strong positive correlation among the DRT-derived parameters, indicating their collective representation of the same underlying charge transfer relaxation process. As the battery ages, these parameters demonstrate consistent trends, reflecting the gradual increase in the charge transfer resistance. In contrast, the correlation between the DRT parameters and SOH exhibits a consistent negative trend, with correlation coefficients ranging from −0.49 to −0.55. This negative relationship highlights the sensitivity of the charge transfer relaxation process to SOH degradation—as the SOH declines, the charge transfer resistance peak becomes more prominent, further emphasizing its importance as a reliable indicator of battery health.

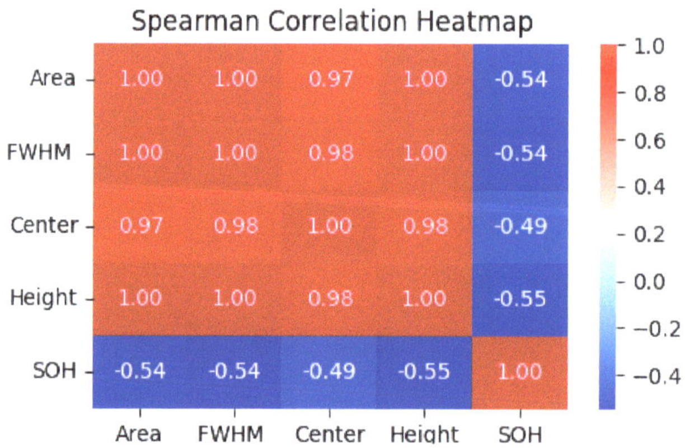

Figure 6. Spearman correlation matrix.

The findings from the Spearman correlation analysis confirm the strong interdependence among the DRT-derived parameters and their inverse relationship with the SOH, establishing them as effective indicators for tracking battery degradation. By incorporating these parameters into machine learning-based models, such as LSTM, for a SOH prediction, the predictive accuracy can be significantly enhanced.

5. LSTM Model for SOH Estimation

5.1. Architecture

The Long Short-Term Memory (LSTM) model is a type of neural network designed for sequential data, with a unique memory feature that enables it to retain and update relevant information over time [21–23]. The architecture of an LSTM cell, as shown in Figure 7, comprises four key gates, the forget gate, input gate, cell state update, and output gate, which work together to manage the flow of information.

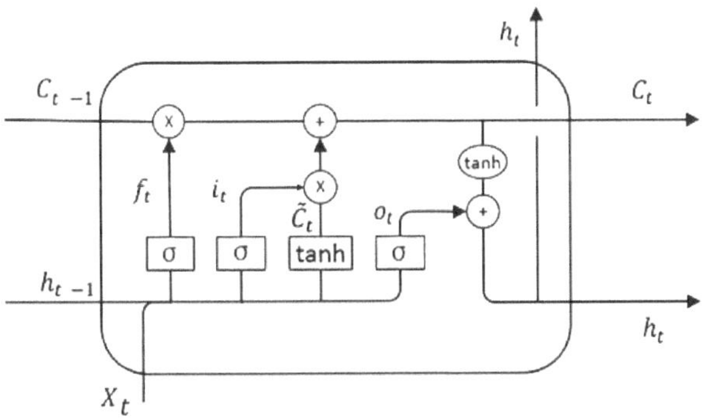

Figure 7. The cell structure of LSTM.

The forget gate decides how much of the previous cell state C_{t-1} should be retained or discarded. It uses the previous hidden state h_{t-1} and the current input x_t, along with a weight matrix W_f and bias b_f, to compute the forget vector f_t using a sigmoid activation function [21]. The sigmoid σ ensures the output ranges between 0 and 1.

$$f_t = \sigma\left(W_f \cdot [h_{t-1}, x_t] + b_f\right) \tag{2}$$

The input gate determines what new information should be added to the cell state. It involves two components: the input gate vector i_t, which determines the proportion of new information to incorporate, and the candidate cell state $\widetilde{C}_t$ which holds potential new values for the cell state [24]. While i_t is computed using a sigmoid function, $\widetilde{C}_t$ is calculated using a *tanh* activation function to scale its output between −1 and 1.

$$i_t = \sigma(W_i \cdot [h_{t-1}, x_t] + b_i) \tag{3}$$

$$\widetilde{C}_t = tanh(W_C \cdot [h_{t-1}, x_t] + b_C) \tag{4}$$

The cell state update combines the outputs from the forget gate and input gate to update the cell state C_t. The forget gate scales down the previous cell state C_{t-1}, while the input gate determines how much of the candidate cell state $\widetilde{C}_t$ should be added. This enables the model to selectively retain old information and incorporate new information effectively.

$$C_t = f_t * C_{t-1} + i_t * \widetilde{C}_t \tag{5}$$

The output gate determines what information from the updated cell state C_t should be passed forward as the current hidden state h_t. The output gate vector o_t is calculated using a sigmoid function, while the updated cell state is scaled using a *tanh* function to produce the final hidden state. This hidden state serves as both the output of the LSTM cell and the input for the next time step [25].

$$o_t = \sigma(W_o[h_{t-1}, x_t] + b_o) \tag{6}$$

$$h_t = o_t * tanh(C_t) \tag{7}$$

5.2. Model Training

In this study, the LSTM model is trained as described in Figure 8. The process begins with the selection of key electrochemical features as input parameters, which include DRT parameters along with the SOC to cover the different state of charge applications. In the data processing stage, the input data undergo two critical preprocessing steps. First, any missing values are handled typically when the Gaussian fitting produced outliers that interrupted the smooth trend parameters, ensuring that the dataset is complete and reliable. Second, the data are normalized to bring all input features to a consistent scale between 0 and 1, which helps improve the convergence of the model during training. After preprocessing, the dataset is split with the ratio of 80% and 20% as a training set to train the model and a testing set to evaluate its performance, respectively.

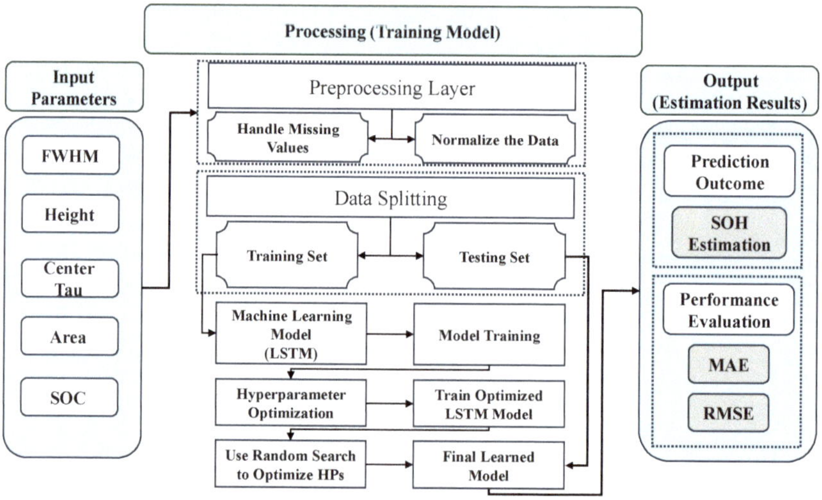

Figure 8. SOH estimation framework based on LSTM model.

The model training process involves an LSTM model with two hidden layers, each consisting of 200 units. To prevent overfitting and improve generalization, L2 regularization and a dropout rate of 0.4 are applied. A dense output layer predicts the SOH values. The model is trained using the Adam optimizer, which efficiently adjusts weights to minimize the prediction error. The Mean Squared Error (MSE) loss function guides the training process over 250 epochs with a batch size of 32 [26]. Additionally, hyperparameter optimization is performed using Random Search to identify the optimal values for the configuration of the model.

5.3. Model Performance

Once the final trained LSTM model is obtained, it is used to predict the SOH values for the testing set. The performance of the model is evaluated using metrics such as the Mean Squared Error (MSE) and Root Mean Squared Error (RMSE), which measure the accuracy of the predictions by quantifying the difference between the predicted and actual SOH values.

The LSTM model was employed to predict the State of Health (SOH) of three lithium-ion batteries across 1000 cycles. The results demonstrate the ability of the model to accurately capture the degradation trends of the batteries using key DRT parameters (FWHM, Height, Center Relaxation Time, and Area) and the SOC as input features. The analysis was performed using four datasets: (a) Battery 1 and Battery 2, (b) Battery 1 and Battery 3, (c) Battery 2 and Battery 3, and (d) a combined dataset including all three batteries.

Figure 9 presents a comparison of the actual SOH and predicted SOH for these datasets. The results indicate that when the LSTM model was trained and tested with datasets from Battery 1 and Battery 3, the prediction accuracy was the highest. This can be attributed to the fact that the initial capacity difference between Battery 1 and Battery 3 is minimal, approximately 1%, resulting in smaller deviations and a closer alignment between the predicted and actual SOH. The performance metrics for this dataset were particularly impressive, achieving a Mean Absolute Error (MAE) of 0.58% and a Root Mean Squared Error (RMSE) of 0.70%.

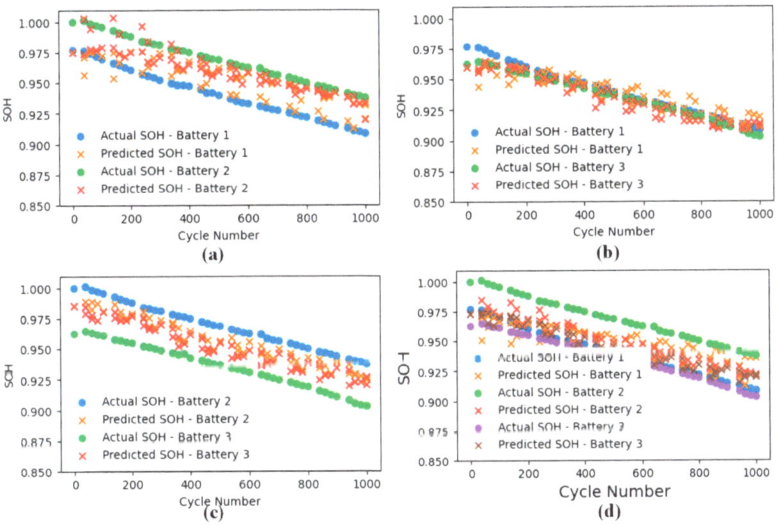

Figure 9. SOH estimation performance of DRT trained LSTM model. (**a**) Results with dataset of battery 1 and 2 (**b**) Results with dataset of battery 1 and 3 (**c**) Results with dataset of battery 2 and 3 (**d**) Results with combined dataset of three batteries.

In contrast, when the model was trained and tested with datasets from Battery 1 and Battery 2 or Battery 2 and Battery 3, the prediction errors were slightly higher. The combined dataset (all three batteries) captured a wider range of capacity variations, leading to a comparatively higher error. However, the combined dataset provides a more realistic representation of real-time scenarios by incorporating diverse aging patterns across different batteries. Despite the slightly increased error, the LSTM model still outperformed traditional EIS-based ECM methods using the same dataset for a predictive battery health estimation using neuro-fuzzy [8]. The Mean Absolute Error (MAE) and Root Mean Squared Error (RMSE), as defined below, were used to evaluate the performance of the model [27].

$$MAE = \frac{1}{N}\sum_{I}^{N}|e_i| \tag{8}$$

$$RMSE = \sqrt{\frac{1}{N}\sum_{i}^{N}(e_i)^2} \quad (9)$$

where N is the total number of data points, and e_i represents the error for each prediction. To validate our method, we compared our proposed DRT-based SOH estimation method with an EIS-based Equivalent Circuit Model (ECM) approach by [8]. The Table 2 below presents the comparative quantitative error metrics, Mean Absolute Error (MAE) and Root Mean Squared Error (RMSE), for the four battery datasets.

Table 2. Comparison of estimation errors.

Battery Dataset	EIS Based ECM Parameters Method [8]		Proposed DRT Parameters Method	
	MAE (%)	RMSE (%)	MAE (%)	RMSE (%)
Battery 1 and Battery 2	1.518	1.676	1.16	1.38
Battery 2 and Battery 3	1.645	1.69	1.37	1.46
Battery 1 and Battery 3	1.023	1.043	0.58	0.70
Combined (Battery 1, 2, 3)	1.43	1.525	1.28	1.41

The results demonstrate that the proposed method achieves lower error metrices across all the datasets, which clearly highlights the improvement and accuracy of the method. The inclusion of data from three cells with different capacities essentially averages out the differences across these varying capacities. It indicates that the LSTM model converges toward stable and reliable predictions by including more data. Although the higher capacity retention of Battery 2 influences the combined dataset, the model shows a better estimation capability when applied to a range of battery behaviors. Since batteries with the same chemistry share similar trends, collecting data across multiple cells ultimately leads to a more generalized model for the SOH estimation. In turn, this kind of broad coverage ensures the model can handle real-world variability and deliver a consistent performance, even when faced with different capacities within the same battery chemistry.

6. Conclusions

This study highlights the effectiveness of integrating a Distribution of Relaxation Times (DRT) analysis with EIS to achieve an advanced and accurate SOH estimation for lithium-ion batteries. By systematically analyzing DRT peak parameters (Area, FWHM, Center Relaxation Time, and Height) over 1000 cycles at six distinct SOC levels, the developed LSTM model demonstrated its capability to capture the complex temporal dependencies associated with battery aging. The model achieved an MAE of 1.28% and an RMSE of 1.41% when tested with combined datasets from three batteries, underscoring its robust prediction capability for real-time SOH estimations across varying cell capacities. The highest predictive accuracy was observed when using datasets from batteries with minimal initial capacity differences, such as Battery 1 and Battery 3, where the model achieved an MAE of 0.58% and an RMSE of 0.70%. These values underline the strength of the proposed methodology in accurately predicting the SOH while accommodating variations in battery aging trends. Comparatively, the LSTM model trained with DRT parameters outperformed the traditional ECM parameters-based estimation model by achieving a higher accuracy across different combinations of datasets.

This work establishes a reliable and scalable framework for enhancing battery health monitoring and prediction across diverse real-world applications, including portable electronics, electric vehicles, and energy storage systems. The integration of DRT parameters with advanced machine learning techniques like LSTM not only improves the accuracy of SOH estimations but also highlights the potential of leveraging electrochemical features for long-term battery performance monitoring. As this research is limited to the standard testing conditions and same kind of battery cells, future work directions will focus on analyzing the dataset of different battery chemistries and various testing scenarios to reflect real-world conditions. It will also focus on incorporating more detailed electrochemical parameters, such as those representing other processes, to provide a more comprehensive understanding of battery degradation mechanisms.

Author Contributions: Conceptualization, A.S.A. and W.C.; Methodology, A.S.A.; Software, A.S.A.; Validation, A.S.A. and M.S.; Formal analysis, A.S.A. and M.S.; Investigation, W.C.; Resources, W.C.; Data curation, M.S.; Writing—original draft, A.S.A.; Writing—review & editing, A.S.A.; Supervision, W.C.; Project administration, W.C. All authors have read and agreed to the published version of the manuscript.

Funding: This research received no external funding.

Data Availability Statement: The original contributions presented in the study are included in the article, further inquiries can be directed to the corresponding author.

Conflicts of Interest: The authors declare no conflicts of interest.

References

1. Rahman, A.; Kim, J.; Hossain, S. Recent advances of energy storage technologies for grid: A comprehensive review. *Energy Storage* **2022**, *4*, e322. [CrossRef]
2. Ahmad, F.; Khalid, M.; Panigrahi, B.K. Development in energy storage system for electric transportation: A comprehensive review. *J. Energy Storage* **2021**, *43*, 103153. [CrossRef]
3. Akram, A.S.; Sohaib, M.; Choi, W. SOH Estimation of the Lithium-ion Batteries using the Parameters of Distribution of Relaxation Times. In Proceedings of the KIPE Autumn Conference, Seoul, Republic of Korea, 22 November 2024; pp. 52–54.
4. Xu, J.; Cai, X.; Cai, S.; Shao, Y.; Hu, C.; Lu, S.; Ding, S. High-energy lithium-ion batteries: Recent progress and a promising future in applications. *Energy Environ. Mater.* **2023**, *6*, e12450. [CrossRef]
5. Lin, Z.; Li, D.; Zou, Y. Energy efficiency of lithium-ion batteries: Influential factors and long-term degradation. *J. Energy Storage* **2023**, *74*, 109386. [CrossRef]
6. Zhang, Z.; John, P.R.; Fang, W. *Safety of Lithium-Ion Batteries. Lithium-Ion Batteries*; Elsevier: Amsterdam, The Netherlands, 2014; pp. 409–435.
7. Chen, T.; Jin, Y.; Lv, H.; Yang, A.; Liu, M.; Chen, B.; Xie, Y.; Chen, Q. Applications of lithium-ion batteries in grid-scale energy storage systems. *Trans. Tianjin Univ.* **2020**, *26*, 208–217. [CrossRef]
8. Kim, M.-S.; Akram, A.S.; Choi, W. A neuro-fuzzy system to evaluate the remaining useful life of the lithium-ion battery using the impedance spectrum in the overall range of SOCs. *J. Power Electron.* **2024**, *25*, 103–114. [CrossRef]
9. Pradyumna, T.K.; Cho, K.; Kim, M.; Choi, W. Capacity estimation of lithium-ion batteries using convolutional neural network and impedance spectra. *J. Power Electron.* **2022**, *22*, 850–858. [CrossRef]
10. Sohaib, M.; Choi, W. Investigation of the Aging Phenomena in Lithium-Ion Batteries Using Distribution of Relaxation Time Analysis. In Proceeding of the KIPE Summer Conference, Seogwipo-si, Republic of Korea, 1–4 July 2024; pp. 187–189.
11. Azizighalehsari, S.; Boj, E.A.; Venugopal, P.; Soeiro, T.B.; Rietveld, G. A Distribution of Relaxation Time Approach on Equivalent Circuit Model Parameterization to Analyse Li-Ion Battery Degradation. *IEEE Trans. Ind. Appl.* **2024**, *60*, 9206–9215. [CrossRef]
12. Kim, K.; Park, K.; Yoon, K.; Moon, H.; Pyeon, H.; Kim, J.; Han, S. A light-weight electrochemical impedance spectroscopy-based SOH estimation method for lithium-ion batteries using the distribution of relaxation times with Grad-CAM analysis. *J. Power Sources* **2025**, *631*, 236169. [CrossRef]
13. Zhang, Q.; Wang, D.; Schaltz, E.; Stroe, D.I.; Gismero, A.; Yang, B. Degradation mechanism analysis and State-of-Health estimation for lithium-ion batteries based on distribution of relaxation times. *J. Energy Storage* **2022**, *55*, 105386. [CrossRef]
14. Chang, Y.-C.; Sohn, H.-J. Electrochemical impedance analysis for lithium-ion intercalation into graphitized carbons. *J. Electrochem. Soc.* **2000**, *147*, 50. [CrossRef]
15. Zhang, D.; Haran, B.; Durairajan, A.; White, R.; Podrazhansky, Y.; Popov, B. Studies on capacity fade of lithium-ion batteries. *J. Power Sources* **2000**, *91*, 122–129. [CrossRef]
16. Suarez-Hernandez, R.; Ramos-Sánchez, G.; Santos-Mendoza, I.O.; Guzmán-González, G.; González, I. A graphical approach for identifying the limiting processes in lithium-ion battery cathode using electrochemical impedance spectroscopy. *J. Electrochem. Soc.* **2020**, *167*, 100529. [CrossRef]

17. Sohaib, M.; Akram, A.S.; Choi, W. Analysis of Aging and Degradation in Lithium Batteries Using Distribution of Relaxation Time. *Batteries* **2025**, *11*, 34. [CrossRef]
18. Iurilli, P.; Brivio, C.; Wood, V. Detection of Lithium-Ion Cells' Degradation through Deconvolution of Electrochemical Impedance Spectroscopy with Distribution of Relaxation Time. *Energy Technol.* **2022**, *10*, 2200547. [CrossRef]
19. Sohaib, M.; Choi, W. A Novel Aging Diagnosis Method Using Distribution of Relaxation Time. In Proceeding of the KIPE Autumn Conference, Seoul, Republic of Korea, 22 November 2024; pp. 46–48.
20. Zhou, Y.; Huang, M.; Chen, Y.; Tao, Y. A novel health indicator for on-line lithium-ion batteries remaining useful life prediction. *J. Power Sources* **2016**, *321*, 1–10. [CrossRef]
21. Park, K.; Choi, Y.; Choi, W.J.; Ryu, H.-Y.; Kim, H. LSTM-based battery remaining useful life prediction with multi-channel charging profiles. *IEEE Access* **2020**, *8*, 20786–20798. [CrossRef]
22. Li, X.; Zhang, L.; Wang, Z.; Dong, P. Remaining useful life prediction for lithium-ion batteries based on a hybrid model combining the long short-term memory and Elman neural networks. *J. Energy Storage* **2019**, *21*, 510–518. [CrossRef]
23. Liu, Y.; Zhao, G.; Peng, X.; Hu, C. Lithium-ion battery remaining useful life prediction with long short-term memory recurrent neural network. *Annu. Conf. PHM Soc.* **2017**, *9*. [CrossRef]
24. Deng, Z.; Lin, X.; Cai, J.; Hu, X. Battery health estimation with degradation pattern recognition and transfer learning. *J. Power Sources* **2022**, *525*, 231027. [CrossRef]
25. Zhang, Y.; Xiong, R.; He, H.; Liu, Z. A LSTM-RNN method for the lithuim-ion battery remaining useful life prediction. In Proceedings of the 2017 Prognostics and System Health Management Conference (PHM-Harbin), Harbin, China, 9–12 July 2017; pp. 1–4.
26. Shi, J.; Rivera, A.; Wu, D. Battery health management using physics-informed machine learning: Online degradation modeling and remaining useful life prediction. *Mech. Syst. Signal Process.* **2022**, *179*, 109347. [CrossRef]
27. Wei, L.; Sun, Y.; Diao, Q.; Xu, H.; Tan, X.; Fan, Y. State of Health Estimation of Lithium-Ion Batteries Based on Stacked-LSTM Transfer Learning with Bayesian Optimization and Multiple Features. *IEEE Sens. J.* **2024**, *24*, 37607–37619. [CrossRef]

batteries

MDPI

Review

Powering the Future Smart Mobility: A European Perspective on Battery Storage

Natascia Andrenacci [1,*], Francesco Vitiello [1], Chiara Boccaletti [2,*] and Francesco Vellucci [1]

1 ENEA, Agenzia Nazionale per le Nuove Tecnologie, l'Energia e lo Sviluppo Economico Sostenibile, Centro Ricerche Casaccia, Via Anguillarese 301, S.M. Galeria, 00123 Rome, Italy; francesco.vitiello@enea.it (F.V.); francesco.vellucci@enea.it (F.V.)

2 Department of Astronautical, Electrical, and Energy Engineering, Sapienza University of Rome, 00184 Rome, Italy

* Correspondence: natascia.andrenacci@enea.it (N.A.); chiara.boccaletti@uniroma1.it (C.B.)

Abstract: Batteries are central to the global energy system and fundamental elements for energy transition and future mobility. In particular, the growth in electric vehicle (EV) sales is pushing up demand for batteries. Most of the battery demand for EVs today can be met with domestic or regional production in China, while the share of imports remains relatively large in Europe and the United States. Boosting the industrial base for battery production is therefore a key task for the EU. To make its battery supply chains secure, resilient, and sustainable, the EU's approach consists of improving cooperation among stakeholders, providing the sector with funding, and establishing a comprehensive regulatory framework. In this paper, an accurate review of the state-of-the-art of automotive batteries is provided, including the performance, safety, sustainability, and costs of the different battery technologies. The significant challenges the EU battery sector must face, such as dependencies on third countries and high energy and labor costs, are discussed. An overview of the present European regulation and of future trends is provided.

Keywords: batteries; electric vehicles; automotive sector; European market; European regulation

1. Introduction

Road transport is responsible for around a quarter of anthropogenic CO_2 emissions in Europe [1]. Vehicle electrification is a viable means to mitigate these emissions, and batteries play a key role in the success of this strategy. Electric mobility is transforming our daily travel experiences, driven by advancements in electric storage systems. Access to batteries is particularly important for ensuring the competitiveness of the automotive sector, as batteries are the most critical component of electric vehicles (EVs) and represent around 25–40% of the total cost of an electric vehicle. In fact, the growth in EV sales is pushing up demand for batteries. According to a Joint Research Center report [2], road transport accounts for 86% of all battery usage, primarily in personal cars, light-duty commercial vehicles, and buses. Automotive batteries need to provide high performance in power, energy storage capacity, and durability, while maintaining low weight, volume, and cost. Additionally, they must comply with automotive safety standards and be recyclable and environmentally sustainable. It is widely believed that an affordable and appropriately sized battery is crucial for the success of electric mobility, and EV manufacturers are actively working toward this goal [3].

In this context, it is not surprising that investments in batteries have been increasing in the last few years. In particular, global investments in EV batteries have increased eightfold since 2018, rising to USD 115 billion in 2023 [4]. Stationary storage accounts for USD 40 billion, for a global battery demand of 850 GWh. China accounts for more than half of this spending, followed by the EU and the USA, while the rest of the world invested only 10% of the global market.

Most venture capital investments focused on innovative battery chemistry and component manufacturers, highlighting a significant interest in alternatives to lithium technologies. An increasing share of investment has also been directed towards battery recycling and reuse, as is evident from Figure 1a. Early-stage investments in lithium technology dominate the market, although non-lithium technologies have been attracting increasing interest in recent times (Figure 1b). These non-lithium batteries include emerging technologies such as metal–hydrogen, solid-state, or sodium-ion, but also more mature solutions, such as redox flow batteries [4].

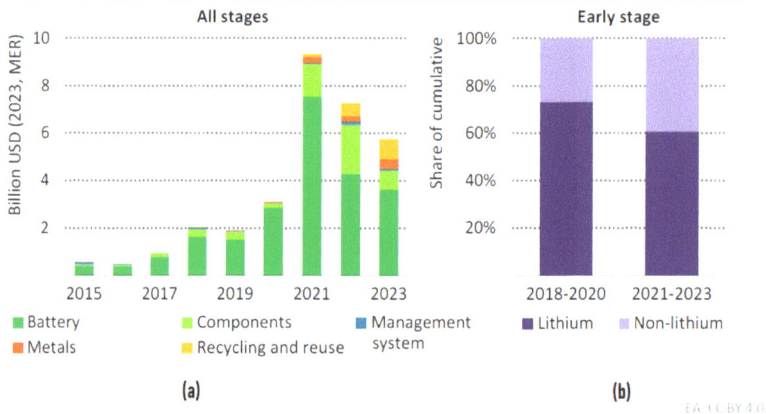

Figure 1. Distribution of venture capital investments per year in battery start-ups (a) by technologies; (b) by chemistry. Reproduced from [4] under CC BY 4.0 terms.

Under current policy settings, battery demand for electromobility is projected to increase 4.5 times by 2030 compared to 750 GWh of LIBs installed in vehicles worldwide in 2023 [5], and more than seven times by 2035, with EV batteries accounting for more than 4.3 TWh [6], as shown in Figure 2. If countries fully meet their announced climate and energy commitments, demand could rise fivefold by 2030 and ninefold by 2035. In a Net Zero Emissions (NZE) scenario by 2050, demand is expected to grow seven times by 2030 and twelve times by 2035 [7].

According to Bloomberg New Energy Finance (BloombergNEF) [8], LIB pack prices dropped 20% from 2023 to a record low of USD 115 per kWh in 2024, although prices may vary significantly across different countries and applications. Battery pack costs in the USA and Europe were 31% and 48% higher than in China, where prices already dipped below USD 100 per kWh in 2024, mainly due to higher manufacturing costs and lower volumes. LIB prices are projected to drop in the next five years, potentially reaching cost parity with internal combustion engine vehicles (ICEVs) in the mid-to-late 2020s [9]. Several factors are contributing to the decline in prices, including overcapacity in cell manufacturing, economies of scale, and reduced prices for metals and components [10]. However, the downward price trend may be halted due to the current very low profit margins of battery producers and possible shortages in the material supply chain. Figure 3 illustrates the price trends of battery packs and cells over the past few years, along with projected prices extending to 2030. These forecasts are averaged from estimates of reference [9], utilizing the average price ratio of packs to cells from the past five years.

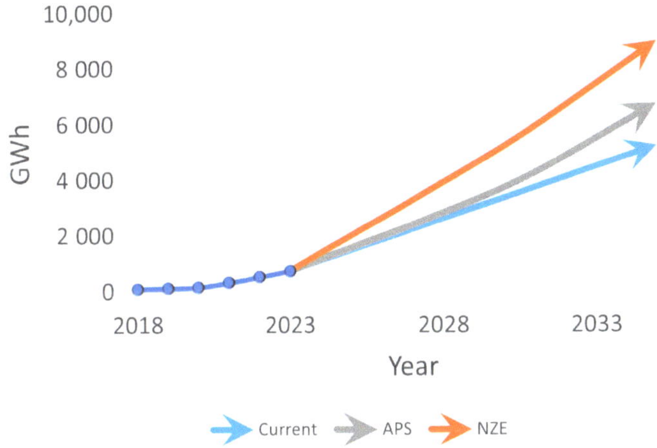

Figure 2. Battery demand growth under different decarbonization scenarios: Current, Announced Pledges Scenario (APS), and Net Zero Emission (NZE).

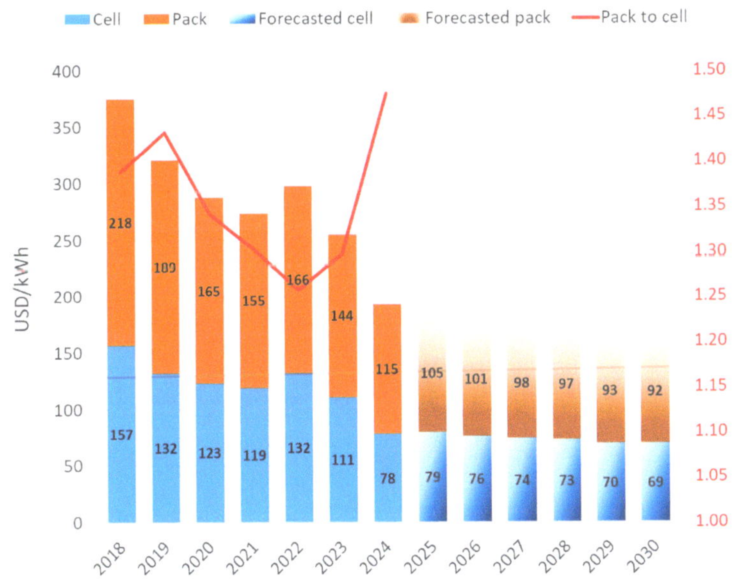

Figure 3. Pack and cell price trend from 2018 to 2024 [8]. The pack-to-cell price ratio is represented by the red line. Forecasted prices are reported for the 2025–2030 period.

Among LIB technologies, lithium iron phosphate (LFP) batteries are expected to become more affordable sooner than nickel–cobalt-based batteries, reinforcing LFP's position in the market [11,12].

Historically, the decrease in prices for cathode materials and improvements in cell energy density have led to lower battery costs. However, in the future, the main driver of cost reductions is likely to shift toward optimizing the manufacturing process and minimizing waste [13]. Transitioning from small-scale battery materials research to large-scale production is crucial, particularly regarding material quality control, raw material procurement, electrode processing, and component design [14]. Advanced characterization tools and AI-based data analytics can significantly enhance quality control, process optimization, and cost reduction [14–20]. Reducing scrap rates will become a crucial factor in improving overall efficiency and affordability in the production of lithium-ion batteries [9]. Indeed, raw materials, such as lithium, cobalt, nickel, and graphite, represent the most relevant cost in batteries. Even though the actual share depends on cell chemistry, Argonne National Laboratories estimated that cell materials represent 63% of the 2024 NMC811-graphite battery pack cost, with cathode and anode materials accounting for 52% and 14% of all materials cost, respectively [21]. The supply and extraction of raw materials are concentrated in specific geographic areas, which poses significant risks to both supply and pricing. In response to this, there are ongoing efforts to decrease the demand for critical raw materials, leading to the development of new battery technologies. Additionally, recycling will be vital in the next decade for reclaiming materials from manufacturing scrap. By 2035, the number of discarded EV batteries will surge, making battery recycling important for reducing critical mineral demand [18,19,22–26]. Effective scaling of recycling could lower raw lithium and nickel demand by 25% and raw cobalt demand by 40% by 2050, while still meeting climate targets [7].

While expected to remain dominant, especially due to the growing BEV market, lithium-ion technology faces potential challenges, mainly due to concerns about lithium supply constraints and changing policy landscapes [6]. This situation highlights the need for research and investment in batteries that use more abundant raw materials. Sodium-ion technology, also known as sodium-ion battery (SIB), is a promising alternative. According to IdTechEx, at present, only pilot plants are operational, along with a few smaller factories that produce limited quantities of SIB batteries [27]. Nonetheless, several manufacturers have disclosed plans that indicate total capacities will surpass 100 GWh by 2030 [27,28]. Their cost and safety advantages make them suitable for specific applications, including two- and three-wheeled vehicles, smaller passenger cars, and certain stationary industrial applications, thus complementing lithium-ion technology. Early commercialization of SIB battery-powered EVs is happening in China [29,30].

In Europe, despite ongoing efforts, the battery industry faces significant challenges that need to be addressed to ensure its long-term success and competitiveness, such as dependencies on third countries, high energy and labor costs, strong competition from Asia, fluctuating demand for EVs, regulatory uncertainties, and difficulties in ramping up production.

Additionally, the battery industry is entangled in geopolitical tensions between China and the USA, which impacts investments in gigafactories [31]. Developing battery recycling or implementing more sustainable value chains could help address some of these challenges. Initiatives like BATTERY 2030+ are set up to promote the development of next-generation batteries. This program encompasses various projects focused on creating ultra-high-performance, durable, safe, sustainable, and affordable batteries for practical applications [32]. Emerging technologies, such as solid-state lithium batteries, promise extended vehicle ranges, shorter charging times, and reduced costs [33].

To make its battery supply chains secure, resilient, and sustainable, the EU adopted three approaches: improve cooperation among stakeholders, provide the sector with funding, and establish a comprehensive regulatory framework.

In this paper, an accurate review of the state-of-the-art of batteries for the automotive sector is provided, including performance, safety, sustainability, and costs, with a focus on the European context. The paper is organized as follows: in Section 2, an overview of the EU regulation framework is given; Section 3 illustrates the state-of-the-art and the future developments of battery technologies for the automotive sector; Section 4 is devoted to the sustainability issue. Promising innovations and regulation improvements are discussed in the Conclusions, in particular from the European perspective. Appendix A illustrates the methods and search results of the literature review.

2. The European Legislative Framework

The European Union has implemented policies to decarbonize industries and the energy system, significantly boosting battery demand. In particular, the global commitment to decarbonizing the transport sector has resulted in continuous growth in the EV battery market. The United States and Europe experienced the fastest growth among major EV markets, reaching more than 40% year-on-year, closely followed by China at about 35%. As a consequence, demand for EV batteries reached more than 750 GWh in 2023, up 40% relative to 2022 [34]. As aforementioned, one of the EU's approaches to support the EU battery sector is to establish a comprehensive legislative framework. Key initiatives include:

- RePowerEU: This initiative seeks to make Europe energy independent from fossil fuels by 2027, prioritizing clean technologies like battery energy storage.
- Fit for 55 Package: This package aims to reduce greenhouse gas emissions by 55% by 2030 and accelerate the electrification of various sectors, increasing battery demand.
- Net-Zero Industrial Act (NZIA): This act aims to increase clean-tech industrial capacity, including battery manufacturing.
- Critical Raw Materials Act (CRMA): This act enhances the collection and recycling of waste products to secure the supply of critical raw materials for batteries.
- EU Battery Regulation: This regulation promotes the circular economy, resource efficiency, and sustainability of batteries throughout their lifecycle. Key points in the Battery Regulation are the following:
    - Mandatory sustainability and safety requirements for the placing of batteries on the European market, including restrictions on certain substances, carbon footprint requirements, performance and durability requirements, etc.;
    - Recycled content requirements;
    - Traceability through labelling, marking, and information requirements, notably with the creation of the digital battery passport;
    - Mandatory implementation of due diligence policies;
    - Extended producer responsibility;
    - Targets for the collection of waste batteries, and provisions regarding the treatment, reuse, and recycling of batteries, notably materials recovery targets;
    - Green public procurement.

The new regulation aligns closely with the European Union's climate neutrality objectives. These policies collectively create a favorable environment for battery innovation, manufacturing, and deployment in Europe, contributing to the EU's goal of achieving climate neutrality.

The regulatory strategies for batteries vary significantly across the EU, China, and the US, reflecting different priorities regarding environmental protection, industrial policy, and consumer safety. The EU's approach emphasizes environmental sustainability, resource efficiency, and developing a circular economy for batteries. The EU aims to establish harmonized standards that apply across all member states. China's regulatory approach prioritizes the safety and reliability of batteries, especially given the rapid growth of the electric vehicle (EV) sector. In contrast, the regulatory landscape in the US is more fragmented. It focuses on transportation safety and on battery recycling through federal guidelines and state-level initiatives. Additionally, there is a growing emphasis on domestic battery production supported by economic incentives. However, the US currently lacks the comprehensive, life-cycle-based approach that is characteristic of the EU. Table 1 summarizes the main regulation aspects for different countries.

Table 1. Main regulatory framework in different countries.

Country	Main Regulations	Main Focus	Main Action	References
EU	Batteries Regulation (2023/1542)	Sustainability; safety; labeling, collection, and recycling of all battery types	Substance restrictions, carbon footprint declarations, recycled content requirements, and battery passports for traceability	[35]
China	EV battery safety standards (GB38031-2025)	Safety; promotion of technological standards	Prevention of fire and explosion after thermal runaway	[36]
US	Mercury-Containing and Rechargeable Battery Management Act; Inflation Reduction Act	Recycling; battery collection and labeling guidelines; safety; domestic battery production	Multi-faceted approach involving federal and state regulations, as well as voluntary standards	[37,38]

3. State-of-the-Art of EV Battery Technologies

The review focuses on battery technologies for automotive applications, from material to battery pack levels. Therefore, the main aspects considered are life duration, fast charge, energy density, cost and market aspects, safety, raw materials, and process sustainability. For this reason, we excluded documents with keywords referring to vehicle and charging technologies, or referring to recycling and post-use applications, or non-automotive purposes. We also excluded reviews focused on battery modeling approaches, as well as those on battery management systems (BMS) and thermal management systems (BMTS). Although these topics are crucial for battery performance and safety, they encompass a broad area of research beyond the scope of this paper and deserve a dedicated review.

Based on the literature review, the current battery technologies for EVs have been identified and are described in the following section, along with the battery requirements for EVs applications and the research developments.

3.1. Lead–Acid Batteries

Lead–acid batteries are a well-established technology that is commonly used in vehicles to power electrical systems. Lead–acid batteries can be configured in large systems without complex management. In automotive applications, six 2 V cells are connected in series to create a 12 V battery. These batteries are known for their low cost per kWh. Available designs include flooded (vented) and enhanced flooded batteries (EFBs), valve-regulated lead–acid (VRLA) batteries, and Absorbent Glass Mat (AGM) batteries [39]. Bipolar lead–acid batteries, a relatively recent development, feature a sandwich construction that reduces internal resistance and increases power density due to their lighter weight, allowing for possible applications in hybrid vehicles [40]. The bipolar electrodes concept has also been extended to other rechargeable battery technologies [41]. Lead–acid batteries excel in recycling, with 99% of active materials being recycled and over 90% efficiency [42]. This supports a circular economy and boosts Europe's reliance on domestic raw materials. Further research is needed to enhance secondary lead utilization and refine the recycling process to reduce impurities. Lead–acid technology is expected to maintain a stable market presence, primarily focused on 12 V starter batteries and industrial applications [43].

3.2. Nickel-Based Technologies

Nickel oxide electrodes serve as the positive plates in various types of rechargeable batteries, including nickel–iron (Ni–Fe), nickel–cadmium (Ni–Cd), nickel–hydrogen (Ni–H2), nickel–metal hydride (Ni–MH), and nickel–zinc (Ni–Zn) batteries. All nickel-based batteries are recognized for their long lifespan and their ability to operate effectively across a wide temperature range [44]. Notably, Ni-Cd batteries are celebrated for their durability; they do not suffer from "sudden death syndrome", and their gradual aging can be monitored effectively. However, they show that the so-called "memory effect", which occurs as soon as a battery is partially charged and discharged, causes their available capacity to degrade suddenly. Furthermore, the cadmium in this battery makes it unfriendly to the environment.

Ni-MH batteries, on the other hand, offer a higher energy density and are generally well-suited for applications with high current consumption [45]. However, this technology requires a BMS for optimal operation, which introduces some vulnerability. Ni-MH batteries are frequently employed as a backup energy source in hybrid EVs like the Toyota Prius [46].

The nominal voltage of Ni-Cd and Ni-MH is 1.2 V, which limits their employment in large battery packs.

3.3. Lithium Batteries

Electrified vehicles with traction motors require higher power and energy capacities than lead–acid batteries can provide. LIBs have proven to be the most suitable alternative for this application. Automotive LIBs demand increased by about 65% to 550 GWh in 2022, from about 330 GWh in 2021, primarily because of growth in electric passenger car sales, with new registrations increasing by 55% in 2022 relative to 2021 [47]. These batteries have a nominal voltage range from 2.3 to 3.85 volts (https://www.grepow.com/lihv-battery/lihv-battery-3-85-v.html accessed on 31 March 2025) and offer significantly high energy density, exceeding 150 Wh/kg [48], reaching 250 Wh/kg [49], surpassing those of lead–acid and Ni-MH batteries, which are 40–60 Wh/kg and 40–110 Wh/kg, respectively. Thanks to their higher energy and power densities, and consequently smaller and lighter cell designs, with respect to other existing battery technologies, LIBs are dominating the automotive battery market.

3.4. Sodium-Based Technologies

High-temperature sodium batteries use liquid sodium electrodes and a solid electrolyte, typically an ion-conducting ceramic. The most common types are sodium–nickel chloride (NaNiCl) and sodium–sulfur (NaS) batteries [50].

NaNiCl, or ZEBRA batteries, operate between 270 °C and 350 °C. This molten state allows high conductivity for sodium ions, providing a specific energy of about 120 and a nominal voltage of 2.3 V to 2.6 V. They are often used in small electric vehicle fleets and stationary energy storage [51,52].

NaS batteries operate in an optimal temperature range between 300 °C and 340 °C. During discharge, sodium oxidizes to Na+, allowing ions to enter the solid electrolyte and release electrons. The voltage during operation is around 2 V. NaS batteries have stable internal resistance, making them ideal for grid storage applications over 1 MWh [50].

NaS and NaNiCl batteries both exceed 4500 cycles and have efficiencies of 75% to 86%. However, they require heating to maintain temperature, which decreases their overall efficiency. NaNiCl batteries have high nickel demand and complex insulation, leading to greater energy and environmental impacts. In contrast, NaS batteries use recyclable materials like steel and aluminum, resulting in a lower environmental footprint [43].

In addition to high-temperature applications, advancements in NaS battery systems for intermediate (100–200 °C) and ambient (25–60 °C) temperatures are promising [53]. Room-temperature sodium-ion batteries (SIBs) are often compared to lithium-ion batteries because both operate on the "rocking chair" principle, where ions move between two host materials at the electrodes [54]. SIB technology is appealing since it is based on an abundant material, its specific energy is comparable to LFP [55], while power density, low temperature performance, and safety are superior [56,57], even though some issues regarding performance stability and safety remain [58,59]. Nevertheless, sodium-ion batteries have reached the market (technological readiness level (TRL) 9) [5]. The producer leader is China, which advances quickly, building the whole supply chain, scaling up production, and testing the first commercial products in real applications [28,29].

3.5. EV Batteries Requirements

Electromobility necessitates advancements in battery technology to compete with conventional vehicles [60]. The focus is on the advancement of LIBs and emerging chemistries. Battery technologies can be categorized into "generations" based on their chemical composition and performance. However, this classification is not officially defined and lacks a single source. In Table 2, we present a widely accepted classification of battery generations, along with their TRL, and typical technologies used for the cathode, anode, and electrolyte. We also add sodium-based technologies, as some batteries based on this chemistry are in TRL 9, and their performance evolution is of interest for automotive applications. However, we did not include them in the battery generation classification, which we use to refer only to LIB technologies.

Key areas for improvement include cell performance advancements and system-level enhancements like standardization, flexible manufacturing, battery swapping capabilities, and vehicle–grid integration for bidirectional charging [61].

At the system level, improving the thermal management of batteries is essential. This involves developing advanced cooling systems and utilizing digital twin models to enhance our understanding of battery performance, which can help to minimize capacity loss and performance degradation. In the short term, research should focus on developing advanced BMS with sensors for diagnostics and failure prediction, supporting remote upgrades while ensuring cybersecurity [62,63]. In the mid-term, the emphasis will be on battery reparability and refurbishment, exploring second-life applications, dismantling and recycling of battery components, and tracking data to meet Battery Regulation and Battery Passport requirements [64].

Table 2. Battery technology maturity. From [2,65,66].

Generation	TRL	Anode	Cathode	Electrolyte
1	9	Carbon/Graphite	LFP, NCA, LCO	Organic liquid
2a	9	Carbon/Graphite	NMC111	Organic liquid
2b	9	Carbon/Graphite	NMC532, NMC622	Organic liquid
3a	9	C/Si (5–10% Si)	NMC622, NMC811	Organic liquid
3b	5–9	Si/C (>10% Si)	HE-NMC	Organic liquid
3b	4	Si/C (>10% Si)	HV-LNMO	Organic liquid
3a	5–6 [1]	Si/C (>10%)	LCO, NMC, LMO, NCA	Solid state

Table 2. *Cont.*

Generation	TRL	Anode	Cathode	Electrolyte
4b	5–6 [2]	Li metal	LCO, NMC, LMO, NCA	Solid state
4	4	Li metal	Li2-S	Solid state
5	4	Li metal	O2	Different possibilities
-	9	Hard-carbon	PBA [3]	Organic liquid
-	4	Na-metal, Tin alloys	layered oxides, polyanion compounds	Different possibilities

[1] TRL 9 is achieved only in specific market segments (e.g., earbuds, wearables) or special operation conditions. [2] TRL 9 is achieved for some niche applications (e.g., space, military). [3] Prussian blue analog.

High gravimetric energy densities are crucial for increased payload and cost-efficiency, especially for commercial and heavy-duty vehicles, alongside considerations of cycle life and total cost of ownership under diverse operating conditions [67]. Given that battery materials comprise a significant portion of cell cost, research and innovation in this area are essential to achieve cost-effective batteries with high energy density, long cycle life, and rapid charging [67].

Figure 4 reports the evolutionary roadmap for lithium-based chemistries from 2023 to 2030 with reference to both gravimetric and volumetric energy density. Actual values can vary depending on specific cell design, operating conditions, and manufacturing processes. Projected values can vary substantially depending on ongoing research and development.

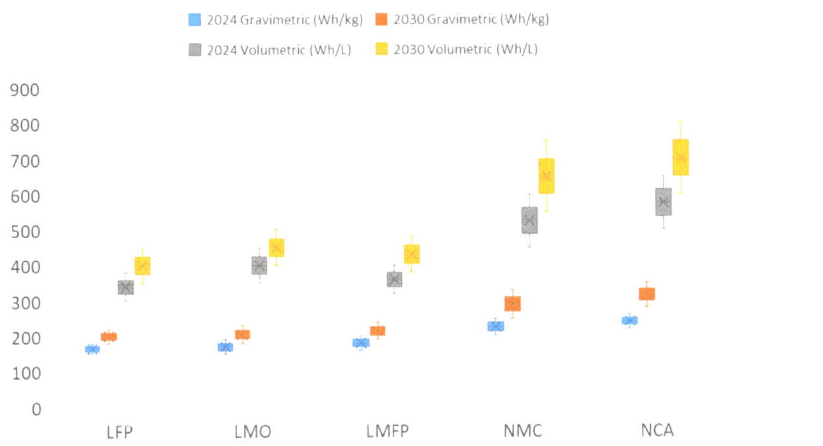

Figure 4. Evolutionary roadmap of lithium-based chemistries 2023–2030.

At the cell level, the objective in the short term is to enhance the energy density performance of cells while eliminating the use of critical materials like cobalt. This approach aims to reduce costs and concentrate on Generation 3b technology [68]. In the mid-term, solid-state Generation 4 LIBs are anticipated to become increasingly significant [65,69]. It is essential to ensure automotive-grade safety, including crash safety and the adoption of flame-retardant materials, as well as to focus on cyclability, lifespan, sustainability, and recyclability [64]. Addressing challenges related to safety, sustainability, and raw material criticality is also paramount for EU competitiveness [70]. This necessitates exploring sustainable materials, including bio-based and bio-mimetic options, and considering raw material abundance and cost in developing new battery chemistries [64]. Current research focuses on advanced materials for higher energy/power density, including high-nickel NMC cathodes, silicon-based anodes, and stabilized electrolytes. Efforts are also focused on reducing inactive materials. "Design-to-cost" approaches, like manganese-rich high lithium manganese (HLM) and improved LFP cathode, are also being pursued [71–73].

While LIBs have seen significant progress in energy/power density and cost reduction, further improvements are needed in material cost competitiveness, fast charging capabilities, ecological and social footprint reduction, high-energy anode and cathode synthesis, supply security, and cycle life enhancement to meet the demands of the automotive market. The industry has identified key battery targets for essential cell and pack characteristics to be achieved in a short-term period to meet mobility needs. In Table 3, we report some parameters' target values elaborated from the European Council for Automotive R&D (EUCAR) [74] and the United States Advanced Battery Consortium (USABC) [75] for cell performance expected goals. We also included the requirements identified in the Batteries Europe Roadmap [64], as well as from the literature [76]. These parameters depend on the application; thus, a range of values has been given instead, which includes high-power applications, such as plug-in hybrid EVs (PHEVs), and high-energy implementations, such as BEVs.

Table 3. Battery cell requirements for automotive applications. The value intervals are determined, including BEV and PHEV future needs and projections. Data from USABC have been estimated for the BOL assuming EOL condition is 80% of the initial capacity.

Parameters at Cell Level	Current 2020–2025	2030–2035	Source
Specific energy	160–290	275–450	[64,74–76]
Energy density Wh/l	450–730	750–1000	[55–57]
Continuous specific power—discharge W/kg	340–750	800–1750	[74,76]
Continuous power density—discharge W/L	1000–1500	2000–3850	[74,76]
Charging rate C (1/h)	2–3	6–3.5	[55–57]
Cost EUR/kWh	60–200	40–100	[74,76]
Hazard level	<=4	<=3	[74,76]

At the pack level, it is anticipated that the cell volume per battery pack will increase from 60% to 75%. Additionally, the cell weight per battery pack is expected to rise from 70% to 80%. The lifetime expectation should be comparable to that of a car, reaching up to 150,000 km. Furthermore, the cost of the battery pack should be approximately 15% to 20% higher than the cost of the individual cells. For PHEVs, the specific energy and density goals are somewhat less stringent than for BEVs, but peak power requests are generally higher. Key parameters are summarized in Table 4.

It should be noted that all USABC goals are written in terms of end-of-life (EOL), battery-pack-level values, whereas researchers and cell developers often highlight beginning-of-life (BOL), cell-level performance. A comparison of the USABC and EUCAR goals, reported on EOL values, on specific energy, calendar life, and cycle life, shows a good degree of alignment.

Table 4. Battery pack requirements for automotive applications. The value intervals are determined, including BEV and PHEV future needs and projections. Data from EUCAR have been estimated based on their actual and foreseen cell-to-pack integration.

Parameters at Pack Level	Current 2020–2025	2030–2035	Source
Specific energy Wh/kg	90–180	190–360	[74,76]
Energy density Wh/l	250–400	450–750	[74,76]
Continuous specific power—discharge W/kg	525	800–1400	[74]
Continuous power density—discharge W/L	900	1650–2600	[74]
Cost EUR/kWh	90–286	65–120	[74,76]

3.6. Life Duration

Battery durability is a critical requirement for EVs. Many vehicle original equipment manufacturers (OEMs) offer warranty periods that extend for many years and hundreds of thousands of miles. A study by P3 and Aviloo [77], analyzing data from over 7000 EVs, reveals promising insights into battery longevity. It shows that while battery capacity initially declines, it stabilizes over time. Specifically, capacity drops to about 95% after the first 30,000 km and around 90% at 100,000 km. Even after 300,000 km, most batteries retain about 87% of their original capacity. These findings exceed many original OEMs' predictions, which often anticipate a decline to 70–80% for the same distance range.

Efforts to improve battery life involve different levels of the storage system, from materials and cells to battery pack management. The cell's main components are the electrodes, where redox reactions occur during charge and discharge. The cathode—the electrode with the higher potential—is usually made of a lithium transition metal (TM) oxide material. The anode is typically an intercalation material like graphite or a graphite hybrid material, or sometimes lithium titanate oxide (LTO). Between the electrodes is a porous separator that prevents short-circuiting while allowing ion migration. The electrolyte plays a fundamental role in the battery, ensuring the transmission of ions and maintaining the reaction balance of the battery. During battery operation, or even during storage, the components may undergo physical and chemical changes that can alter the battery's performance [78]. These degradation mechanisms are not directly observable during normal battery operation, but they affect the battery performance, specifically with the reduction in the usable capacity of the cell (capacity fade) and the decrease in the cell's deliverable power (power fade) [79].

Degradation mechanisms can be categorized into distinct degradation modes, which reflect their impact on cell performance [80–82]. A commonly used classification identifies three main degradation modes:

1. Loss of Lithium Inventory (LLI): This mode includes mechanisms that reduce the amount of cyclable lithium available for transport between the electrodes.
2. Loss of Active Material (LAM): This encompasses mechanisms that lead to a decrease in the material available for electrochemical activity. LAM is often further divided into losses at the anode and losses at the cathode.
3. Conductivity Loss (CL): Also known as impedance change, this mode groups the mechanisms that affect the kinetics of the cell.

Figure 5 provides a simplified graphical representation of the relationships between battery components, degradation mechanisms, and degradation modes. The correlations between these phenomena are much more complex because all battery components are closely interconnected, and the details of specific degradation mechanisms depend on their composition and structure.

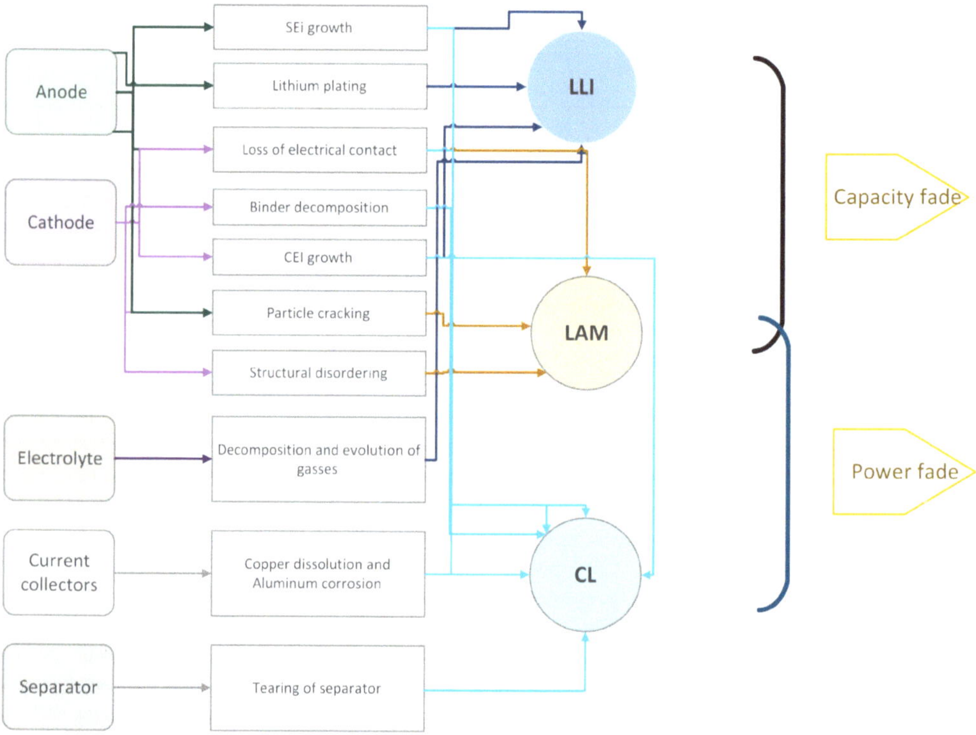

Figure 5. Schematic of degradation mechanisms and modes for different battery components.

Despite the oversimplification, Figure 5 highlights some of the issues the researchers are focusing on to improve battery life performance. Reducing the solid electrolyte interface (SEI) modification or growth during cycling is of particular interest. The SEI is a layer that forms on the electrode–electrolyte interface during the first few cycles. This layer protects the electrolyte material from further depletion and prevents corrosion of the anode. However, during cycling, irreversible side reactions can lead to unwanted growth of the SEI layer on the anode, which significantly reduces the capacity and power rate [83]. This mechanism has a particular impact on intercalation anodes, such as graphite and silicon-based ones, which undergo large volume changes during lithiation and delithiation [84], but it also occurs with other anode materials and at the cathode, where it is usually referred to as cathode electrolyte interface (CEI). The addition of binders, coating, and doping at the electrode can help reduce this degradation mechanism [20]. However, the mechanisms behind SEI growth are still not well understood, and further research is needed to address the issue [85].

In Table 5, we summarize various degradation mechanisms and mitigation strategies for several cathode and anode materials that are currently attracting research interest.

External factors such as operating current, voltage conditions, state of charge, and temperature all play complex roles in influencing the battery degradation mechanisms [86–89].

Table 5. Degradation mechanisms and mitigation strategies for some advanced electrodes.

Type of Electrode	Degradation Mechanisms	Mitigation Strategies	Source
Ni-rich layered cathode	Microcracks, lithium–nickel hybridization and irreversible phase transitions, anisotropic lattice deformation, and surface degradation	Elemental doping, coating modification, electrolyte modification, construction of radial concentration gradients in polycrystalline secondary particles, fabrication of rod-shaped primary particles, single-crystal high-nickel cathodes.	[90–93]
Lithium-rich manganese oxide cathode	Irreversible oxygen loss, structural degradation of the material, particle fragmentation, and transition metal migration	Surface coating, ion doping, component regulation, single crystal structures.	[94,95]
Li-metal anode	Dendrite growth	Coating artificial protective films, surface morphology control, high electrolyte concentration, electrolyte additives.	[85]

The commercially available LIB temperature operating range is usually around −20 to 55 °C, even though the optimal range is generally 15–35 °C [96]. While using the batteries outside their operating range can lead to safety issues, operating outside the optimal range accelerates aging. Generally, low temperatures favor the deposition of metallic lithium on the anode, high temperatures enhance the rate of side reactions and the kinetics of the battery, accelerating the decomposition of SEI and electrolyte consumption [81,82].

Batteries have a specific operating voltage range that depends on the potential of the electrodes. The voltage of a battery is related to its state of charge (SOC); higher voltage indicates a higher SOC, while the low voltage cut-off marks a depleted capacity. It is important to note that the voltage operating windows remain stable throughout the aging process, whereas the SOC values can change, as the available capacity decreases over time relative to the nominal initial capacity.

When cells are overcharged, lithium deposition can occur. An increased cutoff voltage can lead to electrolyte decomposition and uneven degradation of the negative electrode [80–82], and the battery may experience significant volume expansion, as well as increases in temperature and internal resistance, which raises the risk of thermal runaway. Conversely, discharging at low voltages can result in the dissolution of the copper current collector, increasing charge transfer resistance, and leading to micro-short circuits [81,82,96]. Low voltages accelerate the growth of SEI and CEI films, causing changes in the microstructure of the cathode and resulting in faster capacity fade [80,90,96,97].

Table 6 presents an overview of the primary degradation mechanisms caused by various usage-related stresses. It is important to note that, in practical applications, these stress factors often occur simultaneously, amplifying their effects on the battery.

To accurately predict battery aging, it is essential to consider the intricate relationship between external factors related to working or storage conditions, chemical and material-level degradation modes, and aging mechanisms [98–100]. To effectively predict the degradation curve up to the EOL conditions, it is necessary to continue testing until the batteries reach EOL or rely on advanced diagnostic models able to extrapolate early-stage data into predictions for long-term degradation based on observed processes [99,101]. The results are used in BMS to maximize battery performance and duration through diagnostic and prognostic analysis [63,102–105]. Diagnostics assesses how much a battery's performance deviates from its optimal or reference state, a concept known as state of health (SOH). SOH is not linked to a single index, and its definition can vary depending on the perspective from which the battery's performance is evaluated [106–108]. Prognostics estimate how long the device can continue to operate based on its current SOH and usage patterns [99]. Their implementations can be broadly classified into model-based and data-based approaches. Model-based methods, such as physics-based, equivalent circuit, or empirical models, are effective when battery behavior is well understood and require extensive laboratory tests for the accurate identification of model parameters [96,108–110]. In contrast, data-driven approaches are increasingly popular, especially for applications where battery behavior is highly nonlinear and difficult to model precisely [86,96,111,112]. These methods can use real-world usage data to gain insights without relying on specific mathematical equations [87,113]. However, they demand high-quality training data that reflects a wide range of battery behaviors [99,114].

Table 6. Overview of the main degradation mechanisms induced by operating conditions.

External Stress	Induced Degradation Mechanism	Aging Effect
Low temperature	Lithium plating and dendrites formation	Conductivity loss/capacity fade/short circuit
High temperature	Electrolyte and binder decomposition SEI film growth and decomposition	Capacity/power fade
Low voltage/SOC	Corrosion of current collectors Transition metal dissolution Loss of electric contact Lithium plating and dendrites formation SEI and CEI growth	Conductivity loss/capacity fade/power fade
High voltage/SOC	Electrolyte and binder decomposition SEI film growth and decomposition Graphite exfoliation Lithium plating	Capacity/power fade
High current	SEI film growth and decomposition Graphite exfoliation Structural disordering and particle cracking Loss of electric contact	Conductivity loss/capacity fade/power fade

Digital twins and AI analytics can be effectively integrated into a BMS to enable advanced predictive maintenance, moving beyond traditional rule-based or statistical methods [115]. A digital twin is a dynamic virtual representation of a physical battery or battery pack. This model incorporates various elements, including the battery's design specifications, material properties, electrochemical characteristics, thermal behavior, and operational history [116]. The BMS continuously streams real-time data from various sensors embedded in the physical battery to the digital twin [117]. AI algorithms within the digital twin framework continuously process this real-time data to update the virtual model, accurately reflecting the current state and operational conditions of the physical battery [118,119]. AI algorithms, especially machine learning techniques like neural networks, clustering, and statistical process control, can analyze real-time data to identify deviations from typical operating patterns. These anomalies may serve as early indicators of potential faults or degradation. When an anomaly is detected, AI can analyze historical data, failure modes, and the current context within the digital twin to diagnose the potential underlying cause of the issue [120]. By learning from historical degradation patterns, usage profiles, and environmental conditions of similar batteries—both real and simulated within the digital twin—AI models can predict the RUL of individual cells and the entire battery pack. This capability allows for proactive maintenance planning. Furthermore, the digital twin, enhanced by AI, can simulate various future operating conditions (such as different charging profiles, temperature extremes, and load cycles) to predict their impact on battery health and performance. This enables the optimization of battery usage and the evaluation of different maintenance strategies. However, implementing AI and digital twins in BMS faces challenges such as poor data quality, integration issues, and high initial costs. AI struggles with data context and reliability, while digital twins require continuous updates and deal with model complexity and scalability. Both technologies also raise concerns about data security and organizational readiness [121–123]

To estimate the SOH of batteries, methods can be categorized as either direct or indirect [124]. Direct methods involve measuring indicators such as capacity and impedance, while indirect methods analyze data to gain insights into battery aging mechanisms [105,112,125,126]. Examples of direct methods include Coulomb counting, open circuit voltage (OCV) measurements, impedance assessments, and the acoustic emission technique [127–130]. Advanced sensors help to determine direct SOH measurements [131–133]. Indirect methods rely on voltage and current values, which are much easier to measure than battery capacity. They derive the degradation state from models based on the number of cycles performed or through numerical approaches [130,134–137].

Battery thermal models describe the generation and transfer of heat in a battery to predict its temperature. A crucial step in ensuring the efficiency and accuracy of these models is determining and extracting the model parameters [138]. Thermal models are particularly important for ensuring the safe and reliable use of batteries, as temperature levels outside of operating limits can shorten battery life, degrade performance, and increase safety risks [114,139–143]. These models are integrated into the BMS to predict and mitigate thermal risks. Battery thermal management system (BTMS) includes running additional components such as liquid cooling systems or phase-change materials that are employed to keep the temperature settings [142,144–148].

3.7. Fast Charge

Charging time is a significant concern for customers who frequently take long-distance trips and for those who do not have access to private charging. Direct current fast charging (DCFC) equipment can charge a BEV to 80% in just 20 min to 1 h. However, there is still a desire for charging speeds that can approach those of internal combustion engine vehicles (ICEVs).

Key areas for improving fast charging include enhancing the diffusion rate of lithium ions (Li+) in the electrodes, optimizing transport in the electrolyte, and improving charge transfer kinetics at the electrode/electrolyte interface [149–153]. Research opportunities exist at both the cell and battery levels [150,154–156], focusing on the areas described in the following.

3.7.1. Electrode

Optimizing the design and structure of electrodes to enhance their conductivity and surface area significantly improves charge transfer rates. Reducing electrode thickness can improve ion diffusion but lower energy storage capacity. To balance capacity and fast charging with thicker electrodes, it is vital to optimize porosity and design, such as adding electrolyte channels or using a 3D grid porous electrode with vertically aligned pores for better ion transport [156].

Fast charging capability depends on the cathode's crystal structure. Layered oxides allow for high ionic mobility and quick charging due to minimal diffusion barriers. However, they can become structurally unstable during deep delithiation, but this can be contrasted with element doping or synthesizing layered Li transition-metal oxides with single-crystal (SC) [157,158]. As to polyanionic oxides, surface modification and coating improve their poor ionic conductivity [153,159,160]. Conversion-type cathode materials show poor performance upon cycling [161–163]. However, a composite cathode of FeF_2 nanoparticles embedded into a polymer-derived carbon (PDC) matrix showed a high capacity retention after 500 cycles at a high rate of 60C [164].

Carbon-based materials are the most widely used anode materials, but high current densities impact their performance. Graphite materials with porous structures, expanded interlayer spacing, or ordered alignment of graphite flakes perpendicular to the current collector are among the strategies proposed to enhance fast charging performance [153,165–167]. Transition-metal oxide anodes, like LTO, are better suited for fast charging but struggle with low intrinsic conductivity and slow Li+ diffusion kinetics. These challenges can be mitigated through surface modification, doping, and morphology control [149,168,169]. Silicon anodes face issues of large volume changes and low conductivity. To improve their rate and cycling performance while capitalizing on their high specific capacity, techniques such as particle size reduction, new microstructure design, and surface coatings are utilized [151,170–173].

3.7.2. Electrolytes

Important characteristics of electrolytes for fast charging include high conductivity, a high Li+ transference number, low desolvation activation energy, strong reductive and oxidative stability, good electrode compatibility, and a stable SEI layer [150,156,174]. Additives in electrolytes can contribute to the formation of a stable SEI layer, while the use of low-viscosity co-solvents can enhance ion mobility [175,176]. Flame-retardant additives can help reduce flammability and minimize side reactions [177–179]. In contrast, solid electrolytes are intrinsically non-flammable but tend to exhibit low ionic conductivity and high interfacial resistance. Recent advancements in garnet-type and sulfide-based solid electrolytes have demonstrated ionic conductivities that are comparable to those of liquid electrolytes [180,181].

3.7.3. Battery Engineering

The design of the battery cell is essential for ensuring mechanical stability and thermal safety. For instance, a high impedance ratio between the cathode and anode can help minimize lithium plating on the graphite anode and improve the distribution of current density on the collector, thus reducing the risks associated with fast charging [182,183]. At a broader system level, an effective thermal management system is vital for maintaining the performance, longevity, and safety of LIBs, especially in extreme conditions [184–186]. Additionally, implementing safer and more efficient charging strategies can significantly enhance battery performance, stability, and prolong battery lifespan [153,187–189].

3.8. Energy Density

There is an increasing market demand for batteries with higher energy density, driven by the need to extend the range of EVs and boost their adoption rates. Enhancements in battery energy density can be achieved at both the system level and the cell level [48,190,191], e.g., by improving the capacity of electrode materials and raising the operating voltage [192].

3.8.1. Anode

The dominant chemistry used nowadays in EV applications for the anode is graphite, thanks to its relatively low cost, abundance, high energy density, power density, and long cycle life [193]. Graphite consists of layers of single atomic hexagonal carbon atoms that provide enough interstitial sites to store lithium, with a 10% volume deformation during charge and discharge. However, limited capacity, degradation during cycle life, and low potential of graphite versus Li require the search for new anode materials [194]. For carbon anodes, many solutions have been investigated to meet the demand for stable and high-energy anodes. Anodes with different dimensionalities, such as nanotubes (1D), graphene nanosheets (2D), and various spherical structures (3D), have been developed to mitigate the effects of volume change during lithiation [194–196]. Fullerene derivatives (0D) have been investigated for their high charge capacity (861 mAh/g for carboxyl C60) and stable structure at a molecular level. Fullerene proved also useful to suppress dendrite formation on li-metal anodes [197].

Lithium titanate (LTO) is another alternative for the commercialized cells' anode. LTO-anode batteries show a lower nominal voltage (around 2.4 V) and lower specific energy compared to graphite anode batteries. On the other hand, LTO supports fast-charging and shows a remarkable lifespan [198,199]. However, degradation mechanisms are still under investigation, with gassing phenomena representing a potential hazard for safety [200,201]. Along with higher cost, these characteristics prevented the broad adoption of LTO lithium batteries in automotive applications, even though niche market applications are viable [198,202]. A series of titanium- and niobium-based intercalation materials have been developed as potential replacements for LTO, showing excellent electrochemical performance in terms of reversible capacity and electric conductivity [203]. However, most have high operating potentials above 1.5 V relative to Li/Li+. New titanium- and vanadium-based materials have emerged, capable of stable lithium storage below 1 V relative to Li/Li+ [204,205].

Both graphite and LTO work on the principle of intercalation: this process involves the storage of mobile ions in vacant sites or within interlayer spaces of a host lattice [206]. Among intercalation-type materials, MXenes—a new family of two-dimensional metal carbides or carbonitrides—have gained significant attention as anodes [207,208]. MXenes consist of graphene-like transition metal–carbon or nitrogen compounds. Their high conductivity and excellent surface chemistry make them particularly promising [209]. Unfortunately, the intercalation-type anode has a low rate capacity and a low specific capacity [206].

Alloy-based anodes are of great interest for their theoretically high specific capacity and increased energy density compared to graphite. Silicon anodes are particularly interesting alternatives to graphite anodes due to their high theoretical capacity of nearly 4200 mAh/g [210–212]. Silicon is appealing due to its abundance in the Earth's crust and the capacity of a single silicon atom to bond with four lithium ions. However, like all anode alloy materials, silicon anodes suffer from severe deformation during lithiation and de-lithiation, with a change of 300% of volume, which causes problems such as buffering, pulverization of the material, electrolyte consumption, and loss of contact [194,213]. The combined use of Si and graphite has become an appealing choice for high-energy anodes in LIBs [212,214,215]. This integration can also be achieved through the use of Si nanoparticles, Si suboxides, and Si-Graphite composites [216]. Silicon nanocrystals and nanotubes can reduce volume fluctuations during charging and discharging, improving the diffusion and conductivity of electrons and lithium ions [217]. However, cycling can cause agglomeration, increasing surface area and potentially leading to side reactions and reduced Coulombic efficiency [218]. Silicon suboxides (Si-Ox) materials have less volume expansion during lithium insertion compared to crystalline silicon. However, Si-Ox electrodes typically achieve a Coulombic efficiency of around 70% or lower, despite having a reversible capacity of about 1500 mAh/g [218]. The addition of specific binders and the use of tailored liquid electrolytes can help overcome those critical aspects [213,219].

Besides silicon, alloy metals such as tin (Sn), aluminum (Al), and antimony (Sb) have been intensively investigated. Sn-alloy anodes have a high onset voltage above Li/Li+, which can help prevent lithium deposition and dendrite formations while reducing interfacial resistance [211,220]. However, they usually suffer from high-volume deformation during cycling. Nanoscale synthesis of metals, such as nanoparticles and nanotubes, holds promise for addressing volume change during lithiation [194,221].

Transition metal oxides (TMOs) are of significant interest as anode materials for LIBs due to their favorable properties, including being non-toxic, having high power density, and being inexpensive [222–224]. They are based on the conversion reaction, which involves the formation and breaking of chemical bonds during lithiation and delithiation [206]. In LIBs, they can suppress the dendrite growth, increasing the battery's safety [194]. Notably, iron oxide (Fe_3O_4) has a theoretical capacity of about 926 mAh/g, making it a promising anode material. However, it suffers from rapid capacity loss and poor cycling stability due to the pulverization of active materials during operation, along with lower electrical conductivity compared to graphite [222–224].

Table 7 reports a comparison of some characteristics of these anode materials.

Other conversion-type transition metal compounds, such as transition metal chalcogenides, oxalates, carbides, nitrides, aluminum niobates, phosphides, and hydroxides, are also being explored as alternative anode materials in lithium-ion batteries [162].

Metal–organic frameworks (MOFs) are an emerging area of research and are considered promising materials for LIBs [225]. Their advantages include a high surface area, well-defined pore structures, and controllable chemical compositions. The unique porous structures of MOFs enhance electrolyte penetration and ion transport, support active materials, and selectively screen ions, making them ideal for use in battery separators, electrolytes, and electrodes [226]. Additionally, MOFs have several beneficial properties for battery applications, such as significant hygroscopic adsorption capabilities, high thermal stability, excellent electrochemical stability, and substantial mechanical robustness [227].

Table 7. Comparison of some electrochemical characteristics for some anode materials.

Composition	Gravimetric Capacity (mAh/g)	V vs. Li/Li+	Volume Change	Reference
C (graphite)	372	0.3 V	10%	[194,228]
$Li_4Ti_5O_{12}$	175	0.87 V	1%	[194,199]
$Li_{22}Si_5$ ($Li_{4.4}Si$)	4200	0.1 V	310%	[210,211]
$Li_{15}Si_4$	3570	50–60 mV	280%	[179]
Porous carbon–iron oxide (PC–Fe_3O_4)	926	0.8V	200%	[222–224,229]

Other emerging candidates for anodes are metal hydrides, like MgH_2 and TiH_2, which are made from abundant elements and offer low voltage hysteresis and high specific capacity. However, their capacity degrades after around 1000 to 2000 cycles, and they exhibit poor Coulomb efficiency. Metal phosphides are promising anode materials for lithium-ion and sodium-ion batteries due to their high specific capacity, safe operating potential, and excellent thermal stability. Improving long-term cycling performance is essential for hydrides to become viable for battery applications [206].

Finally, lithium metal is a promising anode material due to its high theoretical specific capacity (3860 mAh/g), low electrochemical potential (−3.04 V vs. SHE), and low density (0.59 g cm^{-3}). However, it faces challenges like lithium dendrite formation and continuous solid electrolyte interface (SEI) development, which can cause safety hazards, low Coulombic efficiency (CE), and reduced cycle life. Another challenge is the price of lithium metal itself and the expenses associated with refining the preparation process [222].

3.8.2. Cathode

Lithium battery cathodes use various technologies, each with specific advantages and disadvantages related to cost, safety, and performance. Lithium cobalt oxide (LCO) is the main cathode chemistry used in consumer electronics. In the past, lithium nickel oxide (LNO) was the second most common choice due to its lower cost and higher capacity, but its thermal instability limits its applications.

Recently, the automotive sector has moved to alternative cathode chemistries like lithium nickel cobalt aluminum oxide (NCA), lithium iron phosphate (LFP), lithium nickel manganese cobalt oxide (NMC), and lithium manganese oxide spinel (LMO) due to their better performance and safety compared to LCO [193]. According to the IEA [47], in 2022, NMC market share was 60%, followed by LFP (about 30%), and NCA with a share of about 8%. Cobalt-based technologies (NCA and NMC) are among the favorites for automotive applications due to their higher energy density, reaching 260 Wh/kg, and specific energy compared to LMO and LFP batteries. However, they suffer from thermal instability, which leads to a shorter cycle life than NMC [10]. NMC was originally formulated in the 1:1:1 ratio, which resulted in high energy density, power density, durability, and safety. However, NMC111 cost and value chain sustainability become critical, due to the growing demand for cobalt and its poor ecological and political sustainability [230]. NMC532 and NMC622 have recently become the dominant battery in EV applications, with NMC811 gaining market share [2], and higher-nickel content chemistry such as NMC955 emerging. However, increasing nickel content requires more complex production processes [4]. Moreover, cobalt guarantees the stability of the cathode.

On the other hand, LFP technology is expected to gain traction in automotive applications due to its stable and safe performance. LFP batteries have life cycles exceeding 2000 cycles, a wide operating temperature range, and costs that are over 20% lower than NMC batteries [4,13,231]. However, LFP batteries typically have an energy density that is 20–30% lower than that of high-nickel chemistries. A growing number of EV manufacturers are adopting LFP technology, including Tesla, Ford, Opel, and Citroën [232,233]. This trend is bolstered by researchers' successes in increasing the energy and power density of LFP batteries. In 2024, Zeekr announced an upgrade to its electric vehicle LFP battery, which can support ultra-fast charging at a rate of 5.5 C [234]. Meanwhile, SAIC-GM revealed the production of a 6 C charging LFP battery in collaboration with CATL [235]. Additionally, Geely introduced its latest generation of the self-developed Aegis LFP blade battery, which boasts an energy density of 192 Wh/kg [236].

Researchers have also developed a new iron and manganese form of LFP, termed LMFP, which was commercialized in 2024 [12] According to producer Gotion, the LMFP can reach an energy density of 240 Wh/Kg, has a volumetric energy density of 525 Wh/L, and a lifespan of 4000 cycles at room temperature and 1800 cycles at high temperatures [237]. A volumetric cell to pack (CPT) ratio of 76% allows for an LMFP battery pack energy density of 190 Wh/kg, surpassing the commercial NMC performance. The Chinese OEM BYD reached a 40% increase in cell-to-pack integration efficiency, reaching an energy density pack of 140 Wh/kg, using elongated LFP battery cells (blade cells), which are as long (600–2500 mm) as the pack [238].

In Table 8, we compare different lithium-ion technologies in terms of some electrical parameters and performance parameters.

Table 8. Comparison among some lithium-ion battery technologies for automotive applications. Data were processed from [100,194,199,203,238–244].

Cathode	Anode	Operating Voltage (V)	Energy Density (Wh/kg)	Power Density (W/kg)	Fast-Charging	Lifespan (Cycles)
LFP	C or Si-C	2.5–3.6	90–190	247	3 C	2000–4000
NMC	C or Si-C	3–4.2	130–280	300–800	0.7–1 C	1000–2000
NMC	LTO	1.5–2.8	70–90	2200	10 C	3000–10,000
LMO	C or Si-C	3–4.2	100–185	925	0.7–1 C	300–1000
LMO	LTO	1.5–2.8	70–90	3600	5 C	3000–7000
NCA	C or Si-C	3–4.2	175–300	670	0.7–2 C	500–1000

3.9. Safety

Battery safety encompasses a broad range of investigations aimed at understanding and mitigating the hazards associated with lithium-based batteries. It is of paramount importance, since the ageing process, abusive use, or battery defects can compromise performance and lead to dangerous consequences like fires or explosions [245–247].

Among the objectives of fault diagnosis, there is understanding the failure mechanisms by investigating the root causes of battery failures, including thermal runaway, internal short circuits, over-charging, over-discharging, and mechanical damage, and analyzing the chemical and physical processes that lead to these failures [248]. The risk assessment comprises determining the likelihood and severity of potential hazards, such as fires, explosions, and the release of toxic gases, and quantifying the risks associated with different battery chemistries, designs, and operating conditions. Reliability assessment employs experiments (accelerated aging, in situ diagnostics, failure analysis), computational models, and model or data-driven methods [109]. Experimental techniques include accelerated aging to simulate long-term degradation and in situ diagnostics for real-time monitoring [249]. Understanding thermal runaway is critical, especially in relation to operating conditions such as extreme temperatures, overcharging/discharging, and fast charging [250]. Thermal runaway is a critical failure mode where internal battery temperature rapidly increases due to exothermic reactions, leading to potential fires or explosions [248,251–254]. External factors like mechanical stress [249,255–257] and environmental conditions [84,258] can damage batteries, necessitating robust designs for diverse real-world use.

Risk management and safety assessment must take a comprehensive approach that considers both material and cell risks to evaluate the safety of battery systems effectively [247]. Indeed, the material choices for cathode, anode, electrolyte, and separator directly impact thermal stability and safety, with stable SEI layer formation and separator integrity being crucial [85,259–261]. Ongoing research is devoted to improving the intrinsic safety of battery components. Fire retardants, tailored solvents, and solid or quasi-solid state electrolytes are possible solutions to enhance battery safety [177,178,262,263]. Element doping, surface engineering, nanostructure, single-crystal design, and concentration gradient structure are strategies proposed to stabilize high-nickel cathodes [264–268].

Safety concerns are particularly critical for batteries used in automotive applications. As power and energy levels rise, the risk of vehicle fires and potential harm to people and property increases. For instance, in NCM-type batteries, a higher nickel (Ni) content results in increased specific capacity but reduced thermal stability.

The thermal runaway process is a complex chain reaction that consists of multiple steps, with its outcome influenced by various factors. The different components of the cells—such as electrolytes, electrodes, binders, and separators—affect both the sensitivity to thermal runaway and the severity of its consequences. LIB electrolytes typically consist of organic solvents and participate in several exothermic reactions during thermal runaway.

One critical reaction is the breakdown of the SEI, which is exothermic and starts at temperatures between 51 °C and 69 °C, depending on the electrolyte, particularly in the presence of carbonaceous anodes like graphite. Research has demonstrated that LTO anodes exhibit better safety performance due to their higher thermal stability [269].

The decomposition of the cathode, aside from being influenced by its chemistry, also depends on the electrolyte employed. For example, in an ethylene carbonate/diethyl carbonate (EC/DEC) solvent, the onset temperatures for the decomposition of LCO, NCM, and LFP are approximately 150 °C, 220 °C, and 310 °C, respectively.

Another crucial factor in assessing the safety of the cathode is the heat released during its decomposition and the rate of that heat release. Generally, the thermal reactivity of the cathodes can be ranked as follows: LCO > NCA > NMC > LMO > LFP. For more detailed information on the safety performance of various lithium-ion batteries (LIBs), readers are encouraged to refer to the analyses presented in specific research papers [270–272].

The unique nonlinear characteristics of lithium-ion batteries can render traditional reliability models ineffective [273]. Early fault diagnosis is vital for preventing component breakdown and improving overall EV reliability and safety [274]. Designing and evaluating safety features, such as thermal management systems, protection circuits, and flame-retardant materials, is fundamental to ensure safety [246]. Finally, manufacturing quality must be consistent to prevent defects like contamination and misalignment, requiring rigorous testing [249]. Utilizing data from real-world applications can be extremely beneficial. Recognizing this, many large companies have developed cloud-based platforms for monitoring and analyzing EV batteries in real-world settings [109,111,275].

Policies are essential for improving battery safety throughout their lifecycle, from manufacturing to disposal [276]. Regulations can enforce quality control measures, such as mandatory testing for defects and thermal stability, and require the use of safer materials. They also establish guidelines for the safe transport and storage of batteries, particularly lithium-ion ones, which present fire risks, and mandate safety features in battery-powered devices, like thermal management and overcharge protection. Effective recycling and disposal systems are crucial to prevent environmental contamination and fire hazards from damaged batteries, with regulations promoting responsible practices and traceability in production to identify defects and their origins.

The EU Battery Regulation [35], which includes the battery passport, aims to enhance sustainability and safety in battery use. The battery passport provides transparency and traceability by recording essential information about each battery's composition, performance, and lifecycle. This system allows for better tracking of batteries throughout their lifespan, making it easier to identify and recall defective batteries. Additionally, the battery passport offers detailed information about a battery's condition and history, which can help prevent misuse and promote safe handling practices.

Table 9 reports a résumé of the main performance characteristics reported for LIB technologies with respect to automotive applications.

Table 9. Summary of the main performance index for LIB batteries.

Battery Chemistry	Cost	Energy Density	Specific Power	Lifespan	Safety	Thermal Stability	Overall Suitability for EVs
LFP	Low to Med	Moderate	Good	Excellent	Excellent	Very Good	Good for mass-market EVs, especially where cost and longevity are prioritized.
LMFP	Medium	Moderate to High	Good to Very Good	Very Good to Excellent	Good	Good to Very Good	Promising for mid-range EVs, potentially bridging the gap between LFP and NMC.
LMO	Low to Med	Low to Moderate	High	Moderate	Good	Good	Niche applications (e.g., some hybrids, power-focused EVs if lifespan is acceptable).
LMNO	High	High	Good to Very Good	Potentially Good	Moderate	Moderate	Potential for future high-performance EVs if safety and cost challenges are overcome.
LCO	High	High	Moderate	Moderate	Poor to Med	Poor to Moderate	Limited suitability for EVs due to cost, safety concerns, and lifespan.
NCA	Medium to High	Very High	High	Good	Moderate	Moderate	Well-suited for premium, long-range EVs where performance is a key factor.

3.10. Current Trends and Future Developments in EV Batteries Research

Lithium-ion batteries (LIBs) alone cannot meet the increasing demand for batteries in the coming years, creating opportunities for more affordable and sustainable alternatives. Continuous advancements in Li-ion technology raise the performance standards for new technologies. However, improvements in one area, like energy density, can sometimes compromise safety or drive up costs. Therefore, emerging technologies must balance enhancements in range, safety, charging time, and sustainability to offer transformative solutions for EVs. Below, we highlight several promising battery technologies at various TRLs.

Research is focused on many other technologies, such as aqueous batteries [277–279] and dual-ion batteries [278,280]. Even though these technologies show promising performances, it is not yet proven that these batteries can be competitive in transportation applications, as they are still in the early stages of development.

3.10.1. Solid State LIBs

Lithium-ion solid-state batteries (SSBs) are among the most promising technologies for the near future. They are composed of the same cathode and anode of today's LIBs, but instead of a separator soaked with liquid electrolytes, they use a solid-state electrolyte (SSE). Since liquid electrolytes constitute one of the main safety concerns of LIBs due to the high flammability of the organic solvents they are made of, solid-state LIBs may reduce the risk of venting and fire [20]. In addition to improved safety, SSEs broaden the electrochemical window up to 6 V compared to lithium metal, allowing the use of high-voltage cathode materials [281]. Furthermore, SSEs could facilitate the adoption of lithium metal as anode resulting in an increased energy density thanks to its high theoretical specific capacity (3860 mA h/g), its low electrochemical potential (−3.04 V vs. the standard hydrogen electrode, SHE) and its low density (0.53 g cm^{-3} at room temperature) [282,283].

In comparison with traditional organic liquid electrolytes (OLEs), SPEs are lighter, non-flammable, and have lower cost, superior mechanical and processing properties, good flexibility, and uniform lithium deposition. Despite the considerable advantages offered by solid-state batteries, several challenges and limitations currently hinder their widespread adoption in the automotive industry. One of the primary hurdles is the increased electrical resistance of many solid electrolytes compared to liquid electrolytes [284]. This higher resistance can lead to difficulties in achieving the fast charging speeds that are highly desired for EVs. It can also contribute to a gradual degradation of the battery's performance over extended periods of use [285]. To be of interest, solid electrolytes must meet the following requirements [286]:

(1) Room temperature conductivity $\geq 10^{-4}$ S cm^{-1};
(2) Electronic insulation <10^{-10} S cm^{-1} (Li+ migration number is approximately 1);
(3) Wide electrochemical window (>5.5 V vs. Li/Li+);
(4) Good compatibility with the selected electrode material;
(5) Good thermal stability and mechanical properties, wet environment resistance;
(6) Low cost and low environmental impact raw materials;
(7) A simple synthesis method.

SSEs have excellent elasticity and flexibility that result in good interface contact properties [287]. Especially in the presence of electrodes with high volume expansion and contraction during charging and discharging, they can maintain good contact, reducing the interface impedance and assuring the stability of the SSB [288].

The SSEs investigated in the last few decades are generally grouped in three categories: solid polymer electrolytes (SPEs), inorganic solid electrolytes (ISEs), and organic–inorganic composite solid electrolytes (CSEs) [289,290]. Common polymer matrix materials used for SPEs include poly-ethylene oxide (PEO), polyacrylonitrile (PAN), poly-vinylidene fluoride (PVDF), and poly-methyl methacrylate (PMMA) [286,291], while the following lithium salts are generally used: lithium hexafluorophosphate ($LiPF_6$), lithium bis(fluorosulfonyl)imide (LiFSI), [292] and Lithium Bis(Trifluoromethanesulfonyl)Imide (LiTFSI) [293]. Generally, the ionic conductivity for SPEs varies from 10^{-4} and 10^{-5} S cm^{-1}. Sun et al. [294] developed a polymer bi-phase SSE with an excellent ionic conductivity of 1.9 mS cm^{-1} at room-temperature, a high oxidation potential of 4.9 V (vs Li/Li$^+$) and a lithium transference number (t_{Li+}) of 0.56, by using in situ thermal cross-linking of 2-ethyl cyanoacrylate (CA), polyethylene glycol methyl ether acrylate (PEGMEA), succinonitrile (SN), and fluoroethylene carbonate (FEC) additives. They coupled an 11 μm bi-phase SPE with an NCM811 cathode and a 30 μm lithium metal anode, obtaining a discharge capacity of 208 mAh/g with a cut-off voltage of 4.5 V, resulting in an energy density of 400 Wh/kg.

ISEs, on the other hand, show the highest ionic conductivity, reaching 10^{-2} S cm^{-1}, which is comparable to that of liquid electrolytes at the operating temperature [295]. The ISEs can be grouped into three types: oxide, sulfide, and halide electrolytes.

Solid oxide electrolytes have a high electrochemical oxidation voltage and chemical and mechanical stability, while the disadvantages are mainly the high resistance at the electrode–electrolyte interface and the high synthesis temperature [284]. Among them, the garnet-type SSEs have the highest ionic conductivity (10^{-4}–10^{-3} S cm^{-1}), a good oxidation stability (6 V vs. Li/Li+), and a high stability allowing compatibility with lithium metal anodes. The most promising garnet-type SSE is the $Li_7La_3Zr_2O_{12}$ (LLZO) with a Li+ conductivity of 1 mS cm^{-1} [296].

Sulfide ISEs have the highest ionic conductivity, comparable to that of organic liquid electrolytes [297]. Most sulfide ISEs have a conductivity that exceeds 1 mS cm^{-1}, and some can reach 20 mS cm^{-1}. They also have good flexibility that allows good contact with the electrodes. On the other hand, sulfide ISEs have low oxidation stability, and they are sensitive to air and decompose in the presence of toxic gases in the air, constituting a safety problem [298].

The third type of ISEs, halide solid electrolytes, are not inherently stable with lithium metal anodes, but they can be stabilized through various methods [299]. They offer high mechanical strength and good cycling performance in high voltage windows, but they suffer from low ionic conductivity, which in a few cases can reach 10^{-3} S cm^{-1}, and poor air stability [300,301].

To overcome the low ionic conductivity of SPEs and the high interfacial resistance of ISEs, organic–inorganic composite solid electrolytes (CSEs) were developed. They are composed of both a polymeric matrix with lithium salt and an inorganic filler. The polymeric matrix reduces the interface resistance with the electrode materials, bringing toughness and elasticity, while the inorganic filler improves the ionic conductivity by reducing the crystallinity of the polymer matrix while bringing mechanical strength [302]. Several types of polymeric matrix and lithium salt described before are used in CSEs, as well as several types of ISEs described before are used too. The resulting ionic conductivity for the different CSEs lies between 10^{-4} and 10^{-3} S cm^{-1} [303,304].

Maintaining a stable and effective interface between the solid electrolyte and the solid electrodes (anode and cathode) is crucial for efficient ion transport. However, achieving good and consistent contact at these solid-to-solid interfaces, while also ensuring electronic insulation to prevent short circuits, has been a persistent technical difficulty, due to high interfacial resistance, poor interface contacts, interface instability, and side reactions between solid-state electrolytes and electrodes. To address these challenges, surface coating is often used [305,306]. Another way to improve the electrode–electrolyte interface is the adoption of bilayer heterostructure composite electrolytes where poly-vinylidene fluoride-hexafluoropropylene (PVDF-HFP) and oxidation-resistant PAN were used for the contact with the cathode ($LiFePO_4$), and PVDF-HFP and reduction-resistant PEO were used for the contact with the anode (metal lithium) [305,307]. Another strategy is the adoption of laser-ablated geometrical structures [308].

Some types of solid electrolytes, particularly ceramics, can exhibit sensitivity to temperature and pressure. Their performance might degrade at low temperatures, and some designs require the application of high pressure to maintain adequate contact with the electrodes [309]. Additionally, ceramic electrolytes can be brittle, making them susceptible to cracking under mechanical stress [310], which is a concern in the demanding environment of an electric vehicle. While solid electrolytes are generally expected to suppress the formation of lithium dendrites, this problem can still occur in solid-state batteries that utilize lithium metal anodes [311], potentially leading to short circuits and battery failure.

The current state-of-the-art of these SSBs, developed by enterprises worldwide, has achieved a specific energy of over 250 Wh/kg, generally about 370 Wh/kg (peaking at 417 Wh/kg), an energy density of over 900 Wh/L, a capability of 4 C fast charge, and cycle life greater than 500 cycles and even up to 4000 cycles [193,286,298].

Based on the current state of development and the timelines announced by various companies, the initial commercialization of solid-state batteries in electric vehicles is anticipated in the late 2020s, with some manufacturers targeting the period between 2026 and 2029 for their introduction in higher-end or limited production models [312]. However, widespread adoption across a broader range of electric vehicle segments is more likely to occur in the early to mid-2030s as the technological and manufacturing challenges are effectively addressed and the production costs are significantly reduced to make the technology economically viable for the mass market [313]. Semi-solid-state batteries, where liquid and solid electrolytes are used together to enhance electrode–electrolyte interface performances, may serve as a transitional technology in the interim.

3.10.2. Sodium-Ion Batteries

Non-lithium metal-ion batteries have attracted increasing interest in the last four years, as shown in Figure 6.

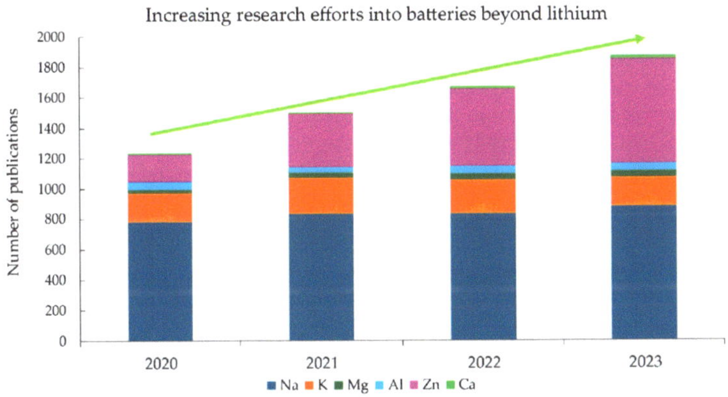

Figure 6. Trend in number of publications regarding non-lithium batteries in the last years. Reproduced from [314] under CC BY 4.0 terms.

The main requirement that alternative chemistries must satisfy is the use of more abundant and easier-to-recycle raw materials. In fact, besides comparable (or at least not significantly worse) performance in terms of energy density, power density, and cycle life, alternative chemistries should have improved safety and environmental sustainability.

Sodium has a low cost and is highly abundant. SIBs function on a very similar principle to LIBs but utilize sodium ions (Na+) as the charge. The fundamental construction of a SIB closely mirrors that of a LIB, making it a promising alternative [315,316]. The main disadvantage of sodium compared to lithium is the higher density, which, combined with a less negative standard electrode potential (−2.71 V of sodium versus −3.040 V of lithium), results in a lower energy and power density than LIBs, making SIBs less attractive for the transportation sector.

The anode is often made of a disordered carbon material known as hard carbon (HC), which has been shown to effectively intercalate sodium ions. Graphite-based anodes are not compatible with sodium-ion, mainly because of the larger ionic diameter of sodium (1.06 Å) versus lithium (0.67 Å) that results in a less ionic diffusivity and causes structural distortion of the active electrode material during intercalation. The use of HC materials in SIBs has already been commercialized, and extensive research is focused on enhancing their rate capability and overall performance. Specifically, research groups are exploring strategies such as heteroatom doping to create porosity and introduce defects on the surface, which can increase the number of active sites [317,318].

Other intercalation materials suitable as anodes in SIBs are titanium-based materials. Among them, titanium-based oxides exhibit considerable potential, due to the various crystalline phases, significant structural stability, and high abundance [319].

Transition metal chalcogenides (TMCs) are also explored as potential negative electrodes for SIBs, since they can provide high capacity [320].

Regarding electrolytes, the most common for SIBs are organic carbonate-based electrolytes and ether-based electrolytes. Carbonates generally use fluoroethyl carbonate (FEC) as an additive to improve the SEI formation [321], and are suitable for high-voltage cathodes since they are more stable than ethers at high voltage. However, in cells with hard carbon as an anode, ethers show better long-term cycling stability. The latter are also more suitable for low-temperature applications [314,322].

A promising solution to increase the energy density of sodium-ion batteries (SIBs) involves replacing the insertion-type anodes with ultra-thin metallic sodium, which has a density of 1166 mAh/g [323]. The metallic sodium anode can be used in the form of a foil deposited on the electrode. However, this foil may contain excess metal, which can decrease the overall energy density. Additionally, metallic sodium is highly reactive to air, and its softness and stickiness complicate the manufacturing process. In an alternative method, the anode is formed during the initial charging process using sodium sourced from the cathode material. This approach benefits from the absence of air during production, eliminating concerns about excess sodium [324]. However, this method may reduce the total capacity that the cathode can provide due to the formation of SEI. Li et al. [325] developed AFNBs that used a layered oxide-based cathode ($Na[Cu_{1/9}Ni_{2/9}Fe_{1/3}Mn_{1/3}]O_2$) and an aluminum current collector coated with graphitic carbon on the anode side, reaching an energy density of 200 Wh/kg with a cycle life of 260.

Typical SIB cathode materials are layered transition metal oxide (LTMO), polyanionic compounds, Prussian blue analogs (PBAs), and organic polymer.

In LTMO, Na+ ions intercalate between the layers through the trigonal prismatic or the octahedral vacancies, and their formula is generally represented as Na_xTMO_2, where $0 < x < 1$ and TM represents 3D transition metals [314]. The use of these cathode materials, already developed for LIBs, could reduce the initial investment required for SIBs production. However, the large size of sodium-ions leads to a great volume expansion during cycling, resulting in reduced stability and lower cycle life, especially at high C-rate. Strategies such as structural modifications can improve LTMO performance [326]. Wang et al. [327] synthesized a transition-metal oxide cathode based on P2-Na_xMnO_2 doped with potassium: $Na_{0.612}K_{0.056}MnO_2$ to reinforce the Mn-O bonds and reduce the phase transition during cycling. When coupled with a presodiated hard carbon, they reached a full cell energy density of 314.4 Wh/kg (based on the total mass of active anode and cathode materials), with a specific capacity of 230.6 mAh/g. Sathiya et al. [328] gradually replaced manganese with a non-transitional metal ion, Sn^{4+}, into a $NaNi_{0.5}Mn_{0.5-\gamma}Sn_\gamma O_2$ cathode (γ = 0–0.5) to reduce cathode phase transition and increase cycling performance. When coupled with a hard carbon anode, the energy density of materials decreases from 335 Wh/kg for the lightest $NaNi_{0.5}Mn_{0.5}O_2$ to 270 Wh/kg when substituting all the Mn with Sn. However, with a γ between 0.3 and 0.5, they improved the cycling stability of the cell. Zheng et al. [329] synthesized a Mn-based layered oxide cathode with a high-temperature thermal shock strategy (NMO-HTS) to suppress Mn ion vacancy within transition material layers. The average discharge is 2.65 V, and the rate capability is 180 mAh/g at 1C and from 28 to 67 mAh/g at 20 C. The full cell, with NMO-HTS cathode and hard carbon anode, reached an energy density of 248 Wh/kg based on the weight of both electrodes.

Among polyanionic compounds, the sodium superionic conductors (NASICON) are among the most promising materials due to their significant structural stability and high ionic conductivity [330,331]. The NASICON molecular formula is generally represented as $Na_xMM'(XO_4)_3$, where M or M′ = V, Fe, etc., X = P or S, x = 0–4. Deng et al. [332] optimized a $Na_3V_2(PO_4)_2F_3$ (NVPF) cathode, obtaining an electrode energy density of 446.4 Wh /kg at 1 C with a high reversible capacity (120.8 mAh/g at 1 C), good rate capability (89 mAh/g at 30 C) and long cycle life (81.3%@10 C after 1700 cycles). The presence of vanadium can pose a challenge, since this metal is scarce in the EU, with China being its primary producer. Zhang et al. [333] developed another NASICON-type $Na_4MnCr(PO_4)_3$ cathode with an excellent energy density of 566.5 Wh/kg (greater than that of $LiFePO_4 \approx 530$ Wh/kg), a reversible capacity of 160.5 mAh/g.

PBA compounds are of interest for their three-dimensional open framework, adjustable structures, and chemical composition, and recent advancements have demonstrated promising enhancements in electrochemical performance and synthesis efficiency. PBAs have a typical structure of a double perovskite framework with $(C{\equiv}N)^-$ anions bridging MN_6 (M = Fe, Mn, Co, Ni, Cu, Zn, and Ti) and FeC_6 octahedra, while Na^+ ions and H_2O occupy the interstitial sites [334]. Within PBAs, sodium manganese hexacyanoferrate ($Na_xMnFe(CN)_6$) is one of the most promising, thanks to its high energy density, non-toxicity, and abundance of its metal components, Mn and Fe. Tang et al. [334] synthesized the cubic and monoclinic structures of ($Na_xMnFe(CN)_6$). The cubic structure achieves better electrochemical performances, with a specific capacity of about 120 mAh/g at 3.5 V vs. Na^+/Na, and a capacity retention of 70% over 500 cycles. PBAs are already used in production SIB cells fabricated by the company Novasis Energies, Inc., providing cell-specific energy between 100 and 130 Wh/kg and volumetric energy density of 150–210 Wh/L [335]. The interstitial water presence is one of the main drawbacks since it compromises the cell performance, and it is difficult to remove [325].

Organic polymers are seen as potential cathode materials for SIBs due to their lightweight nature, flexibility in design, and potentially lower cost compared to traditional inorganic materials [336]. Examples of aromatic functional groups that can be used include imides, quinones, conjugated polymers, and aromatic carboxylates. The design flexibility includes the possibility to modulate the material's physical properties, such as flexibility and electronic conductivity [337]. These materials can achieve a higher capacity than metal oxides and polyanions, but this comes at the cost of a lower voltage range and reduced stability [316].

Table 10 presents the specific energy and capacity of various non-commercial SIBs based on the composition of the anode and cathode, demonstrating performance comparable to LIBs.

It is noteworthy that several studies confirmed that the construction of high-performance SIBs shows a low dependence on cobalt [338]. The main technical challenge with SIBs is their limited reversibility, which results from unstable electrode materials during cycling and uncontrollable side reactions at the electrode/electrolyte interface. Research shows that capacity fading and poor cycling stability are linked to these issues [339]. Additionally, the formation of the SEI layer is still not fully understood. The dynamic process of SEI formation and its chemical composition are critical for ion mobility and the stability of the electrolyte and electrodes, significantly influencing the cells' electrochemical performance [340,341].

The safety performance of SIBs is still under investigation. If future research confirms that sodium batteries are indeed safer than LIBs, their commercialization could increase significantly since one of the key motivations for developing batteries beyond lithium is the potential to enhance safety features. However, sodium, like other alternative ions, is unstable in ambient air and reacts vigorously with water.

Among the components contributing to the overall stability of the cell, electrolytes play a particularly important role in ensuring cell safety. Since organic electrolytes commonly used in SIBs are inherently flammable, it is crucial to thoroughly evaluate the safety aspects associated with them. Yue et al. [342] compared the risk of thermal runaway in SIBs and LIBs, finding that the hazard associated with sodium-ion batteries using transition metal cathodes was lower than that of NCM LIBs but higher than that of LFP batteries. Buthia et al. [343] reported in their review the onset and maximum temperatures related to thermal runaway in both LIBs and SIBs with various chemistries. They noted that all SIB chemistries exhibit a lower maximum temperature compared to LIBs, although their onset temperatures are similar. Additionally, the authors emphasize that the risks of thermal runaway in SIBs can vary depending on the cathode material used. Specifically, cathodes made from layered oxides present a greater risk than those made from Prussian blue analogs, while polyanionic compounds are considered less risky.

Table 10. Electrochemical properties of some non-commercial SIBs.

Cathode	Anode	Specific Energy [Wh/kg]	Specific Capacity [mAh/g]	Source
$Na_{0.612}K_{0.056}MnO$	Presodiated HC	314.4	230.6	[327]
$NaNi_{0.5}Mn_{0.5-\gamma}Sn_{\gamma}O_2$	HC	335–270 with γ = 0–0.5	-	[328]
NMO-HTS	HC	248	180	[329]
$Na_3V_{1.8}(CrMnFenAl)_{0.2}(PO_4)_3$	HC	202	-	[344]
$Na_xMnFe(CN)_6$	TiO_2	111	120	[334]
PBA	HC	100–130	-	[335]
$Na[Cu_{1/9}Ni_{2/9}Fe_{1/3}Mn_{1/3}]O_2$	Anode Free	200	-	[325]

3.10.3. Potassium-Ion Batteries

Potassium has garnered significant scientific interest due to its membership in the same group as lithium and sodium. It exhibits a standard electrode potential of −2.931 V, which is more negative than that of sodium (−2.71 V). Additionally, potassium is one of the most abundant elements in the Earth's crust. However, the diffusion of potassium ion batteries (PIBs) is hindered by the larger ionic radius compared to sodium, which leads to an increased volume expansion during ion intercalation in the electrode materials.

The larger radius of K^+ implies a smaller charge density, which corresponds to a weaker solvent shell, also compared to Na^+, and consequently to a minor solvation energy. Since in the intercalation mechanism, the ion desolvation is a key feature, PIBs could show high-rate capability and high power [314].

A common cathode used in PIB full cells is the organic compound perylenetetracarboxylic dianhydride (PTCDA), which is compatible with common organic electrolytes. Other cathodes include LTMO, polyanionic compounds, and PBA.

Regarding the PBAs, similarly to sodium ion, the interstitial water presence gives rise to unwanted reactions such as thermal instability and capacity decay during cycling. Liao et al. [345] focused on a cathode of metal hexacyanoferrates (HCFs), because of their low cost and high energy density, and developed a method to obtain nanoplate of (100) face-oriented potassium magnesium hexacyanoferrates (KMgHCF) with low $[Fe(CN)_6]$ vacancies, high crystallinity and no crystal water. When coupled with an anode composed of graphite and dipotassium terephthalate (K2TP), they obtained a full cell with a 3.45 V average discharge voltage, with a specific capacity of 83 mAh/g, and an energy density of 214.8 Wh/kg.

Research on anodes for PIBs primarily focuses on carbon-based materials, including graphitic carbon, soft carbon, and hard carbon. Although Na^+ cannot intercalate in graphite, K^+ can. This phenomenon is related to the enthalpy of formation rather than simply the size of the ions [346].

In general, electrolytes used in PIBs include carbonate or ether-based electrolytes, while aqueous electrolytes are not common due to the limited electrochemical stability window.

Zhang et al. [347] reported the carbonaceous anodes as a bottleneck of PIBs capacity, so they focused on the development of a carbon anode, doped with nitrogen, that possesses a 3D framework (3D-NTC), to enhance the electrode capacity. Also, Wang et al. [348] focused on the anode development, using a facet growth mechanism with $(BiO_2)CO_3$ nanocrystal and amorphous iron oxide obtaining outstanding results. Ji et al. [349] found that a magnet, the $KFeC_2O_4F$, can be used as a cathode in PIBs, with a discharge capacity of 112 mAh/g and an excellent capacity retention of 94% after 2000 cycles. The reported results are summarized in Table 11 for the full cells.

Table 11. Electrochemical properties of some PIBs.

Cathode	Anode	Specific Energy [Wh/kg]	Specific Capacity [mAh/g]	Authors
PTCDA	3D-NTC	187	241	[347]
PTCDA	$(BiO_2)CO_3$	732.8	-	[348]
$KFeC_2O_4F$	Soft Carbon	235	112	[349]
KMgHCF	Graphite and K2TP	214.8	83	[345]

3.10.4. Magnesium-Ion Batteries

Magnesium is one of the ten most abundant elements on Earth (1000 times more abundant than lithium), and it is the third most used metal in industry. It has a good specific density (2205 mAh/g of Mg vs. 3861 mAh/g of Li), a high density per unit of volume (3833 mAh/mL of Mg vs. 2062 mAh/mL of Li), and a discrete redox potential (−2.372 V of Mg/Mg^{2+} vs. SHE). These advantages, together with the reputation of dendrite-free plating of Mg onto Mg metal [350], make magnesium ion batteries (MIBs) of interest. Nevertheless, many challenges must be faced to achieve a satisfactory performance, and research is still at an early stage.

The main drawbacks of MIBs are the sluggish kinetics of Mg ions and the difficulty in finding electrolytes compatible with both electrodes and SEI formation when using common solvents. The magnesium ion, Mg^{2+}, has a smaller ionic radius than Li+ and a double valence state that results in a higher charge density and therefore in slow diffusion inside the electrodes and high desolvation energy in the electrolytes [322,351].

As a result, the chemistry of the magnesium ion battery is different and more complicated than that of other alkali metals already used for secondary batteries, and the TRL is still low [352]. Uncommon electrolytes are used in MIBs, the two best performing electrolytes are the "All Phenyl Complex" (APC) from Aurbach et al. [353] and the "Magnesium Aluminum Chloride Complex" (MACC) by Doe et al. [354]. Also aqueous electrolytes are used, but, similar to the common organic electrolytes, they tend to passivate pure Mg anodes [355].

The most common cathodes are inorganic oxides, polyanionic, and sulfide compounds. Metal oxides can store Mg^{2+} ions both by intercalation and by conversion [316]. Within the sulfide compounds, the benchmark is the Chevrel-phase Mo_6S_8 [356,357].

For anodes, the principal materials under investigation are metallic magnesium, alloy-based, and carbon-based materials. Metallic magnesium could allow the high energy density that MIBs promise. While alloy-based materials, mainly bismuth, tin, and antimony, are used to avoid the passivation layer formed from Mg metal in the presence of a common electrolyte [358,359]. In Table 12, we summarize some electrochemical properties of MIBs.

Table 12. Electrochemical properties of some MIBs.

Cathode	Anode	Specific Energy [Wh/kg]	Specific Capacity [mAh/g]	Source
$NaV_2O_2(PO_4)_2F$ @rGO	$Mg_{0.79}NaTi_2(PO_4)_3/C$	76	-	[360]
Te @CSs	Mg metal	337	387	[361]
$Mg_{0.75}V_{10}O_{24}\cdot 4H_2O$	PTCDA	67	-	[362]

Dominko et al. [353] calculated the performance between various cathodes, considering as anode either a Mg foil or an alloy (Mg_2Sn) and taking into account many parameters such as conductive additive, binder, current collectors, porosity of electrodes and separator, etc. All combinations based on insertion cathodes have an operating voltage below 2 V, and their performance is lower than that of LIBs (~685 Wh/L and 180 Wh/kg). Organic materials can increase gravimetric density up to 250 Wh/kg, but only the sulfur conversion cathode can have an energy density comparable to LIBs. Wang et al. [361] developed an intercalation cathode based on tetragonal $NaV_2O_2(PO_4)_2F$ reduced graphene oxide (rGO) obtained by a electrochemical desodiated method, that shows an average working voltage of 3.3 V vs. Mg^{2+}/Mg and 30.3 mAh/g at 5 mA/g. When coupled with a $Mg_{0.79}NaTi_2(PO_4)_3/C$ anode material to build a full cell, they obtained a 76 Wh/kg energy density with a power density of 1300 W/kg, with a working voltage between 0.3 and 2.2 V. Chen et al. [362] developed a conversion-type cathode based on Tellurium (Te) encapsulated on carbon spheres (CSs). When coupled with Mg metal anode, the electrode capacity is up to 387 mAh/g with an energy density of 337 Wh/kg. Ma et al. [363] developed a lamellar $Mg_{0.75}V_{10}O_{24}{\cdot}4H_2O$ (denoted as MVOH), which contains abundant lattice H_2O that led to fast Mg^{2+} migration through a complex shuttle mechanism. When coupling this cathode with a perylene-3,4,9,10-tetracarboxylic dianhydride (PTCDA) anode in an aqueous solution electrolyte, they achieved a 67 Wh/kg energy density with a cycle life of 5000 cycles and a maximal power density of 2 kW/kg.

3.10.5. Lithium–Sulfur Batteries

Lithium–sulfur (LiS) batteries represent another promising emerging technology that utilizes a lithium metal anode and a sulfur-based cathode. LiSs potential for exceptionally high energy density, which can be two to five times greater than that of current LIBs, makes them a remarkable candidate for automotive applications [363]. Another major benefit is the potential for lower costs. Sulfur, a primary component of Li-S batteries, is an abundant and inexpensive material compared to the cobalt and nickel used in many lithium-ion batteries. Furthermore, Li-S batteries are considered to be potentially more environmentally friendly due to the abundance of sulfur and the less resource-intensive processes associated with its extraction compared to some of the metals used in traditional batteries [364].

Like conventional lithium-ion batteries, they typically employ a liquid organic electrolyte to facilitate the movement of ions. The generation of electrical energy in an LiS battery occurs through an electrochemical reaction where lithium at the anode is oxidized, releasing lithium ions that then travel through the electrolyte to the sulfur cathode. At the cathode, elemental sulfur undergoes a complex, multi-step reduction process. This process involves the formation of various intermediate compounds known as lithium polysulfides (Li_2Sn), which have different chain lengths (with n ranging from 3 to 8). The discharge process continues until the sulfur is fully reduced to lithium sulfide (Li_2S). During charging, this reaction is reversed, with lithium sulfide being oxidized back to elemental sulfur, and lithium ions returning to the anode [365]. This chemical transformation involves a multi-electron transfer process, resulting in a significantly high theoretical energy density estimated to be between 2500 and 2600 Wh/kg [366]. Intermediate lithium polysulfides are soluble in the electrolyte, allowing them to diffuse between the cathode and anode. This polysulfide shuttle creates performance challenges, leading to a decrease in battery capacity and Coulombic efficiency [367]. Additionally, the cycle stability of LiS batteries is compromised by the poisoning of the lithium metal negative electrode, which hampers their practical application [368]. Other issues include the low conductivity of active materials, significant volume changes during redox cycling, and the formation of dendrites [365].

Table 13 summarizes the performance metrics reported in several recent research articles and company announcements concerning LiS batteries in 2023 and 2024. It highlights the advancements in energy density, capacity density, and voltage achieved through various materials and experimental conditions.

Table 13. Electrochemical properties of some LiS batteries.

Energy Density (Wh/kg)	Capacity Density (mAh/g)	Average Discharge Voltage (V)	Cell Characteristics	Source
395	-	-	9.5 Ah pouch cell	[369]
>550 (pack), up to 700 (cell)	-	-	Low electrolyte-to-sulfur ratio pouch cells	[370]
-	>250 (pouch cell), >3 mAh/cm^2	-	Catholyte, sulfur-free carbon nanofiber cathode	[371]
2500 (theoretical)	1672 (theoretical)	~2.1	Theoretical values	[372]
~542.7 (estimated All-solid-state)	-	~2.0	All-solid-state, 90% sulfur utilization, 60 wt% sulfur in cathode	[373]
2400 (theoretical with Li anode)	1675 (theoretical cathode), 3860 (theoretical anode)	-	Theoretical values	[374]

Analysis of the data reveals a clear trend towards achieving higher energy densities in practical LiS battery cells. These advancements are often associated with specific material choices and cell designs, such as the use of pouch cell formats and optimized electrolyte-to-sulfur ratios. While theoretical values for energy and capacity density remain much higher, the gap is gradually being narrowed by ongoing research. The average discharge voltage for LiS batteries remains consistently around 2.0–2.1 V, as dictated by fundamental chemistry. The table underscores the dynamic nature of the field, with continuous improvements being reported through innovative approaches.

To improve LiS performance in terms of cell cyclability, rate capability, safety, and lifespan, electrolyte additives can be used [375]. Material engineering is actively studied to prolong the lifespan of LiS [365], along with different protection strategies for the anode to avoid dendrite formation [376]. Other research areas include the development of nanostructured sulfur cathodes [377,378], the creation of protective coatings [379], the use of MXene to anchor polysulfides [380], and the advancement of all-solid-state LiS batteries [368].

While lithium–sulfur batteries hold significant promise, their widespread commercialization in electric vehicles is still some years away.

3.10.6. Zinc-Based Batteries

Zinc-based batteries (ZBs) represent a diverse category of electrochemical energy storage systems that utilize metallic zinc as the primary negative electrode material. Zinc is an abundant, inexpensive, and non-toxic material, making it a more sustainable and cost-effective option compared to lithium and other critical battery materials [381].

Unlike lithium-ion batteries, which primarily rely on the intercalation of lithium ions into electrode materials, zinc-based batteries often operate through different mechanisms, such as the dissolution and deposition of zinc metal or oxygen reduction reactions, depending on the specific type of zinc battery. There are several main types of zinc-based batteries being explored for various applications, including zinc-ion batteries (ZIBs), which typically use aqueous electrolytes and involve the reversible shuttling of zinc ions between the zinc anode and a cathode made of materials like manganese oxide or vanadium oxide [382]. Aqueous electrolytes are non-flammable and eliminate the risk of thermal runaway associated with the organic electrolytes used in LIBs.

Another prominent type is zinc–air batteries (ZABs), which are open-system batteries that utilize zinc as the anode and oxygen from the ambient air as the cathode. These batteries generate energy through the reaction of zinc with oxygen and can be recharged either electrically or by physically replacing the spent zinc anode [383,384]. They have a very high theoretical energy density, potentially exceeding that of LIBS.

Despite their promising attributes, zinc-based batteries face several challenges that currently limit their widespread use in electric vehicles. To enable large-scale practical applications, several challenges must be overcome, including limited discharging capacity, low operating voltage, low energy density, short cycle life, and complex energy storage mechanisms, which require methodologies that depend on the specific redox mechanism [385,386]. ZIBs generally have a lower energy density compared to high-performance lithium-ion batteries, which could restrict the driving range of EVs powered solely by ZIBs. A significant technical hurdle for both ZIBs and ZABs is the zinc anode, which is susceptible to corrosion, the formation of dendrites (needle-like zinc structures that can cause short circuits), and hydrogen evolution reactions, all of which can severely compromise the battery's Coulombic efficiency and cycling stability, hindering their broader adoption [382,387]. ZABs have faced challenges related to their rechargeability, often requiring physical replacement of the zinc anode [388], and can suffer from slow discharge rates [389]. For the same energy capacity, ZB packs can also be bulkier compared to lithium-ion packs [381]. Additionally, some ZBs chemistries have a lower nominal voltage compared to lithium-ion batteries [390].

ZBs are already utilized in various niche applications, including hearing aids and some stationary storage systems. Commercialization for electric vehicles, particularly zinc–air batteries, is expected to begin within this decade, potentially targeting smaller vehicle segments like two- and three-wheelers initially [391].

Figure 7 depicts a maturity roadmap of lithium-ion alternative battery chemistries presented in this section.

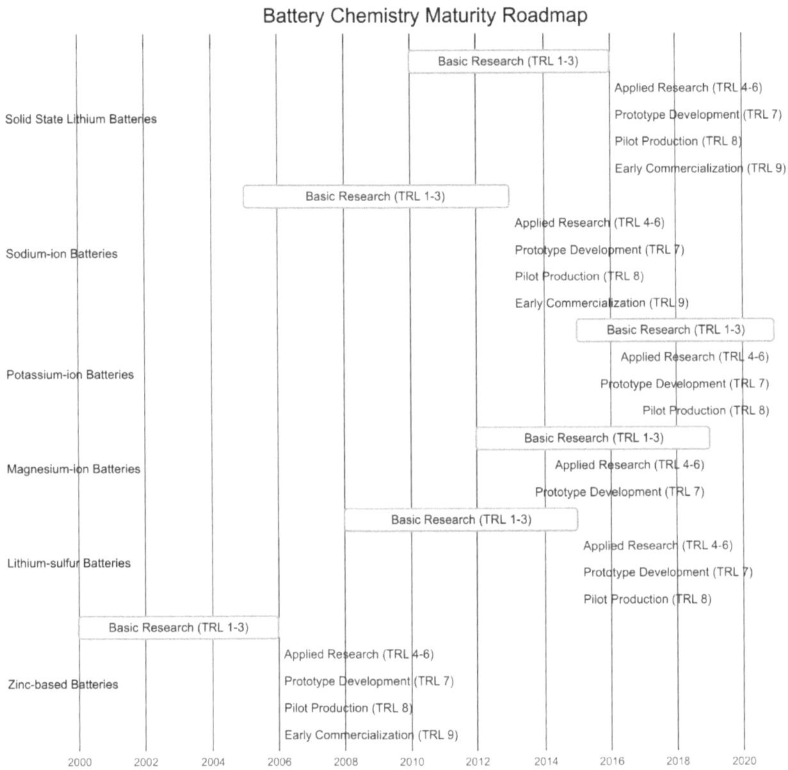

Figure 7. Maturity roadmap for some lithium-ion alternative battery chemistries.

4. The Sustainability Issue

Batteries are one of the key enablers for sustainable development, green mobility, clean energy, and climate neutrality. For the widespread adoption of electric power across industries like transportation, sustainable battery practices are vital. To truly benefit from a low-carbon economy driven by electrification, we must tackle the environmental burdens of battery manufacturing, specifically CO_2 output, resource scarcity, and ethical material procurement. Achieving sustainable consumption and production necessitates a swift transition to well-engineered batteries that are clean, circular, and long-lasting [392]. Accordingly, a harmonized regulatory framework governing the entire lifecycle of batteries placed on the market ought to incorporate the following sustainability stipulations:

- Minimizing the carbon footprint across the entire value and production chain;
- Ensuring ethical and responsible raw material acquisition;
- Fostering a circular design that incorporates recycled content and enables reuse, repurposing, remanufacturing, and final recycling;
- Providing transparent communication and tracking of performance and material/chemical contents for end users and supply chain stakeholders.

From the perspective of EU environmental policy, batteries are considered highly valuable and strategic products. According to the recent Battery Regulation [35], batteries must be designed and manufactured to optimize performance, durability, and safety while minimizing their environmental impact. Specific sustainability requirements have been established for EV batteries, as this market segment is expected to grow significantly in the coming years.

A "battery passport" is mandated by the Regulation to increase transparency throughout battery supply and value chains for all stakeholders. This passport will monitor and trace batteries and disclose vital information such as the carbon footprint of their production, the origin of their components, potential for repair and repurpose, and dismantling instructions. It will also detail the treatment, recycling, and recovery pathways for batteries at the end of their usable life.

The battery passport also presents challenges that could lead to drawbacks if left unmitigated [393,394]. The regulation requires actors to collect, integrate, and certify extensive data throughout the battery lifecycle. This requires transparent and efficient communication between actors to ensure everybody along the supply chain can provide the information needed. Passport issuers are expected to face mainly technical and system challenges, thus requiring industry collaboration, investment in emerging technologies, and support from authorities to ensure standard enforcement. Conversely, small- and medium-sized enterprises are anticipated to encounter primarily capability and resource challenges, necessitating early internal alignment, standardized requirements, and financial aid.

The involvement of numerous supply chain actors makes it difficult to access data on aspects like the metallic composition of batteries and the specifics of their manufacturing processes. Moreover, determining the state of health of batteries can also be complicated. Adding to these challenges, companies are unwilling to disclose data unless they have assurances through non-disclosure agreements. Although these unresolved issues might lessen the overall effectiveness of the passport, its benefits are still expected to be greater than its drawbacks.

4.1. Life Cycle Emissions

Encompassing all greenhouse gas (GHG) emissions produced across the supply chain's diverse phases, the life cycle emissions of batteries originate with the mining and processing of ores. This initial step yields precursor materials crucial for creating the battery's active components: cathode and anode materials. Subsequently, these active materials, along with other elements like the electrolyte, separator, current collector, and casing, are utilized in battery cell production. Ultimately, individual cells are assembled into battery modules or packs, which also incorporate vital components such as battery management systems, electronics, and cooling systems. Notably, the carbon footprint of batteries (CFB) stands as one of the sustainability metrics that must be calculated and communicated, as detailed in Article 7 of the EU Regulation [35].

Quantifying the overall greenhouse gas impact of a battery over its anticipated service life, the CFB is expressed in g CO_2 equivalent per kWh of total energy output. The Regulation mandates CFB declaration for specific battery types placed on the Union market: rechargeable industrial batteries with a capacity over 2 kWh, light means of transport (LMT) batteries, and EV batteries. Subsequently, these CFB declarations will be utilized to establish CFB performance classes and the CFB thresholds that batteries must adhere to when entering the European market. A CFB model is constructed from processes that combine elementary flows with their associated characterization factors, along with activity data connected to the respective life cycle inventory or carbon footprint of the process. Notably, both activity data and elementary flows can originate from processes and their constituent sub-processes, with information potentially being company-specific or derived from secondary sources like databases.

Figure 8 represents the system boundaries of the carbon footprint of a generic EV battery. Each square represents a process, while each arrow represents activity data. The different colors indicate in which life cycle stage each process belongs, while red arrows and red borders indicate if a process/activity data shall be company-specific.

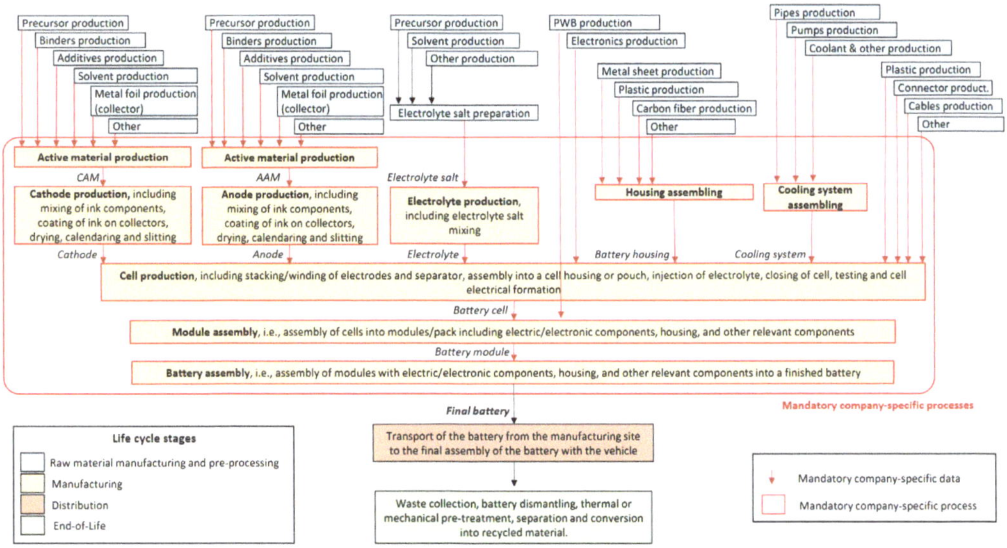

Figure 8. System boundaries of the carbon footprint of a generic EV battery. PWB: printed wiring. CAM: cathode active material, AAM: anode active material. Reproduced from [395] under CC BY 4.0.

Resulting from extensive discussions and representing a consensus among numerous stakeholders on how to calculate the carbon footprint of EV batteries, the CFB rules aim to be broadly applicable. Although methodological guidelines for determining the CFB of EVBs have been proposed, aligning with the Product Environmental Footprint Categories Rules for Batteries (PEFCR) [395], these guidelines also maintain a general approach. This generality would enable the application of the CFB rules not only to batteries currently available but also to those potentially entering the market in the near future.

According to the PEFCR for batteries, the benchmark value for climate change (measured in kg CO_2eq) for e-mobility lithium-ion batteries (excluding the use phase) is 0.42 kg CO_2eq/kWh [396].

Life cycle emissions from batteries are significantly determined by their chemical composition. In the case of NMC batteries, the prevalent chemistry in the US and EU, the processing of critical minerals is the largest emission contributor, accounting for 55% of the total. This contrasts with LFP batteries, the primary type in China, where critical mineral processing represents about 35% of emissions. This difference results in NMC batteries having greater overall life cycle emissions than LFP batteries, primarily due to the higher emissions from critical mineral processing. A key factor in the elevated emissions of NMC batteries is the use of nickel, an element absent in LFP batteries. Notably, half of all nickel is currently mined in Indonesia, and about two-thirds undergoes refining in Indonesia and China, utilizing energy-intensive methods that predominantly rely on coal.

Given the diversity of battery chemistries, effective emission reduction in the supply chain necessitates targeted strategies. For NMC batteries, the most impactful approach would be to minimize emissions from critical mineral processing. This could involve several actions, such as sourcing less carbon-intensive nickel, improving the carbon efficiency of existing nickel processing, increasing the proportion of recycled materials in batteries, or even reducing the nickel content in NMC batteries while preserving high energy density through the adoption of high-manganese NMC chemistries.

While LFP battery cell manufacturing currently contributes the most to their lifecycle emissions—nearly 50% compared to just 15% for NMC—future trends suggest a decrease in overall battery emissions. This anticipated reduction stems from progress in areas like increased battery pack energy density, the ongoing decarbonization of electricity grids, and a greater supply of cathode active materials derived from recycling. Notably, strategies for reducing LFP battery emissions should therefore focus on lowering the carbon footprint of cell production through enhanced electrification and the use of low-emission power. Additionally, the steam utilized in many cell manufacturing plants presents an opportunity for emissions reduction by sourcing it from low-carbon alternatives, such as waste heat from nuclear or other industrial facilities, or renewable fuels like biogenic waste.

Furthermore, as the volume of end-of-life EV batteries suitable for recycling increases after 2035, recycling is expected to become an even more crucial factor in decreasing battery emissions, offering advantages for material-intensive chemistries such as NMC.

4.2. Recycling

The increasing demand for batteries is driving a heightened need for critical raw materials. Specifically, the global need for raw materials used in batteries—such as nickel, graphite, and lithium—is projected to rise significantly by 2040, with anticipated increases of 20 times for nickel, 19 times for graphite, and 14 times for lithium, compared to 2020 levels, as shown in Figure 9. This rising demand is primarily due to the rapid growth of EVs and, to a lesser extent, energy storage applications. Therefore, as the demand for batteries grows, so does the need for effective battery recycling to ensure a sustainable and competitive industry. In this regard, the battery passport introduced by the EU Regulation [35] constitutes a key instrument.

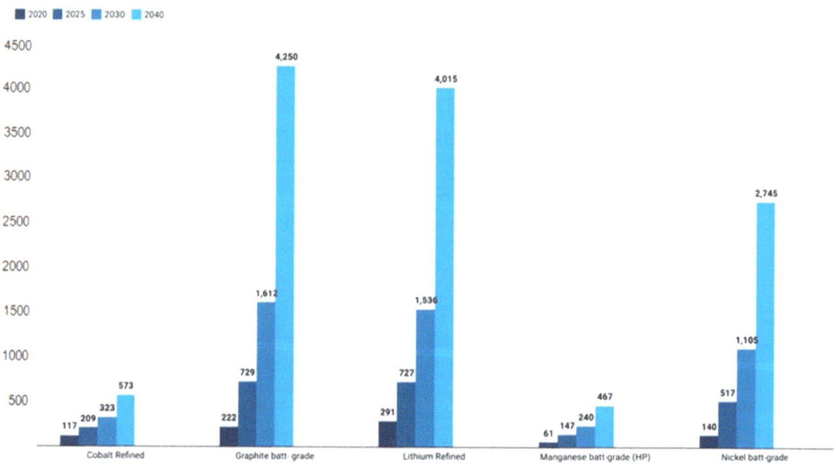

Figure 9. Forecast of battery demand globally from processed raw materials (kt) [397].

Globally, according to a recent study of the Battery Pass consortium [398], the recycling pre-processing and treatment costs could be reduced by ~10–20%, as a result of the combination of the following factors:

- Access to detailed battery composition and chemistry information eliminates costly sampling procedures, allowing for more efficient and lower-cost sorting while minimizing contamination risks. Sampling costs could potentially decrease by 50% to 80%.
- A detailed dismantling manual can reduce disassembly time and costs of battery packs by 20–40%.

- Additionally, dismantling manuals can facilitate the automation of parts of the dismantling process, particularly for heavy and hazardous operations, resulting in a further 20–30% reduction in dismantling costs.
- Optimizing the recycling treatment process and potentially reducing material and processing costs by 10–20% could be achieved through a homogenous battery recycling feedstock. This consistent input, pre-processed to eliminate unwanted materials, would streamline the feed-in process (batch sequencing) and allow for finer control over process parameters.

Improved recycling efficiency offers a dual advantage: not only does it reduce costs associated with pre-processing and treatment, but it also enables the retrieval of additional active materials, thereby lowering carbon emissions.

In particular, the production of cathodes is identified as a significant contributor to the overall environmental impact of LIBs [399].

The proportion of mined lithium allocated for EV batteries has quadrupled in the past five years, reaching 60% in 2022. That same year, demand for lithium exceeded supply, even with a remarkable 180% increase in production since 2017. Similarly, the shares of cobalt and nickel used for electric vehicle batteries have also risen in the same period, increasing from 10% and 2% to 30% and 10%, respectively [47].

Essential for the clean energy transition, both cobalt and lithium are unfortunately vulnerable to supply shortages and present considerable risks to their supply, although these risks are rooted in somewhat different circumstances. Figure 10 offers a geographical perspective on the reserves and supply chain of both cobalt and lithium, encompassing all stages [400].

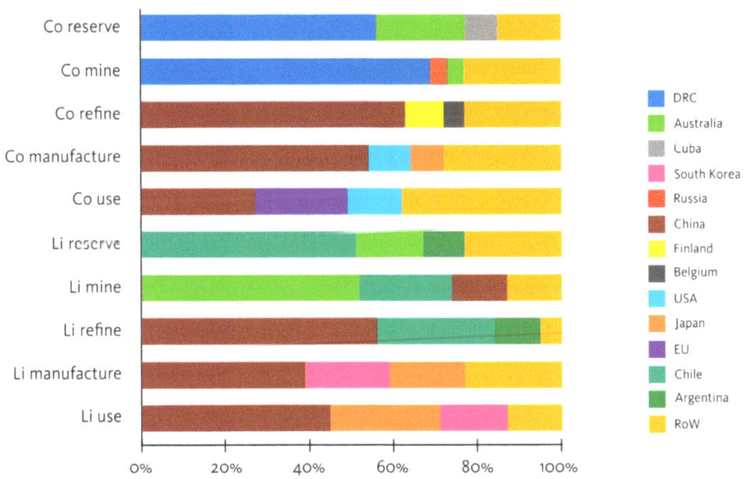

Figure 10. An overview of the geographical location of cobalt and lithium reserves and supply.

While secondary supply through recycling offers an alternative source of materials, its impact on overall supply is expected to be quite small by 2030. Looking ahead, the recycling of LIBs, particularly from electric vehicles, will be crucial for unlocking substantial volumes of recycled cobalt and lithium. Figure 11 presents the projected annual production capacity for lithium and cobalt, factoring in this secondary supply stream.

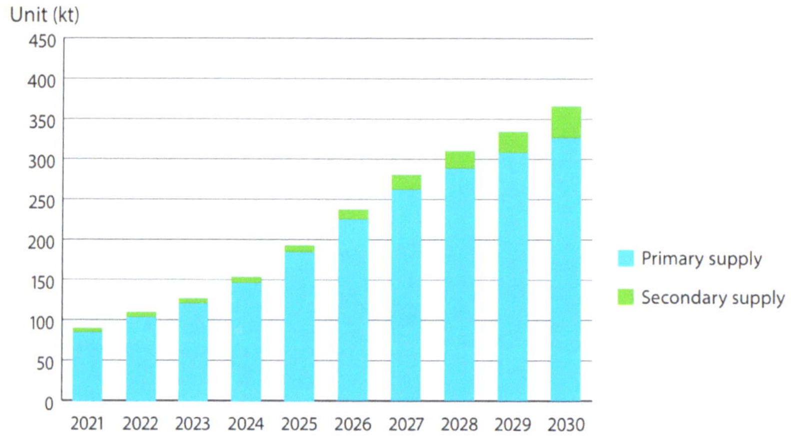

(a)

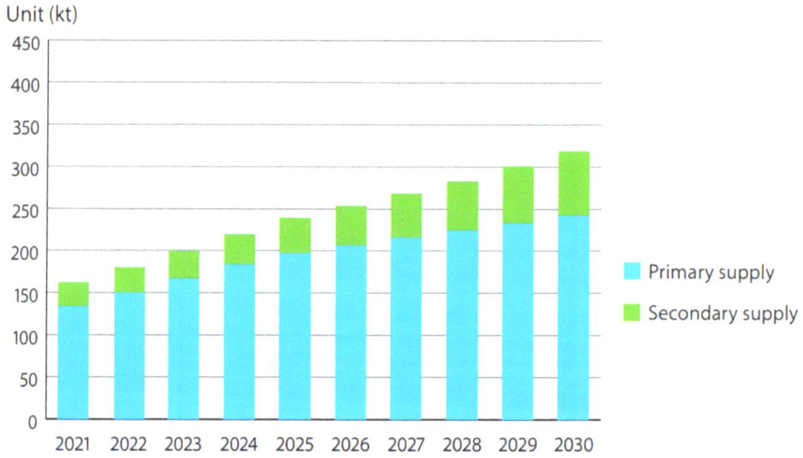

(b)

Figure 11. Annual lithium (**a**) and cobalt (**b**) supply growth by 2030. Source: Benchmark Mineral Intelligence (https://www.benchmarkminerals.com/ accessed on 31 March 2025).

Given the complexity of the battery ecosystem, to analyze the actual economic and environmental impacts derived from the implementation of the battery passport digital framework requires advanced simulation tools, for instance, based on Agent-Based Modeling. The latter is a relatively contemporary approach when compared to System Dynamics and Discrete Event Modeling. By means of these advanced tools, the entire lifecycle of LIBs, including end-of-life strategies such as reuse, remanufacturing, and repurposing, can be explored and evaluated.

As a result of the introduction of the battery passport, it is estimated that European recyclers could recover between ~4 and 8 kilotons of additional cathode active materials each year, starting in 2045 [398].

Compared to other battery technologies, such as lead–acid or sodium–nickel chloride batteries, LIBs generally have a lower environmental impact over their life cycle, especially for their superior performance in practical applications [401,402] although the results depend on the life-cycle assessment method applied [403,404].

Among LIBs, LFP batteries stand out because they use iron and phosphorus instead of the nickel, manganese, and cobalt found in NCA and NMC batteries. This shift also offers both cost and long life, although the overall environmental impact could be higher than for other LIB technologies [405]. Additionally, the presence of phosphorus in LFPs, which is also used in food production, could lead to competing demands for this resource as battery demand continues to rise.

Adopting batteries that require fewer critical minerals may be beneficial to mitigate the demand for critical raw materials and enhance sustainability, resilience, and supply chain security. Additionally, supporting the development of vehicle designs with optimized battery sizes and battery recycling practices could help address these challenges [22,406]. Therefore, there is a rapidly growing focus on the necessary technological and knowledge advancements in the field of EV battery recycling. The total recycling amount of copper, lithium, nickel, and cobalt by 2040 could decrease the demand for the extraction of mentioned minerals from their primary resources by approximately 10% [61]. Moreover, enhancing circularity along the battery value chains has the potential to decrease the EU's supply dependency. It is estimated that by 2040, recycling could contribute to up to 51% and 42% of cobalt and nickel EU demand, respectively [397].

Since the volume of EOL batteries, in particular LIBs, is projected to increase dramatically, robust recycling solutions are needed. China is at the forefront of global battery recycling, holding approximately 80% of pretreatment capacity and nearly 85% of material recovery. By 2030, if all planned initiatives are realized, North America and Europe are expected to contribute 10% and 5% to global recycling capacity, respectively. South Korea aims for a recovery capacity of nearly 600 kilotons, which would account for over 5% of global material recovery.

The trend of reusing EOL batteries significantly helps reduce their environmental impact [61]. This process involves disassembling and repurposing spent EV batteries, which still retain 80–85% of their original energy capacity, for use in renewable energy technologies. After the reuse phase, these batteries can ultimately be recycled.

There are various types and combinations of recycling processes for LIBs. Currently, hydro- and pyrometallurgical processes dominate the global industrial landscape for recovering valuable metals [407]. Hydro-metallurgical processes are complex and involve high costs for chemicals and wastewater treatment, although they achieve high recovery efficiency. On the other hand, pyrometallurgical processes have disadvantages, such as high-energy consumption and toxic gas emissions, but they are the most used methods. Novel technologies are being developed, such as the direct repair and regeneration of cathode and functional materials. However, their actual costs and environmental impacts still require thorough evaluation.

5. Conclusions

While LIBs have improved energy density, the driving range of current EVs remains a concern for consumers. Many potential buyers experience "range anxiety", fearing their vehicle will run out of charge before reaching a destination, especially during long trips or in areas with limited charging infrastructure, and recharging an EV battery still takes significantly longer than refueling a gasoline car. The maximum theoretical energy density of conventional LIBs using graphite anodes is nearing its limits, indicating the need for alternative technologies for further range improvements.

The high cost of LIBs contributes significantly to the price of electric vehicles. This is primarily due to expensive materials like cobalt and nickel, as well as the complex manufacturing processes involved. Additionally, conventional LIBs use flammable liquid electrolytes, posing risks of thermal runaway and associated safety concerns. These issues necessitate complex safety systems that increase the overall cost and weight of EVs.

Moreover, the production of LIBs relies on critical materials that are geographically concentrated, raising ethical and environmental concerns. This creates vulnerabilities in the supply chain, potentially leading to price fluctuations and geopolitical risks. Therefore, there is a pressing need to diversify battery chemistries and reduce reliance on these essential materials. For emerging technologies to gain traction in the market, they must offer clear and significant advantages in key areas such as cost efficiency, improved safety, and superior performance.

Transitioning from an established technology like lithium-ion requires substantial investment in new manufacturing processes and the development of robust supply chains for new materials. Therefore, any emerging battery technology aiming to capture a significant share of the automotive market must provide a compelling value proposition that surpasses the existing advantages and infrastructure associated with lithium-ion batteries.

Based on the current state of development and the projected timelines from various companies and research institutions, the future of electric vehicle (EV) batteries is likely to feature a diverse range of technologies. Lithium iron phosphate (LFP) batteries, a cobalt-free variant of lithium-ion batteries, are gaining significant traction due to their cost-effectiveness and safety benefits.

In the near to medium term, silicon anode technology is expected to be integrated into existing lithium-ion batteries, which could incrementally improve their energy density and charging speeds. However, addressing the challenge of silicon's volume expansion during charging and discharging cycles is crucial.

Solid-state batteries hold significant promises for improved energy density, safety, and charging speeds in the long term, particularly for high-performance EVs. The potential lower cost, improved sustainability, and lower lifespan compared to LIBs can help the EU achieve its EU2030 targets [408]. Nonetheless, further breakthroughs in cost-effective manufacturing and solutions to technical challenges such as scalability and interfacial resistance are necessary. While LIBs will likely dominate the EV market in 2030, SSBs are expected to move beyond the pilot and demonstration phases. Market forecasts suggest that the solid-state battery market could reach a significant value by 2030 [409]. If European companies can become leaders in SSB technology development and production, it could significantly boost the EU's competitiveness in the global battery market and contribute to job creation within the EU.

Sodium-ion batteries are also emerging as a vital option for the lower-cost and standard-range EV segments, especially as their energy density improves. Additionally, these batteries utilize abundant and sustainable materials, which helps reduce the European Union's dependence on critical raw materials with concentrated supply chains. This aligns well with the goals of the Critical Raw Materials Act (CRMA) and promotes a secure and resilient battery value chain. The potentially lower cost and enhanced safety can boost the adoption of EVs. The commercialization of SIB technology is still in its early stages compared to LIBs. However, given the increasing focus on supply chain diversification and cost reduction, SIBs are expected to gain a notable foothold in the EU market by 2030 [410]. The development of a strong domestic SIB manufacturing industry within Europe will be key to realizing the full potential of this technology [411].

Table 14 reports a short comparison of relevant parameters for emerging technologies.

The overall outlook for the adoption of these emerging technologies in the automotive sector is positive, though it requires ongoing investment in research and development. This investment should focus on addressing the key limitations of each technology to accelerate their commercialization.

Table 14. Comparison among emerging battery technologies.

Technology	Energy Density (Wh/kg)	Energy Density (Wh/L)	Power Density (W/kg)	Charging Speed (Typical)	Cycle Life (Typical)	Safety Characteristics	Estimated Cost (per kWh)	Estimated Commercialization Timeline for Automotive
Solid-State	200–500+	400–800+	Up to 1000+	10–30 min	1000–5000+	Non-flammable solid electrolyte, reduced thermal runaway	Higher	Late 2020s–Mid 2030s
Silicon Anode (Li-ion)	250–400+	500–800+	Up to 1000+	10–30 min	300–1000+	Similar to Li-ion	Similar	Late 2020s onwards
Sodium-Ion	100–160	200–300	Up to 500	30–60 min	2000–5000+	Lower thermal runaway risk, better low-temp perf.	Lower	2025 onwards
Lithium-Sulfur	300–600+	300–500+	Up to 500	Varies	100–500+	Lithium metal anode poses dendrite risk	Lower	2030 and beyond

Government funding for battery technology development has increased across all major global economies, though their strategic approaches and specific goals differ significantly [412]. China initially focused on its domestic EV market demand, but has shifted towards a supply side strategy to solidify its global market leadership. Japan, a former battery technology leader, is now expanding lithium-ion production while exploring advanced technologies like fluorine shuttle batteries. South Korea aims to spearhead the international battery industry through substantial R&D investment focused on near-term commercialization.

The United States employs a balanced strategy combining supply- and demand-side measures with an open technological approach, aiming for R&D leadership and independence from China by targeting cost and sustainability performance. In contrast, the EU prioritizes supply side policies with ambitious sustainability and recycling targets to become a leading provider of sustainable battery technologies and ensure supply chain resilience. Initiatives like the European Battery Alliance and the European Green Deal aim to establish a competitive and sustainable European battery ecosystem. The Automotive Skills Alliance supports the reskilling of European automotive workers for the sector's future needs.

Developing a robust European battery supply chain is crucial for reducing reliance on external suppliers and securing supply for future mobility and energy demands. Addressing the sustainable, ethical, and diversified sourcing of raw materials is critical due to increasing global demand and competition, where Europe faces strong Asian competition and higher production costs. Europe's nascent battery production sector offers a competitive edge, allowing for the immediate adoption of advanced technologies to build a state-of-the-art supply chain without extensive industrial retrofitting.

Europe could achieve battery cell self-sufficiency by 2026 and largely meet its demand for key components and lithium by 2030. However, over half of European gigafactory plans still face delay or cancellation, necessitating stronger government intervention. Key industrial policy should ensure investment, provide EU-level funding, and favor best-in-class "made in EU" projects. Finland, the UK, Norway, and Spain have the most at-risk capacity, while France, Germany, and Hungary have shown the most progress in securing it [413].

Maintaining competitiveness requires emphasizing efficiency and innovation, alongside training a skilled workforce across the battery value chain, including material chemistry, production, recycling, and data management. Continuous investment in R&D is essential for the rapid integration of innovations into European production processes.

Establishing common European standards is vital to ensure battery quality, safety, and sustainability, while also facilitating recycling, reuse, and repair processes to promote a circular economy. In particular, the battery passport, recently introduced by EU Regulation, supports the collection and sharing of product-related data among supply chain actors, addressing existing information gaps for products and components throughout global supply chains, thus becoming a key enabler for circular business models. Europe's leadership in setting high environmental and social standards can provide a competitive advantage in developing more sustainable batteries and recycling methods.

Funding: This research received no external funding.

Conflicts of Interest: The authors declare no conflict of interest.

Abbreviations

The following abbreviations are used in this manuscript:

3D-NTC	Three-dimensional nitrogen-doped turbostratic carbon
AFNB	Anode free sodium batteries
APC	All phenyl complex
BEV	Battery electric vehicle
BMS	Battery management system
BOL	Begin of life
BTMS	Battery thermal management system
CA	Cyanoacrylate
CS	Carbon sphere
CSE	Composite solid electrolytes
EOL	End of life
EUCAR	European Council for Automotive R&D
EV	Electric vehicle
FEC	Fluoroethylene carbonate
HC	Hard carbon
HCF	Hexacyanoferrates
HE-NMC	High-energy lithium nickel manganese cobalt oxide
HLM	High lithium manganese oxide
HTS	High temperature thermal shock
HV-LNMO	High-voltage lithium nickel manganese oxide
ISE	Inorganic solid electrolyte
KMgHCF	Potassium magnesium hexacyanoferrates
K2TP	Dipotassium terephthalate
LCO	Lithium cobalt oxides
Li+	Lithium ion
LIB	Lithium-ion battery
LiS	Lithium–sulfur
LMO	Lithium manganese oxide
LFP	Lithium iron phosphate
LLZO	$Li_7La_3Zr_2O_{12}$
LTO	Lithium titanate oxide
MACC	Magnesium Aluminum Chloride Complex
MIB	Magnesium ion battery
MF	Muffle furnace-sintered
MOF	Metal–organic framework
MVOH	$Mg_{0.75}V_{10}O_{24}\cdot 4H_2O$

NaNiCl	Sodium–nickel chloride
NaS	Sodium–sulfur
NASICON	Sodium superionic conductors
NaxMnFe(CN)6	Sodium manganese hexacyanoferrate
NCA	Lithium nickel cobalt aluminum oxide
Ni-Cd	Nickel–cadmium
Ni–Fe	Nickel–iron
Ni-H2	Nickel–hydrogen
Ni–MH	Nickel–metal hydride
Ni–Zn	Nickel–zinc
NMB	Sodium metal batteries
NMC	Lithium nickel manganese cobalt oxide
NMO	Sodium manganese oxide
NVP	$Na_3V_2(PO_4)_3$
OEM	Original equipment manufacturer
OLE	Organic liquid electrolyte
PAN	Polyacrylonitrile
PBA	Prussian blue analog
PEGMEA	Poly-ethylene glycol methyl ether acrylate
PEO	Poly-ethylene oxide
PHEV	Plug-in hybrid electric vehicle
PIB	Potassium ion battery
PTCDA	Perylenetetracarboxylic dianhydride
PMMA	Poly-methyl methacrylate
PVDF	Poly-vinylidene fluoride
PVDF-HFP	Poly-vinylidene fluoride-hexafluoropropylene
rGO	Reduced graphene oxide
SC	Single crystal
SE	Solid electrolyte
SEI	Solid electrolyte interphase
SN	Succinonitrile
SHE	Standard hydrogen electrode
SIB	Sodium-ion battery
SOC	State of charge
SOH	State of health
SPE	solid polymer electrolyte
SSB	Solid state battery
SSE	Solid state electrolyte
TiO_2	Titanium dioxide
tLi+	Lithium transference number
TM	Transition metal
TMC	Transition metal chalcogenides
TMO	Transition metal oxide
TRL	Technological readiness level
USABC	United States Advanced Battery Consortium
ZB	Zinc-based battery

Appendix A

This appendix details the review searching criteria and analyzes the results.

Appendix A.1. Review Methods

Searching on Scopus with the string "review AND battery AND (automotive OR "electric vehicle" OR "electric vehicles")" in the fields of "article, title, and keywords" yields 3730 documents published between 2019 and 2025 (as of 12 February 2025). The distribution of the publications year by year is shown in Figure A1. In contrast, using the same search criteria in Google Scholar returns 18,600 papers when searching "everywhere in the article". However, limiting the search to the title results in only 176 records. Given that the search options in Google Scholar are either too broad or too narrow for our needs, we will focus on the results from Scopus. We refined our search to focus on the specific topics relevant to our current work, adding new criteria to our previous search string (see Table A1 and Figure A2).

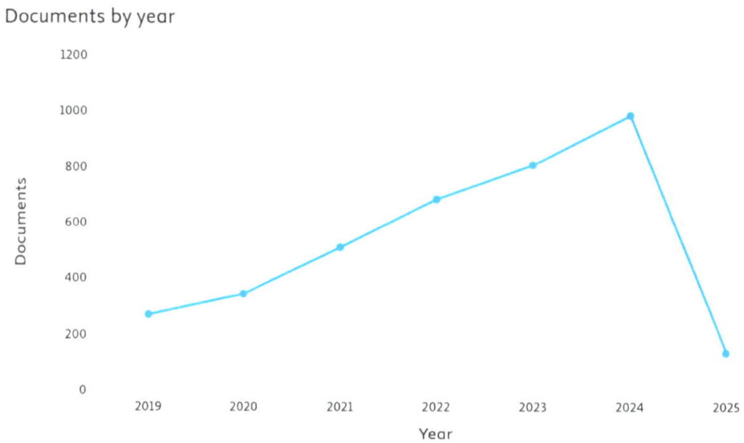

Figure A1. Articles on EV batteries published between 2019 and February 2025 and registered in Scopus.

Table A1. Literature search results with additional criteria.

Additional Criteria	Safety	Life OR Lifespan OR Cyclelife	Cost OR Market	Energy OR Power	Sustainability
No. of articles from 2019	883	984	1331	3039	256
No. of articles from 2023	487	536	672	1550	178

Appendix A.2. Search Results

From Figure A1, it can be concluded that most of the research reviews concentrated on energy and power performance. The second most common topic was battery cost and market aspects, followed by battery life performance and safety. Sustainability is the least explored aspect.

By limiting the search from 2023 up to today, we found 1924 reviews related to battery and automotive topics on Scopus. The trend of topic interests, illustrated in Figure A2, remains mostly the same over the two periods.

Among the results from Scopus, 156 articles discuss the "state of the art", while 269 records meet the search criteria "review AND battery AND automotive AND (emerging OR 'future development')". When we combined the two criteria, we retrieved a total of 402 documents. The percentage distribution of subject areas and the number of articles per country are illustrated in Figures A3 and A4, respectively.

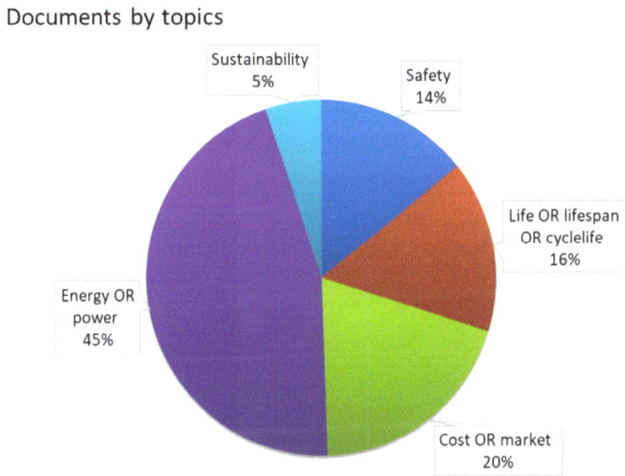

Figure A2. Percentage distribution of articles by topics since 2023.

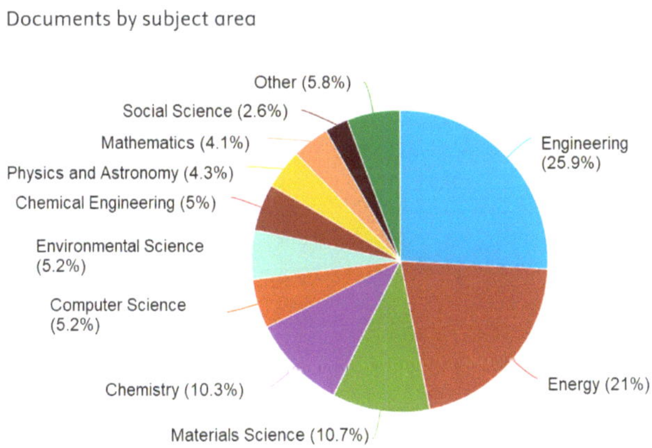

Figure A3. Percentage distribution of article subject areas in the last year (until February 2025).

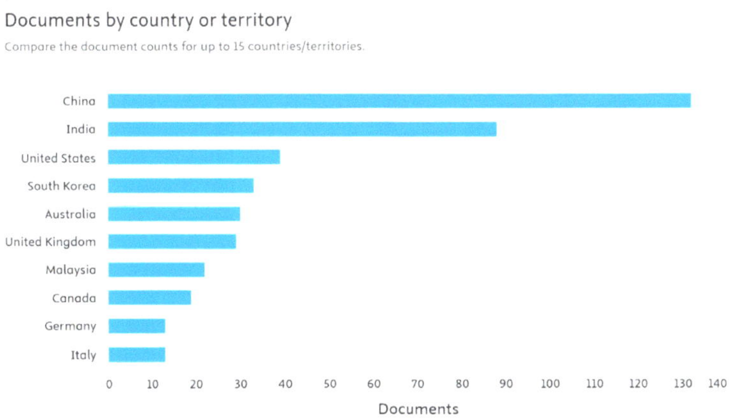

Figure A4. Distribution of articles among countries in the last year (until February 2025).

Most review articles are categorized under engineering and energy, reflecting the focus on automotive applications (Figure A3). The authors of these studies predominantly come from China and India (Figure A4), while the contributions from all European countries exceed those from the United States.

References

1. European Environment Agency. *Transport and Environment Report 2022: Digitalisation in the Mobility System: Challenges and Opportunities*; Publications Office: Luxembourg, 2022.
2. Bielewski, M.A.P.; Bobba, S.; Kronberga, A.; Georgakaki, A.; Letout, S.; Kuokkanen, A.; Mountraki, A.; Ince, E.; Shtjefni, D.; Joanny, G.; et al. *Clean Energy Technology Observatory: Batteries for Energy Storage in the European Union—2022 Status Report on Technology Development, Trends, Value Chains and Markets*; Joint Research Centre: Luxemburg, 2022.
3. Sanguesa, J.A.; Torres-Sanz, V.; Garrido, P.; Martinez, F.J.; Marquez-Barja, J.M. A Review on Electric Vehicles: Technologies and Challenges. *Smart Cities* **2021**, *4*, 372–404. [CrossRef]
4. *International Energy Agency World Energy Outlook Special Report Batteries and Secure Energy Transitions*; International Energy Agency (IEA): Paris, France, 2024.
5. Bielewski, M.; Pfrang, A.; Quintero Pulido, D.; Bobba, S.; Schade, B.; Georgakaki, A.; Letout, S.; Mountraki, A.; Ince, E. *Clean Energy Technology Observatory Battery Technology-KJ0124064ENN*; Joint Research Centre: Luxemburg, 2024.
6. Fleischmann, J.; Hanicke, M.; Horetsky, E.; Ibrahim, D.; Jautelat, S.; Linder, M.; Schaufuss, P.; Torscht, L.; van de Rijt, A. *Battery 2030: Resilient, Sustainable, and Circular*; McKinsey & Company: New York, NY, USA, 2023.
7. IEA. *EV Battery Supply Chain Sustainability*; International Energy Agency (IEA): Paris, France, 2024.
8. Bloomberg NEF. Lithium-Ion Battery Pack Prices See Largest Drop Since 2017, Falling to $115 per Kilowatt-Hour: Bloomberg NEF. Available online: https://about.bnef.com/blog/lithium-ion-battery-pack-prices-see-largest-drop-since-2017-falling-to-115-per-kilowatt-hour-bloombergnef/ (accessed on 3 March 2025).
9. Orangi, S.; Manjong, N.; Clos, D.P.; Usai, L.; Burheim, O.S.; Strømman, A.H. Historical and Prospective Lithium-Ion Battery Cost Trajectories from a Bottom-up Production Modeling Perspective. *J. Energy Storage* **2024**, *76*, 109800. [CrossRef]
10. Houache, M.; Yim, C.-H.; Karkar, Z.; Abu-Lebdeh, Y. On the Current and Future Outlook of Battery Chemistries for Electric Vehicles—Mini Review. *Batteries* **2022**, *8*, 70. [CrossRef]
11. Paul Lienert For EV Batteries, Lithium Iron Phosphate Narrows the Gap with Nickel, Cobalt. Available online: https://www.reuters.com/business/autos-transportation/ev-batteries-lithium-iron-phosphate-narrows-gap-with-nickel-cobalt-2023-06-22/ (accessed on 5 February 2025).
12. Timo Möller, C.C.; Azevedo, M.; Campagnol, N.; Kinzel, Y. *The Battery Chemistries Powering the Future of Electric Vehicles*; McKinsey & Company: New York, NY, USA, 2024.
13. Lu, Y.; Zhu, T. Status and Prospects of Lithium Iron Phosphate Manufacturing in the Lithium Battery Industry. *MRS Commun.* **2024**, *14*, 888–899. [CrossRef]
14. Xiao, J.; Shi, F.; Glossmann, T.; Burnett, C.; Liu, Z. From Laboratory Innovations to Materials Manufacturing for Lithium-Based Batteries. *Nat. Energy* **2023**, *8*, 329–339. [CrossRef]
15. Ghasemi Yeklangi, A.; Ghafari, A.; Asgari Sima, F.; Akbari, S. Advancing Lithium-Ion Battery Manufacturing: Novel Technologies and Emerging Trends. *J. Appl. Electrochem.* **2024**, *54*, 2653–2682. [CrossRef]
16. Fonseca, N.; Thummalapalli, S.V.; Jambhulkar, S.; Ravichandran, D.; Zhu, Y.; Patil, D.; Thippanna, V.; Ramanathan, A.; Xu, W.; Guo, S.; et al. 3D Printing-Enabled Design and Manufacturing Strategies for Batteries: A Review. *Small* **2023**, *19*, e2302718. [CrossRef]
17. Maksimovna Vakhrusheva, D.; Xu, J. Model-Driven Manufacturing of High-Energy-Density Batteries: A Review. *Batter. Supercaps* **2024**, *8*, e202400539. [CrossRef]
18. Gutwald, B.; Baumann, N.; Funk, F.; Reichenstein, T.; Albayrak, B.; Franke, J. Sustainable Manufacturing Practices: A Systematic Analysis and Guideline for Assessing the Industrial Product Carbon Footprint. In Proceedings of the 2024 1st International Conference on Production Technologies and Systems for E-Mobility, EPTS 2024, Bamberg, Germany, 5–6 June 2024.
19. Wang, Y.; Kang, X.; Chen, Z. A Survey of Digital Twin Techniques in Smart Manufacturing and Management of Energy Applications. *Green Energy Intell. Transp.* **2022**, *1*, 100014. [CrossRef]
20. Ralls, A.M.; Leong, K.; Clayton, J.; Fuelling, P.; Mercer, C.; Navarro, V.; Menezes, P.L. The Role of Lithium-Ion Batteries in the Growing Trend of Electric Vehicles. *Materials* **2023**, *16*, 6063. [CrossRef]
21. Knehr, K.; Kubal, J.; Ahmed, S. Estimated Cost of EV Batteries 2019–2024. Chemical Sciences and Engineering division of Argonne National Laboratory. 2024. Available online: https://www.anl.gov/cse/reference/estimated-cost-of-ev-batteries (accessed on 31 March 2025).

22. Hertel, D.; Bräunig, G.; Thürer, M. Towards a Green Electromobility Transition: A Systematic Review of the State of the Art on Electric Vehicle Battery Systems Disassembly. *J. Manuf. Syst.* **2024**, *74*, 387–396. [CrossRef]
23. Kaarlela, T.; Villagrossi, E.; Rastegarpanah, A.; San-Miguel-Tello, A.; Pitkäaho, T. Robotised Disassembly of Electric Vehicle Batteries: A Systematic Literature Review. *J. Manuf. Syst.* **2024**, *74*, 901–921. [CrossRef]
24. Li, W.; Peng, Y.; Zhu, Y.; Pham, D.T.; Nee, A.Y.C.; Ong, S.K. End-of-Life Electric Vehicle Battery Disassembly Enabled by Intelligent and Human-Robot Collaboration Technologies: A Review. *Robot. Comput. -Integr. Manuf.* **2024**, *89*, 102758. [CrossRef]
25. Neri, A.; Butturi, M.A.; Gamberini, R. Sustainable Management of Electric Vehicle Battery Remanufacturing: A Systematic Literature Review and Future Directions. *J. Manuf. Syst.* **2024**, *77*, 859–874. [CrossRef]
26. Qi, L.; Wang, Y.; Kong, L.; Yi, M.; Song, J.; Hao, D.; Zhou, X.; Zhang, Z.; Yan, J. Manufacturing Processes and Recycling Technology of Automotive Lithium-Ion Battery: A Review. *J. Energy Storage* **2023**, *67*, 107533. [CrossRef]
27. Siddiqi, S. Sodium-Ion Batteries 2024–2034: Technology, Players, Markets, and Forecasts. IDTechEX. 2025. Available online: https://www.idtechex.com/en/research-report/sodium-ion-batteries/978 (accessed on 27 February 2025).
28. Sanchez, J. BYD Breaks Ground on World's Largest Sodium-Ion Battery Plant. Available online: https://www.drivencarguide.co.nz/news/byd-breaks-ground-on-worlds-largest-sodium-ion-battery-plant/ (accessed on 26 February 2025).
29. Farasis The World's First EV Powered by Farasis Energy's Sodium-Ion Batteries Rolls off the Assembly Line. Available online: https://www.farasis-energy.com/en/the-worlds-first-ev-powered-by-farasis-energys-sodium-ion-batteries-rolls-off-the-assembly-line/ (accessed on 26 February 2025).
30. Prnewswire Yadea Launches Sodium Battery Electric Two-Wheelers, Leading a Revolution in the Electric Mobility Industry. Available online: https://www.prnewswire.com/news-releases/yadea-launches-sodium-battery-electric-two-wheelers-leading-a-revolution-in-the-electric-mobility-industry-302345708.html#:~:text=8,%202025%20/PRNewswire/%20--,unveiling%20its%20latest%20electric%20two- (accessed on 26 February 2025).
31. Ragonnaud, G. *Powering the EU's Future: Strengthening the EU Battery Industry*; European Parliamentary Research Service: Luxembourg, 2025.
32. Battery2030+ Roadmap. Available online: https://battery2030.eu/research/roadmap/ (accessed on 21 February 2025).
33. Shah, R.; Mittal, V.; Precilla, A.M. Challenges and Advancements in All-Solid-State Battery Technology for Electric Vehicles. *J* **2024**, *7*, 204–217. [CrossRef]
34. IEA. *Global EV Outlook 2024*; IEA: Paris, France, 2024.
35. European Parliament. *Regulation (EU) 2023/1542 of the European Parliament and of the Council of 12 July 2023 Concerning Batteries and Waste Batteries, Amending Directive 2008/98/EC and Regulation (EU) 2019/1020 and Repealing Directive 2006/66/EC*; European Parliament: Strasbourg, France, 2023.
36. GB 38031-2025 Electric Vehicles Traction Battery Safety Requirements (English Version). Available online: https://www.codeofchina.com/standard/GB38031-2025.html (accessed on 4 April 2025).
37. H.R. 2024—Mercury-Containing and Rechargeable Battery Management Act. Available online: https://www.congress.gov/bill/104th-congress/house-bill/2024 (accessed on 20 February 2025).
38. Inflation Reduction Act of 2022. Available online: https://www.irs.gov/inflation-reduction-act-of-2022 (accessed on 5 April 2025).
39. Garche, J.; Moseley, P.T.; Karden, E. Lead–Acid Batteries for Hybrid Electric Vehicles and Battery Electric Vehicles. In *Advances in Battery Technologies for Electric Vehicles*; Elsevier: Amsterdam, The Netherlands, 2015; pp. 75–101. ISBN 978-1-78242-377-5.
40. Saakes, M.; Woortmeijer, R.; Schmal, D. Bipolar Lead–Acid Battery for Hybrid Vehicles. *J. Power Sources* **2005**, *144*, 536–545. [CrossRef]
41. Liu, T.; Yuan, Y.; Tao, X.; Lin, Z.; Lu, J. Bipolar Electrodes for Next-Generation Rechargeable Batteries. *Adv. Sci.* **2020**, *7*, 2001207. [CrossRef]
42. Battery Council International New Study Confirms U.S.' Most Recycled Consumer Product—Lead Batteries—Maintains Remarkable Milestone: 99% Recycling Rate. Available online: https://batterycouncil.org/news/press-release/new-study-confirms-lead-batteries-maintain-remarkable-99-recycling-rate/ (accessed on 14 February 2025).
43. EUROBAT TF Innovation. *Technical Annex to the EUROBAT White Paper Battery Innovation Roadmap V3.0*; EUROBAT: Brussels, Belgium, 2024.
44. Tsais, P.-J.; Chan, L.I. Nickel-Based Batteries: Materials and Chemistry. In *Electricity Transmission, Distribution and Storage Systems*; Elsevier: Amsterdam, The Netherlands, 2013; pp. 309–397. ISBN 978-1-84569-784-6.
45. Nei, J.; Wang, M. Hydrogen Storage Alloy Development for Wide Operating Temperature Nickel-Metal Hydride Battery Applications. *Int. J. Hydrog. Energy* **2024**, *49*, 19–38. [CrossRef]
46. Arya, S.; Verma, S. Nickel-Metal Hydride (Ni-MH) Batteries. In *Rechargeable Batteries*; Wiley: Hoboken, NJ, USA, 2020; pp. 131–175.
47. International Energy Agency (IEA). *Global EV Outlook 2023*; International Energy Agency (IEA): Paris, France, 2023.
48. Khan, F.M.N.U.; Rasul, M.G.; Sayem, A.S.M.; Mandal, N. Maximizing Energy Density of Lithium-Ion Batteries for Electric Vehicles: A Critical Review. *Energy Rep.* **2023**, *9*, 11–21. [CrossRef]

49. Tian, Y.; Zeng, G.; Rutt, A.; Shi, T.; Kim, H.; Wang, J.; Koettgen, J.; Sun, Y.; Ouyang, B.; Chen, T.; et al. Promises and Challenges of Next-Generation "beyond Li-Ion" Batteries for Electric Vehicles and Grid Decarbonization. *Chem. Rev.* **2021**, *121*, 1623–1669. [CrossRef]
50. Argyrou, M.C.; Christodoulides, P.; Kalogirou, S.A. Energy Storage for Electricity Generation and Related Processes: Technologies Appraisal and Grid Scale Applications. *Renew. Sustain. Energy Rev.* **2018**, *94*, 804–821. [CrossRef]
51. Moseley, P.T.; Rand, D.A.J. High-Temperature Sodium Batteries for Energy Storage. In *Electrochemical Energy Storage for Renewable Sources and Grid Balancing*; Elsevier: Amsterdam, The Netherlands, 2015; pp. 253–268. ISBN 978-0-444-62616-5.
52. Veneri, O.; Capasso, C.; Patalano, S. Experimental Study on the Performance of a ZEBRA Battery Based Propulsion System for Urban Commercial Vehicles. *Appl. Energy* **2017**, *185*, 2005–2018. [CrossRef]
53. Nikiforidis, G.; Van De Sanden Ac, M.C.M.; Tsampas, M.N. High and Intermediate Temperature Sodium-Sulfur Batteries for Energy Storage: Development, Challenges and Perspectives. *RSC Adv.* **2019**, *9*, 5649–5673. [CrossRef] [PubMed]
54. Deng, C.; Li, X.; Chen, R.; Ye, K.; Lipton, J.; Maclean, S.A.; Wang, H.; Taylor, A.D.; Weng, G.M. Recent Advances in Rocking Chair Batteries and Beyond. *Energy Storage Mater.* **2023**, *60*, 102820. [CrossRef]
55. Abraham, K.M. How Comparable Are Sodium-Ion Batteries to Lithium-Ion Counterparts? *ACS Energy Lett.* **2020**, *5*, 3544–3547. [CrossRef]
56. Dorau, F.A.; Sommer, A.; Koloch, J.; Röß-Ohlenroth, R.; Schreiber, M.; Neuner, M.; Abo Gamra, K.; Lin, Y.; Schöberl, J.; Bilfinger, P.; et al. Comprehensive Analysis of Commercial Sodium-Ion Batteries: Structural and Electrochemical Insights. *J. Electrochem. Soc.* **2024**, *171*, 090521. [CrossRef]
57. Rehm, M.; Fischer, M.; Gomez, M.R.; Schütte, M.; Sauer, D.U.; Jossen, A. Comparing the Electrical Performance of Commercial Sodium-Ion and Lithium-Iron-Phosphate Batteries. *J. Power Sources* **2025**, *633*, 236290. [CrossRef]
58. Yao, K.; Xu, S.; Yang, Y.; Zheng, Y.; Zuraqi, K.; Yang, D.; Liu, J.; Mani, U.; Rui, X. Designing Sodium Alloys for Dendrite-free Sodium-metal Batteries. *Inf. Funct. Mater.* **2024**, *1*, 242–263. [CrossRef]
59. Adelhelm, P.; Hartmann, P.; Bender, C.L.; Busche, M.; Eufinger, C.; Janek, J. From Lithium to Sodium: Cell Chemistry of Room Temperature Sodium–Air and Sodium–Sulfur Batteries. *Beilstein J. Nanotechnol.* **2015**, *6*, 1016–1055. [CrossRef]
60. Burd, J.T.J.; Moore, E.A.; Ezzat, H.; Kirchain, R.; Roth, R. Improvements in Electric Vehicle Battery Technology Influence Vehicle Lightweighting and Material Substitution Decisions. *Appl. Energy* **2021**, *283*, 116269. [CrossRef]
61. Jannesar Niri, A.; Poelzer, G.A.; Zhang, S.E.; Rosenkranz, J.; Pettersson, M.; Ghorbani, Y. Sustainability Challenges throughout the Electric Vehicle Battery Value Chain. *Renew. Sustain. Energy Rev.* **2024**, *191*, 114176. [CrossRef]
62. Naseri, F.; Kazemi, Z.; Larsen, P.G.; Arefi, M.M.; Schaltz, E. Cyber-Physical Cloud Battery Management Systems: Review of Security Aspects. *Batteries* **2023**, *9*, 382. [CrossRef]
63. Rahmani, P.; Chakraborty, S.; Mele, I.; Katrašnik, T.; Bernhard, S.; Pruefling, S.; Wilkins, S.; Hegazy, O. Driving the Future: A Comprehensive Review of Automotive Battery Management System Technologies, and Future Trends. *J. Power Sources* **2025**, *629*, 235827. [CrossRef]
64. *Batteries Europe Secretariat Research and Innovation Roadmap on Battery Technologies Powering Europe's Green Revolution: Paving the Way to a More Resilient and Sustainable Battery Industry 2*; Batteries Europe: Brussels, Belgium, 2023.
65. Itani, K.; De Bernardinis, A. Review on New-Generation Batteries Technologies: Trends and Future Directions. *Energies* **2023**, *16*, 7530. [CrossRef]
66. Tong, Z.; Zhu, X. A Patent Landscape Analysis on the High-Voltage Spinel LiNi0.5Mn1.5O4 for next-Generation Lithium-Ion Batteries. *Next Energy* **2024**, *5*, 100158. [CrossRef]
67. König, A.; Nicoletti, L.; Schröder, D.; Wolff, S.; Waclaw, A.; Lienkamp, M. An Overview of Parameter and Cost for Battery Electric Vehicles. *World Electr. Veh. J.* **2021**, *12*, 21. [CrossRef]
68. Kim, Y.; Seong, W.M.; Manthiram, A. Cobalt-Free, High-Nickel Layered Oxide Cathodes for Lithium-Ion Batteries: Progress, Challenges, and Perspectives. *Energy Storage Mater.* **2021**, *34*, 250–259. [CrossRef]
69. Maisel, F.; Neef, C.; Marscheider-Weidemann, F.; Nissen, N.F. A Forecast on Future Raw Material Demand and Recycling Potential of Lithium-Ion Batteries in Electric Vehicles. *Resour. Conserv. Recycl.* **2023**, *192*, 106920. [CrossRef]
70. Choix, B.; Uhlig, F. *Sustainability Requirements for Batteries in the EU*; European Economic and Social Committee: Brussels, Belgium, 2020.
71. Duffner, F.; Wentker, M.; Greenwood, M.; Leker, J. Battery Cost Modeling: A Review and Directions for Future Research. *Renew. Sustain. Energy Rev.* **2020**, *127*, 109872. [CrossRef]
72. Gutierrez, A.; He, M.; Yonemoto, B.T.; Yang, Z.; Wang, J.; Meyer, H.M.; Thackeray, M.M.; Croy, J.R. Advancing Lithium- and Manganese-Rich Cathodes through a Combined Electrolyte Additive/Surface Treatment Strategy. *J. Electrochem. Soc.* **2019**, *166*, A3896–A3907. [CrossRef]
73. Wentker, M.; Greenwood, M.; Leker, J. A Bottom-Up Approach to Lithium-Ion Battery Cost Modeling with a Focus on Cathode Active Materials. *Energies* **2019**, *12*, 504. [CrossRef]
74. EUCAR. *Battery Requirements for Future Automotive Applications EG BEV&FCEV*; EUCAR: Etterbeek, Belgium, 2019.

75. *USABC Goals for Advanced High-Performance Batteries for Electric Vehicle (EV) Applications*; USCAR: Southfield, MI, USA, 2021.
76. Frith, J.T.; Lacey, M.J.; Ulissi, U. A Non-Academic Perspective on the Future of Lithium-Based Batteries. *Nat. Commun.* **2023**, *14*, 420. [CrossRef] [PubMed]
77. Hackmann, M.; Knörzer, H.; Pfeuffer, J.; Jeckel, P. *Battery Aging in Practice: Analysis of over 7,000 Vehicles Provide Deep Insights into Battery Life and Vehicle Residual Value*; P3 group GmbH: Stuttgart, Germany, 2024.
78. European Commission. *Joint Research Centre. Performance and Durability Requirements in the Batteries Regulation. Part 1, General Assessment and Data Basis*; Publications Office: Luxembourg, 2024.
79. Timilsina, L.; Badr, P.R.; Hoang, P.H.; Ozkan, G.; Papari, B.; Edrington, C.S. Battery Degradation in Electric and Hybrid Electric Vehicles: A Survey Study. *IEEE Access* **2023**, *11*, 42431–42462. [CrossRef]
80. Birkl, C.R.; Roberts, M.R.; McTurk, E.; Bruce, P.G.; Howey, D.A. Degradation Diagnostics for Lithium Ion Cells. *J. Power Sources* **2017**, *341*, 373–386. [CrossRef]
81. Edge, J.S.; O'Kane, S.; Prosser, R.; Kirkaldy, N.D.; Patel, A.N.; Hales, A.; Ghosh, A.; Ai, W.; Chen, J.; Yang, J.; et al. Lithium Ion Battery Degradation: What You Need to Know. *Phys. Chem. Chem. Phys.* **2021**, *23*, 8200–8221. [CrossRef]
82. Menye, J.S.; Camara, M.-B.; Dakyo, B. Lithium Battery Degradation and Failure Mechanisms: A State-of-the-Art Review. *Energies* **2025**, *18*, 342. [CrossRef]
83. Takenaka, N.; Bouibes, A.; Yamada, Y.; Nagaoka, M.; Yamada, A. Frontiers in Theoretical Analysis of Solid Electrolyte Interphase Formation Mechanism. *Adv. Mater.* **2021**, *33*, 2100574. [CrossRef]
84. Zhao, J.; Lv, Z.; Li, D.; Feng, X.; Wang, Z.; Wu, Y.; Shi, D.; Fowler, M.; Burke, A.F. Battery Engineering Safety Technologies (BEST): M5 Framework of Mechanisms, Modes, Metrics, Modeling, and Mitigation. *eTransportation* **2024**, *22*, 100364. [CrossRef]
85. Li, B.; Chao, Y.; Li, M.; Xiao, Y.; Li, R.; Yang, K.; Cui, X.; Xu, G.; Li, L.; Yang, C.; et al. A Review of Solid Electrolyte Interphase (SEI) and Dendrite Formation in Lithium Batteries. *Electrochem. Energy Rev.* **2023**, *6*, 7. [CrossRef]
86. Kim, E.; Kim, M.; Kim, J.; Kim, J.; Park, J.-H.; Kim, K.-T.; Park, J.-H.; Kim, T.; Min, K. Data-Driven Methods for Predicting the State of Health, State of Charge, and Remaining Useful Life of Li-Ion Batteries: A Comprehensive Review. *Int. J. Precis. Eng. Manuf.* **2023**, *24*, 1281–1304. [CrossRef]
87. Qu, X.; Shi, D.; Zhao, J.; Tran, M.-K.; Wang, Z.; Fowler, M.; Lian, Y.; Burke, A.F. Insights and Reviews on Battery Lifetime Prediction from Research to Practice. *J. Energy Chem.* **2024**, *94*, 716–739. [CrossRef]
88. Rezaei Larijani, M.; Hedayati Kia, S.; Zolghadri, M.; El Hajjaji, A. Battery Characterization, Life Cycle, and Modeling. In *Handbook of Power Electronics in Autonomous and Electric Vehicles*; Elsevier: Amsterdam, The Netherlands, 2024; pp. 323–335, ISBN 978-0-323-99545-0.
89. Yang, Y.; Zhao, L.; Zhang, Y.; Yang, Z.; Lai, W.-H.; Liang, Y.; Dou, S.-X.; Liu, M.; Wang, Y.-X. Challenges and Prospects of Low-Temperature Rechargeable Batteries: Electrolytes, Interfaces, and Electrodes. *Adv. Sci.* **2024**, *11*, e2410318. [CrossRef] [PubMed]
90. Chen, K.; Cai, W.; Hu, Z.; Huang, Q.; Wang, A.; Zeng, Z.; Song, J.; Sun, Y.; Kong, Q.; Feng, W.; et al. Damage Mechanisms and Recent Research Advances in Ni-Rich Layered Cathode Materials for Lithium-Ion Batteries. *Electron* **2024**, *2*, e27. [CrossRef]
91. Huang, C.; Zheng, H.; Qin, N.; Wang, C.; Wang, L.; Lu, J. Single-Crystal Nickel-Rich Cathode Materials: Challenges and Strategies. *Wuli Huaxue Xuebao/ Acta Phys. Chim. Sin.* **2024**, *40*, 2308051. [CrossRef]
92. Lu, J.; Xu, C.; Dose, W.; Dey, S.; Wang, X.; Wu, Y.; Li, D.; Ci, L. Microstructures of Layered Ni-Rich Cathodes for Lithium-Ion Batteries. *Chem. Soc. Rev.* **2024**, *53*, 4707–4740. [CrossRef]
93. Xie, H.; Peng, H.; Jiang, D.; Xiao, Z.; Liu, X.; Liang, H.; Wu, M.; Liu, D.; Li, Y.; Sun, Y.; et al. Structures, Issues, and Optimization Strategies of Ni-Rich and Co-Low Cathode Materials for Lithium-Ion Battery. *Chem. Eng. J.* **2023**, *470*, 144051. [CrossRef]
94. Li, H.; Zhang, H.; Liang, Y.; Chen, R.; Cao, Y. Modification of Lithium-Rich Manganese Oxide Materials: Coating, Doping and Single Crystallization. *Batter. Supercaps* **2024**, *8*, e202400443. [CrossRef]
95. Lin, Y.; Li, Y.; Tang, M.; Zhan, L.; Zhai, Y.; Chen, W.; Zhou, M.; Ji, Y.; Wang, P. A Review of High-Capacity Lithium-Rich Manganese-Based Cathode Materials for a New Generation of Lithium Batteries. *Inorganica Chim. Acta* **2024**, *572*, 122239. [CrossRef]
96. Ding, S.; Li, Y.; Dai, H.; Wang, L.; He, X. Accurate Model Parameter Identification to Boost Precise Aging Prediction of Lithium-Ion Batteries: A Review. *Adv. Energy Mater.* **2023**, *13*, 2301452. [CrossRef]
97. Kumar, R. *State Estimation of Lithium-Ion Battery for Electric Vehicle Application: Types, Design Methods, and Future Trends*; Taylor&Francis: London, UK, 2024; ISBN 978-1-040-00910-9.
98. Chidambaram, R.K.; Chatterjee, D.; Barman, B.; Das, P.P.; Taler, D.; Taler, J.; Sobota, T. Effect of Regenerative Braking on Battery Life. *Energies* **2023**, *16*, 5303. [CrossRef]
99. Khan, A.; Naqvi, I.H.; Bhargava, C.; Lin, C.-P.; Boles, S.T.; Kong, L.; Pecht, M. Safety and Reliability Analysis of Lithium-Ion Batteries with Real-Time Health Monitoring. *Renew. Sustain. Energy Rev.* **2025**, *212*, 115408. [CrossRef]
100. Soltani, M.; Vilsen, S.B.; Stroe, A.I.; Knap, V.; Stroe, D.I. Degradation Behaviour Analysis and End-of-Life Prediction of Lithium Titanate Oxide Batteries. *J. Energy Storage* **2023**, *68*, 107745. [CrossRef]

101. Masias, A.; Marcicki, J.; Paxton, W.A. Opportunities and Challenges of Lithium Ion Batteries in Automotive Applications. *ACS Energy Lett.* **2021**, *6*, 621–630. [CrossRef]
102. Abdolrasol, M.G.M.; Ayob, A.; Lipu, M.S.H.; Ansari, S.; Kiong, T.S.; Saad, M.H.M.; Ustun, T.S.; Kalam, A. Advanced Data-Driven Fault Diagnosis in Lithium-Ion Battery Management Systems for Electric Vehicles: Progress, Challenges, and Future Perspectives. *eTransportation* **2024**, *22*, 100374. [CrossRef]
103. Bashir, H.; Yaqoob, A.; Jawaid, I.; Khalid, W.; Javed, M.Y.; Sultan, W. A Review of Battery Management System and Modern State Estimation Approaches in Lithiumion Batteries for Electric Vehicle. In Proceedings of the 2022 5th International Conference on Energy Conservation and Efficiency, ICECE 2022—Proceedings, Lahore, Pakistan, 16–17 March 2022.
104. Challoob, A.F.; Bin Rahmat, N.A.; A/L Ramachandaramurthy, V.K.; Humaidi, A.J. Energy and Battery Management Systems for Electrical Vehicles: A Comprehensive Review & Recommendations. *Energy Explor. Exploit.* **2024**, *42*, 341–372. [CrossRef]
105. Hossain Lipu, M.S.; Ansari, S.; Miah, M.S.; Meraj, S.T.; Hasan, K.; Shihavuddin, A.S.M.; Hannan, M.A.; Muttaqi, K.M.; Hussain, A. Deep Learning Enabled State of Charge, State of Health and Remaining Useful Life Estimation for Smart Battery Management System: Methods, Implementations, Issues and Prospects. *J. Energy Storage* **2022**, *55*, 105752. [CrossRef]
106. Yang, B.; Qian, Y.; Li, Q.; Chen, Q.; Wu, J.; Luo, E.; Xie, R.; Zheng, R.; Yan, Y.; Su, S.; et al. Critical Summary and Perspectives on State-of-Health of Lithium-Ion Battery. *Renew. Sustain. Energy Rev.* **2024**, *190*, 114077. [CrossRef]
107. Park, J.; Jin, Y.; Kam, W.; Han, S. A Practical Semi-Empirical Model for Predicting the SoH of Lithium-Ion Battery: A Novel Perspective on Short-Term Rest. *J. Energy Storage* **2024**, *96*, 112659. [CrossRef]
108. Vasta, E.; Scimone, T.; Nobile, G.; Eberhardt, O.; Dugo, D.; De Benedetti, M.M.; Lanuzza, L.; Scarcella, G.; Patanè, L.; Arena, P.; et al. Models for Battery Health Assessment: A Comparative Evaluation. *Energies* **2023**, *16*, 632. [CrossRef]
109. Omakor, J.; Miah, M.S.; Chaoui, H. Battery Reliability Assessment in Electric Vehicles: A State-of-the-Art. *IEEE Access* **2024**, *12*, 77903–77931. [CrossRef]
110. Vanem, E.; Salucci, C.B.; Bakdi, A.; Alnes, Ø.Å.S. Data-Driven State of Health Modelling—A Review of State of the Art and Reflections on Applications for Maritime Battery Systems. *J. Energy Storage* **2021**, *43*, 103158. [CrossRef]
111. Ali, Z.M.; Calasan, M.; Gandoman, F.H.; Jurado, F.; Abdel Aleem, S.H.E. Review of Batteries Reliability in Electric Vehicle and E-Mobility Applications. *Ain Shams Eng. J.* **2024**, *15*, 102442. [CrossRef]
112. Clemente, C.; Giovanni, C.; Ottorino, V.; Stanislao, P.; Ferdinando, V. Estimation of the State of Charge Health of Electric Vehicle Batteries Through Machine Learning Approach. In Proceedings of the Energy Proceedings, Doha, Qatar, 3–7 December 2023; Volume 45.
113. Massaoudi, M.; Abu-Rub, H.; Ghrayeb, A. Advancing Lithium-Ion Battery Health Prognostics With Deep Learning: A Review and Case Study. *IEEE Open J. Ind. Appl.* **2024**, *5*, 43–62. [CrossRef]
114. Cicconi, P.; Kumar, P. Design Approaches for Li-Ion Battery Packs: A Review. *J. Energy Storage* **2023**, *73*, 109197. [CrossRef]
115. Zhong, D.; Xia, Z.; Zhu, Y.; Duan, J. Overview of Predictive Maintenance Based on Digital Twin Technology. *Heliyon* **2023**, *9*, e14534. [CrossRef]
116. Alfaro-Viquez, D.; Zamora-Hernandez, M.; Fernandez-Vega, M.; Garcia-Rodriguez, J.; Azorin-Lopez, J. A Comprehensive Review of AI-Based Digital Twin Applications in Manufacturing: Integration Across Operator, Product, and Process Dimensions. *Electronics* **2025**, *14*, 646. [CrossRef]
117. Mulpuri, S.K.; Sah, B.; Kumar, P. An Intelligent Battery Management System (BMS) with End-Edge-Cloud Connectivity—A Perspective. *Sustain. Energy Fuels* **2025**, *9*, 1142–1159. [CrossRef]
118. Kreuzer, T.; Papapetrou, P.; Zdravkovic, J. Artificial Intelligence in Digital Twins—A Systematic Literature Review. *Data Knowl. Eng.* **2024**, *151*, 102304. [CrossRef]
119. Es-haghi, M.S.; Anitescu, C.; Rabczuk, T. Methods for Enabling Real-Time Analysis in Digital Twins: A Literature Review. *Comput. Struct.* **2024**, *297*, 107342. [CrossRef]
120. Cavus, M.; Dissanayake, D.; Bell, M. Next Generation of Electric Vehicles: AI-Driven Approaches for Predictive Maintenance and Battery Management. *Energies* **2025**, *18*, 1041. [CrossRef]
121. Zhao, J.; Qu, X.; Wu, Y.; Fowler, M.; Burke, A.F. Artificial Intelligence-Driven Real-World Battery Diagnostics. *Energy AI* **2024**, *18*, 100419. [CrossRef]
122. Iranshahi, K.; Brun, J.; Arnold, T.; Sergi, T.; Müller, U.C. Digital Twins: Recent Advances and Future Directions in Engineering Fields. *Intell. Syst. Appl.* **2025**, *26*, 200516. [CrossRef]
123. Saeed, S.; Altamimi, S.A.; Alkayyal, N.A.; Alshehri, E.; Alabbad, D.A. Digital Transformation and Cybersecurity Challenges for Businesses Resilience: Issues and Recommendations. *Sensors* **2023**, *23*, 6666. [CrossRef]
124. Nuroldayeva, G.; Serik, Y.; Adair, D.; Uzakbaiuly, B.; Bakenov, Z. State of Health Estimation Methods for Lithium-Ion Batteries. *Int. J. Energy Res.* **2023**, *2023*, 1–21. [CrossRef]
125. Wang, Z.; Zhao, X.; Fu, L.; Zhen, D.; Gu, F.; Ball, A.D. A Review on Rapid State of Health Estimation of Lithium-Ion Batteries in Electric Vehicles. *Sustain. Energy Technol. Assess.* **2023**, *60*, 103457. [CrossRef]

126. Che, Y.; Hu, X.; Lin, X.; Guo, J.; Teodorescu, R. Health Prognostics for Lithium-Ion Batteries: Mechanisms, Methods, and Prospects. *Energy Environ. Sci.* **2023**, *16*, 338–371. [CrossRef]
127. Sedláčková, E.; Pražanová, A.; Plachý, Z.; Klusoňová, N.; Knap, V.; Dušek, K. Acoustic Emission Technique for Battery Health Monitoring: Comprehensive Literature Review. *Batteries* **2025**, *11*, 14. [CrossRef]
128. Dini, P.; Colicelli, A.; Saponara, S. Review on Modeling and SOC/SOH Estimation of Batteries for Automotive Applications. *Batteries* **2024**, *10*, 34. [CrossRef]
129. Gaga, A.; Tannouche, A.; Mehdaoui, Y.; El Hadadi, B. Methods for Estimating Lithium-Ion Battery State of Charge for Use in Electric Vehicles: A Review. *Energy Harvest. Syst.* **2022**, *9*, 211–225. [CrossRef]
130. Anekal, L.; Samanta, A.; Williamson, S. Wide-Ranging Parameter Extraction of Lithium-Ion Batteries to Estimate State of Health Using Electrochemical Impedance Spectroscopy. In Proceedings of the 1st IEEE Industrial Electronics Society Annual On-Line Conference, ONCON 2022, kharagpur, India, 9–11 December 2022.
131. Wang, W.; Zhang, Y.; Xie, B.; Huang, L.; Dong, S.; Xu, G.; Cui, G. Deciphering Advanced Sensors for Life and Safety Monitoring of Lithium Batteries. *Adv. Energy Mater.* **2024**, *14*, 2304173. [CrossRef]
132. Cheng, A.; Xin, Y.; Wu, H.; Yang, L.; Deng, B. A Review of Sensor Applications in Electric Vehicle Thermal Management Systems. *Energies* **2023**, *16*, 5139. [CrossRef]
133. Li, J.; Li, Y.; Zeng, W. Gas Sensing Technology as the Key to Safety Warning of Lithium-Ion Battery: Recent Advance. *Sens. Actuators A Phys.* **2024**, *365*, 114890. [CrossRef]
134. Diao, W.; Jiang, J.; Zhang, C.; Liang, H.; Pecht, M. Energy State of Health Estimation for Battery Packs Based on the Degradation and Inconsistency. *Energy Procedia* **2017**, *142*, 3578–3583. [CrossRef]
135. Fan, G.; Zhou, B.; Ye, S.; Shen, H.; Huo, D.; Zhang, X. Unveiling the Secrets behind Physics-Based Modeling of Lithium-Ion Battery Degradation and Its Key Applications. *J. Energy Storage* **2024**, *102*, 114086. [CrossRef]
136. Jaime-Barquero, E.; Bekaert, E.; Olarte, J.; Zulueta, E.; Lopez-Guede, J.M. Artificial Intelligence Opportunities to Diagnose Degradation Modes for Safety Operation in Lithium Batteries. *Batteries* **2023**, *9*, 388. [CrossRef]
137. Nigam, A.; Sharma, H.; Kumar, P.; Kumar, A. *Modeling and Simulation Study for Power Management and Battery Degradation of Smart Electric Vehicles*; Wiley Semiconductor: Hoboken, NJ, USA, 2024; ISBN 978-1-394-22504-0.
138. Shi, J.; Zhang, H.; Yu, H.; Xu, Y.; Xu, S.; Sheng, L.; Feng, X.; Wang, X. Experimental Determinations of Thermophysical Parameters for Lithium-Ion Batteries: A Systematic Review. *eTransportation* **2024**, *20*, 100321. [CrossRef]
139. Abramushkina, E.; Zhaksylyk, A.; Geury, T.; El Baghdadi, M.; Hegazy, O. A Thorough Review of Cooling Concepts and Thermal Management Techniques for Automotive Wbg Inverters: Topology, Technology and Integration Level. *Energies* **2021**, *14*, 4981. [CrossRef]
140. Alkhedher, M.; Al Tahhan, A.B.; Yousaf, J.; Ghazal, M.; Shahbazian-Yassar, R.; Ramadan, M. Electrochemical and Thermal Modeling of Lithium-Ion Batteries: A Review of Coupled Approaches for Improved Thermal Performance and Safety Lithium-Ion Batteries. *J. Energy Storage* **2024**, *86*, 111172. [CrossRef]
141. Behi, H.; Gandoman, F.H.; Karimi, D.; Hosen, M.D.S.; Behi, M.; Jaguemont, J.; Van Mierlo, J.; Berecibar, M. *Thermal Management of the Li-Ion Batteries to Improve the Performance of the Electric Vehicles Applications*; Wiley: Hoboken, NY, USA, 2022; ISBN 978-1-119-80107-8.
142. Chavan, S.; Venkateswarlu, B.; Salman, M.; Liu, J.; Pawar, P.; Joo, S.W.; Choi, G.S.; Kim, S.C. Thermal Management Strategies for Lithium-Ion Batteries in Electric Vehicles: Fundamentals, Recent Advances, Thermal Models, and Cooling Techniques. *Int. J. Heat Mass Transf.* **2024**, *232*, 125918. [CrossRef]
143. Chen, Z.; Jia, L.; Yin, L.; Dang, C.; Ren, H.; Zhang, Z. A Review on Thermal Management of Li-Ion Battery: From Small-Scale Battery Module to Large-Scale Electrochemical Energy Storage Power Station. *J. Therm. Sci.* **2025**, *34*, 1–23. [CrossRef]
144. Garud, K.S.; Tai, L.D.; Hwang, S.-G.; Nguyen, N.-H.; Lee, M.-Y. A Review of Advanced Cooling Strategies for Battery Thermal Management Systems in Electric Vehicles. *Symmetry* **2023**, *15*, 1322. [CrossRef]
145. Khan, A.; Yaqub, S.; Ali, M.; Ahmad, A.W.; Nazir, H.; Khalid, H.A.; Iqbal, N.; Said, Z.; Sopian, K. A State-of-the-Art Review on Heating and Cooling of Lithium-Ion Batteries for Electric Vehicles. *J. Energy Storage* **2024**, *76*, 109852. [CrossRef]
146. Rallabandi, S.; Issac Selvaraj, R.V. Advancements in Battery Cooling Techniques for Enhanced Performance and Safety in Electric Vehicles: A Comprehensive Review. *Energy Technol.* **2024**, *12*, 2301404. [CrossRef]
147. Wu, C.; Sun, Y.; Tang, H.; Zhang, S.; Yuan, W.; Zhu, L.; Tang, Y. A Review on the Liquid Cooling Thermal Management System of Lithium-Ion Batteries. *Appl. Energy* **2024**, *375*, 124173. [CrossRef]
148. Yang, B.; Ji, J.; Zhang, X.; Hua, W. Passive Cooling of Lithium-Ion Batteries Based on Flexible Phase Change Materials: Molecular Structure, Interactions and Mechanistic Aspects. *J. Mol. Liq.* **2023**, *391*, 123340. [CrossRef]
149. He, J.; Meng, J.; Huang, Y. Challenges and Recent Progress in Fast-Charging Lithium-Ion Battery Materials. *J. Power Sources* **2023**, *570*, 232965. [CrossRef]
150. Song, C.; Han, S.H.; Moon, H.; Choi, N.-S. Unlocking Fast-Charging Capabilities of Lithium-Ion Batteries through Liquid Electrolyte Engineering. *EcoMat* **2024**, *6*, e12476. [CrossRef]

151. Li, Z.; Liu, J.; Qin, Y.; Gao, T. Enhancing the Charging Performance of Lithium-Ion Batteries by Reducing SEI and Charge Transfer Resistances. *ACS Appl. Mater. Interfaces* **2022**, *14*, 33004–33012. [CrossRef]
152. Wang, R.; Wang, L.; Liu, R.; Li, X.; Wu, Y.; Ran, F. "Fast-Charging" Anode Materials for Lithium-Ion Batteries from Perspective of Ion Diffusion in Crystal Structure. *ACS Nano* **2024**, *18*, 2611–2648. [CrossRef]
153. Xiao, H.; Zhao, J.; Gao, Q.; Zhang, W.; Cheng, X.; Song, C.; Li, G. Recent Advances in Fast-Charging Lithium-Ion Batteries: Mechanism, Materials, and Future Opportunities. *Chem. Eng. J.* **2025**, *506*, 159927. [CrossRef]
154. Pegel, H.; Wycisk, D.; Sauer, D.U. Influence of Cell Dimensions and Housing Material on the Energy Density and Fast-Charging Performance of Tabless Cylindrical Lithium-Ion Cells. *Energy Storage Mater.* **2023**, *60*, 102796. [CrossRef]
155. Pegel, H.; Wycisk, D.; Scheible, A.; Tendera, L.; Latz, A.; Sauer, D.U. Fast-Charging Performance and Optimal Thermal Management of Large-Format Full-Tab Cylindrical Lithium-Ion Cells under Varying Environmental Conditions. *J. Power Sources* **2023**, *556*, 232408. [CrossRef]
156. Yeganehdoust, F.; Kumar, A.; Reddy, M.R.; Zaghib, K. Cell Architecture Design for Fast-Charging Lithium-Ion Batteries in Electric Vehicles. *Batteries* **2025**, *11*, 20. [CrossRef]
157. Manthiram, A.; Song, B.; Li, W. A Perspective on Nickel-Rich Layered Oxide Cathodes for Lithium-Ion Batteries. *Energy Storage Mater.* **2017**, *6*, 125–139. [CrossRef]
158. Ahangari, M.; Zhou, M.; Luo, H. Review of Layered Transition Metal Oxide Materials for Cathodes in Sodium-Ion Batteries. *Micromachines* **2025**, *16*, 137. [CrossRef]
159. Zhong, J.; Yang, Z.; Yu, Y.; Liu, Y.; Li, J.; Kang, F. Surface Substitution of Polyanion to Improve Structure Stability and Electrochemical Properties of Lithium-Rich Layered Cathode Oxides. *Appl. Surf. Sci.* **2020**, *512*, 145741. [CrossRef]
160. Zhang, S.; Li, Y.; Xing, T.; Liu, H.; Liu, Z.; Li, Z.; Wu, M. Recent Progress in the Use of Polyanions as Solid Electrolytes. *New Carbon Mater.* **2022**, *37*, 358–370. [CrossRef]
161. Huang, Q.; Turcheniuk, K.; Ren, X.; Magasinski, A.; Song, A.-Y.; Xiao, Y.; Kim, D.; Yushin, G. Cycle Stability of Conversion-Type Iron Fluoride Lithium Battery Cathode at Elevated Temperatures in Polymer Electrolyte Composites. *Nat. Mater.* **2019**, *18*, 1343–1349. [CrossRef]
162. Wang, H.; Chen, S.; Fu, C.; Ding, Y.; Liu, G.; Cao, Y.; Chen, Z. Recent Advances in Conversion-Type Electrode Materials for Post Lithium-Ion Batteries. *ACS Mater. Lett.* **2021**, *3*, 956–977. [CrossRef]
163. Wu, F.; Yushin, G. Conversion Cathodes for Rechargeable Lithium and Lithium-Ion Batteries. *Energy Environ. Sci.* **2017**, *10*, 435–459. [CrossRef]
164. Su, Y.; Chen, J.; Li, H.; Sun, H.; Yang, T.; Liu, Q.; Ichikawa, S.; Zhang, X.; Zhu, D.; Zhao, J.; et al. Enabling Long Cycle Life and High Rate Iron Difluoride Based Lithium Batteries by In Situ Cathode Surface Modification. *Adv. Sci.* **2022**, *9*, 2201419. [CrossRef] [PubMed]
165. Pan, Z.; Yang, J.; Kong, J.; Loh, X.J.; Wang, J.; Liu, Z. "Porous and Yet Dense" Electrodes for High-Volumetric-Performance Electrochemical Capacitors: Principles, Advances, and Challenges. *Adv. Sci.* **2022**, *9*, 2103953. [CrossRef] [PubMed]
166. Saju, S.K.; Chattopadhyay, S.; Xu, J.; Alhashim, S.; Pramanik, A.; Ajayan, P.M. Hard Carbon Anode for Lithium-, Sodium-, and Potassium-Ion Batteries: Advancement and Future Perspective. *Cell Rep. Phys. Sci.* **2024**, *5*, 101851. [CrossRef]
167. Li, C.; Li, X.; Yu, W.; Xiao, R.; Huang, F.; Tian, H.; Wang, C.; Chen, X.; Shao, J. Scalable Fabrication of Turbostratic Graphene with High Density and High Ion Conductivity for Compact Capacitive Energy Storage. *Matter* **2023**, *6*, 4032–4049. [CrossRef]
168. Miao, J. Review on Electrode Degradation at Fast Charging of Li-Ion and Li Metal Batteries from a Kinetic Perspective. *Electrochem* **2023**, *4*, 156–180. [CrossRef]
169. Ezhyeh, Z.N.; Khodaei, M.; Torabi, F. Review on Doping Strategy in $Li_4Ti_5O_{12}$ as an Anode Material for Lithium-Ion Batteries. *Ceram. Int.* **2023**, *49*, 7105–7141. [CrossRef]
170. Khan, M.; Yan, S.; Ali, M.; Mahmood, F.; Zheng, Y.; Li, G.; Liu, J.; Song, X.; Wang, Y. Innovative Solutions for High-Performance Silicon Anodes in Lithium-Ion Batteries: Overcoming Challenges and Real-World Applications. *Nano-Micro Lett.* **2024**, *16*, 179. [CrossRef]
171. Asamoah, G.A.; Korsah, M.; Jeyasundar, P.G.S.A.; Ahmed, M.; Lau, S.Y.; Danquah, M.K. Nanotechnology-Based Lithium-Ion Battery Energy Storage Systems. *Sustainability* **2024**, *16*, 9231. [CrossRef]
172. Wang, L.; Guo, Q.; Wang, J.; Li, H.; Wang, G.; Yang, J.; Song, Y.; Qin, Y.; Liu, L. Improved Cycling Performance of a Silicon Anode for Lithium Ion Batteries Using Carbon Nanocoils. *RSC Adv.* **2014**, *4*, 40812–40815. [CrossRef]
173. Gao, Y.; Fan, L.; Zhou, R.; Du, X.; Jiao, Z.; Zhang, B. High-Performance Silicon-Rich Microparticle Anodes for Lithium-Ion Batteries Enabled by Internal Stress Mitigation. *Nano-Micro Lett.* **2023**, *15*, 222. [CrossRef] [PubMed]
174. Lei, S.; Zeng, Z.; Cheng, S.; Xie, J. Fast-charging of Lithium-ion Batteries: A Review of Electrolyte Design Aspects. *Battery Energy* **2023**, *2*, 20230018. [CrossRef]
175. Liu, C.; Sheng, L.; Jiang, L. Research on Performance Constraints and Electrolyte Optimization Strategies for Lithium-Ion Batteries at Low Temperatures. *RSC Adv.* **2025**, *15*, 7995–8018. [CrossRef]

176. Wan, S.; Ma, W.; Wang, Y.; Xiao, Y.; Chen, S. Electrolytes Design for Extending the Temperature Adaptability of Lithium-Ion Batteries: From Fundamentals to Strategies. *Adv. Mater.* **2024**, *36*, 2311912. [CrossRef]
177. Kianfar, M.; Ipakchi, H.; Mohajer, S.; Rasoulifard, M.H.; Seyed Dorraji, M.S.; Louaguef, D.B.; Azat, S.; Saeb, M.R.; Vahabi, H. Flame-Retardant Self-Healing Polymers: A Review. *J. Polym. Sci.* 2024. [CrossRef]
178. Liu, W.; Jiang, Y.; Wang, N.; Fu, W. Recent Progress in Flame Retardant Technology of Battery: A Review. *Resour. Chem. Mater.* **2023**, *2*, 80–99. [CrossRef]
179. Gao, Z.; Rao, S.; Zhang, T.; Li, W.; Yang, X.; Chen, Y.; Zheng, Y.; Ding, Y.; Dong, T.; Li, S. Design Strategies of Flame-Retardant Additives for Lithium Ion Electrolyte. *J. Electrochem. Energy Convers. Storage* **2022**, *19*, 030910. [CrossRef]
180. Han, Y.; Chen, Y.; Huang, Y.; Zhang, M.; Li, Z.; Wang, Y. Recent Progress on Garnet-Type Oxide Electrolytes for All-Solid-State Lithium-Ion Batteries. *Ceram. Int.* **2023**, *49*, 29375–29390. [CrossRef]
181. Sarfraz, N.; Kanwal, N.; Ali, M.; Ali, K.; Hasnain, A.; Ashraf, M.; Ayaz, M.; Ifthikar, J.; Ali, S.; Hendi, A.; et al. Materials Advancements in Solid-State Inorganic Electrolytes for Highly Anticipated All Solid Li-Ion Batteries. *Energy Storage Mater.* **2024**, *71*, 103619. [CrossRef]
182. Lain, M.J.; Kendrick, E. Understanding the Limitations of Lithium Ion Batteries at High Rates. *J. Power Sources* **2021**, *493*, 229690. [CrossRef]
183. Zhou, H.; Alujjage, A.S.; Terese, M.; Fear, C.; Joshi, T.; Rikka, V.R.; Jeevarajan, J.A.; Mukherjee, P.P. Effect of Fast Charging on Degradation and Safety Characteristics of Lithium-Ion Batteries with LiFePO4 Cathodes. *Appl. Energy* **2025**, *377*, 124465. [CrossRef]
184. Dai, N.; Long, J. Research on Fast-Charging Battery Thermal Management System Based on Refrigerant Direct Cooling. *Sci Rep* **2023**, *13*, 11707. [CrossRef] [PubMed]
185. Tai, L.D.; Garud, K.S.; Hwang, S.-G.; Lee, M.-Y. A Review on Advanced Battery Thermal Management Systems for Fast Charging in Electric Vehicles. *Batteries* **2024**, *10*, 372. [CrossRef]
186. Khan, S.A.; Hussain, I.; Thakur, A.K.; Yu, S.; Lau, K.T.; He, S.; Dong, K.; Chen, J.; Xiangrong, L.; Ahmad, M.; et al. Advancements in Battery Thermal Management System for Fast Charging/Discharging Applications. *Energy Storage Mater.* **2024**, *65*, 103144. [CrossRef]
187. Wassiliadis, N.; Schneider, J.; Frank, A.; Wildfeuer, L.; Lin, X.; Jossen, A.; Lienkamp, M. Review of Fast Charging Strategies for Lithium-Ion Battery Systems and Their Applicability for Battery Electric Vehicles. *J. Energy Storage* **2021**, *44*, 103306. [CrossRef]
188. Ji, G.; He, L.; Wu, T.; Cui, G. The Design of Fast Charging Strategy for Lithium-Ion Batteries and Intelligent Application: A Comprehensive Review. *Appl. Energy* **2025**, *377*, 124538. [CrossRef]
189. Wang, Z.; Wu, X.; Bai, Z.; Yang, N.; Guo, G.; Banjoko, O.S. Enhancing Fast Charging Performance of Lithium-Ion Batteries: The Role of Operating Temperature and Charging Rate. *Electrochim. Acta* **2025**, *511*, 145390. [CrossRef]
190. Ghani, F.; An, K.; Lee, D. A Review on Design Parameters for the Full-Cell Lithium-Ion Batteries. *Batteries* **2024**, *10*, 340. [CrossRef]
191. Niu, H.; Zhang, N.; Lu, Y.; Zhang, Z.; Li, M.; Liu, J.; Song, W.; Zhao, Y.; Miao, Z. Strategies toward the Development of High-Energy-Density Lithium Batteries. *J. Energy Storage* **2024**, *88*, 111666. [CrossRef]
192. Xu, J.; Cai, X.; Cai, S.; Shao, Y.; Hu, C.; Lu, S.; Ding, S. High-Energy Lithium Ion Batteries: Recent Progress and a Promising Future in Applications. *Energy Environ. Mater.* **2023**, *6*, e12450. [CrossRef]
193. Salgado, R.M.; Danzi, F.; Oliveira, J.E.; El-Azab, A.; Camanho, P.P.; Braga, M.H. The Latest Trends in Electric Vehicles Batteries. *Molecules* **2021**, *26*, 3188. [CrossRef] [PubMed]
194. Mishra, A.K.; Monika, A.K.; Patial, B.S. A Review on Recent Advances in Anode Materials in Lithium Ion Batteries. *Mater. Today Electron.* **2024**, *7*, 100089. [CrossRef]
195. Yi, X.; Qi, G.; Liu, X.; Depcik, C.; Liu, L. Challenges and Strategies toward Anode Materials with Different Lithium Storage Mechanisms for Rechargeable Lithium Batteries. *J. Energy Storage* **2024**, *95*, 112480. [CrossRef]
196. Joy, R.; Balakrishnan, N.T.M.; Das, A.; Shafeek, S.; Thakur, V.K.; Zaghib, K.; Jaffarali, J.F.M.; Reddy, M.V.V.; Raghavan, P. Graphene: Chemistry and Applications for Lithium-Ion Batteries. *Electrochem* **2022**, *3*, 143–183. [CrossRef]
197. Yu, L.; Chen, X.; Yao, N.; Gao, Y.-C.; Yuan, Y.-H.; Gao, Y.-B.; Tang, C.; Zhang, Q. Advanced Carbon as Emerging Energy Materials in Lithium Batteries: A Theoretical Perspective. *InfoMat* **2025**, e12653. [CrossRef]
198. Dang, G.; Zhang, M.; Min, F.; Zhang, Y.; Zhang, B.; Zhang, Q.; Wang, J.; Zhou, Y.; Liu, W.; Xie, J.; et al. Lithium Titanate Battery System Enables Hybrid Electric Heavy-Duty Vehicles. *J. Energy Storage* **2023**, *74*, 109313. [CrossRef]
199. Chen, H.; Chahbaz, A.; Yang, S.; Zhang, W.; Sauer, D.U.; Li, W. Thermodynamic and Kinetic Degradation of LTO Batteries: Impact of Different SOC Intervals and Discharge Voltages in Electric Train Applications. *eTransportation* **2024**, *21*, 100340. [CrossRef]
200. He, Y.-B.; Li, B.; Liu, M.; Zhang, C.; Lv, W.; Yang, C.; Li, J.; Du, H.; Zhang, B.; Yang, Q.-H.; et al. Gassing in $Li_4Ti_5O_{12}$-Based Batteries and Its Remedy. *Sci. Rep.* **2012**, *2*, 913. [CrossRef]
201. Liu, J.; Bian, P.; Li, J.; Ji, W.; Hao, H.; Yu, A. Gassing Behavior of Lithium Titanate Based Lithium Ion Batteries with Different Types of Electrolytes. *J. Power Sources* **2015**, *286*, 380–387. [CrossRef]

202. Cicconi, P.; Postacchini, L.; Pallotta, E.; Monteriù, A.; Prist, M.; Bevilacqua, M.; Germani, M. A Life Cycle Costing of Compacted Lithium Titanium Oxide Batteries for Industrial Applications. *J. Power Sources* **2019**, *436*, 226837. [CrossRef]
203. Peng, C.; Liang, S.; Yu, Y.; Cao, L.; Yang, C.; Liu, X.; Guo, K.; Müller-Buschbaum, P.; Cheng, Y.; Wang, C. A Chronicle of Titanium Niobium Oxide Materials for High-performance Lithium-ion Batteries: From Laboratory to Industry. *Carbon Neutralization* **2024**, *3*, 1036–1091. [CrossRef]
204. Zhang, Y.; Huang, J.; Saito, N.; Zhang, Z.; Yang, L.; Hirano, S.-I. Search for Stable Host Materials as Low-Voltage Anodes for Lithium-Ion Batteries: A Mini-Review. *Energy Storage Mater.* **2023**, *55*, 364–387. [CrossRef]
205. Wen, X.; Ma, C.; Du, C.; Liu, J.; Zhang, X.; Qu, D.; Tang, Z. Enhanced Electrochemical Properties of Vanadium-Doped Titanium Niobate as a New Anode Material for Lithium-Ion Batteries. *Electrochim. Acta* **2015**, *186*, 58–63. [CrossRef]
206. Kumar Prajapati, A.; Bhatnagar, A. A Review on Anode Materials for Lithium/Sodium-Ion Batteries. *J. Energy Chem.* **2023**, *83*, 509–540. [CrossRef]
207. Ayodhya, D. A Review of Recent Progress in 2D MXenes: Synthesis, Properties, and Applications. *Diam. Relat. Mater.* **2023**, *132*, 109634. [CrossRef]
208. Naguib, M.; Gogotsi, Y.; Barsoum, M.W. Mxenes: A New Family of Two-Dimensional Materials and Its Application As Electrodes for Li and Na-Ion Batteries. *Meet. Abstr.* **2015**, *MA2015-01*, 849. [CrossRef]
209. Awan, H.T.A.; Abdah, M.A.A.M.; Mehar, M.; Walvekar, R.; Chaudhary, V.; Khalid, M.; Khosla, A. MXene-Polymer Hybrid Composites for Advanced Energy Storage: Insights into Supercapacitors and Batteries. *J. Energy Storage* **2024**, *95*, 112449. [CrossRef]
210. Feyzi, E.; M R, A.K.; Li, X.; Deng, S.; Nanda, J.; Zaghib, K. A Comprehensive Review of Silicon Anodes for High-Energy Lithium-Ion Batteries: Challenges, Latest Developments, and Perspectives. *Next Energy* **2024**, *5*, 100176. [CrossRef]
211. Nzereogu, P.U.; Omah, A.D.; Ezema, F.I.; Iwuoha, E.I.; Nwanya, A.C. Anode Materials for Lithium-Ion Batteries: A Review. *Appl. Surf. Sci. Adv.* **2022**, *9*, 100233. [CrossRef]
212. Peter Holderith Silicon Anode EV Batteries Are The Real Deal, But It's Complicated. Available online: https://insideevs.com/news/740255/silicon-anode-batteries-advantages-evs/ (accessed on 25 February 2025).
213. Saidi, N.M.; Abdah, M.A.A.M.; Mustafa, M.N.; Walvekar, R.; Khalid, M.; Khosla, A. Advancements in Silicon Anodes for Enhanced Lithium-Ion Batteries Performance: Innovations Toward Next-Gen Superbatteries. *Battery Energy* **2025**, e20240048. [CrossRef]
214. Prachi Patel The Age of Silicon Is Here…for Batteries. Available online: https://spectrum.ieee.org/silicon-anode-battery (accessed on 24 February 2025).
215. Saddique, J.; Wu, M.; Ali, W.; Xu, X.; Jiang, Z.-G.; Tong, L.; Zheng, H.; Hu, W. Opportunities and Challenges of Nano Si/C Composites in Lithium Ion Battery: A Mini Review. *J. Alloys Compd.* **2024**, *978*, 173507. [CrossRef]
216. Choi, M.; Lee, E.; Sung, J.; Kim, N.; Ko, M. Comparison of Commercial Silicon-Based Anode Materials for the Design of a High-Energy Lithium-Ion Battery. *Nano Res.* **2024**, *17*, 5270–5277. [CrossRef]
217. Li, L.; Deng, Y.; Hu, K.; Xu, B.; Wang, N.; Bai, Z.; Xu, X.; Yang, J. Nanostructure Designing and Hybridizing of High-Capacity Silicon-Based Anode for Lithium-Ion Batteries. *Prog. Nat. Sci. Mater. Int.* **2023**, *33*, 16–36. [CrossRef]
218. Hossain, M.A.M.; Tiong, S.K.; Hannan, M.A.; Ker, P.J.; Fattah, I.M.R.; Mahlia, T.M.I. Recent Advances in Silicon Nanomaterials for Lithium-Ion Batteries: Synthesis Approaches, Emerging Trends, Challenges, and Opportunities. *Sustain. Mater. Technol.* **2024**, *40*, e00964. [CrossRef]
219. Toki, G.F.I.; Hossain, M.K.; Rehman, W.U.; Manj, R.Z.A.; Wang, L.; Yang, J. Recent Progress and Challenges in Silicon-Based Anode Materials for Lithium-Ion Batteries. *Ind. Chem. Mater.* **2024**, *2*, 226–269. [CrossRef]
220. Amardeep, A.; Freschi, D.J.; Wang, J.; Liu, J. Fundamentals, Preparation, and Mechanism Understanding of Li/Na/Mg-Sn Alloy Anodes for Liquid and Solid-State Lithium Batteries and Beyond. *Nano Res.* **2023**, *16*, 8191–8218. [CrossRef]
221. Guo, S.; Feng, Y.; Wang, L.; Jiang, Y.; Yu, Y.; Hu, X. Architectural Engineering Achieves High-Performance Alloying Anodes for Lithium and Sodium Ion Batteries. *Small* **2021**, *17*, 2005248. [CrossRef]
222. Xu, H.; Li, H.; Wang, X. The Anode Materials for Lithium-Ion and Sodium-Ion Batteries Based on Conversion Reactions: A Review. *ChemElectroChem* **2023**, *10*, e202201151. [CrossRef]
223. Ming, J.; Kwak, W.J.; Youn, S.J.; Ming, H.; Hassoun, J.; Sun, Y. Lithiation of an Iron Oxide-Based Anode for Stable, High-Capacity Lithium-Ion Batteries of Porous Carbon–Fe_3O_4/Li[$Ni_{0.59}Co_{0.16}Mn_{0.25}$]$O_2$. *Energy Technol.* **2014**, *2*, 778–785. [CrossRef]
224. Wu, Q.; Jiang, R.; Mu, L.; Xu, S. Fe_3O_4 Anodes for Lithium Batteries: Production Techniques and General Applications. *Comptes Rendus Chim.* **2019**, *22*, 96–102. [CrossRef]
225. Li, Q.; Han, N.; Chai, J.; Zhang, W.; Du, J.; Tian, H.; Liu, H.; Wang, G.; Tang, B. Strategies to Improve Metal-Organic Frameworks and Their Derived Oxides as Lithium Storage Anode Materials. *Energy* **2023**, *282*, 128378. [CrossRef]
226. Liu, Z.; Huang, B.; Li, C.; Zhu, H.; Liu, G. Review of Progress in the Application of Polytetrafluoroethylene-Based Battery Separators. *ACS Appl. Mater. Interfaces* **2024**, *16*, 63109–63128. [CrossRef] [PubMed]

227. Mu, A.U.; Cai, G.; Chen, Z. Metal-Organic Frameworks for the Enhancement of Lithium-Based Batteries: A Mini Review on Emerging Functional Designs. *Adv. Sci.* **2024**, *11*, 2305280. [CrossRef] [PubMed]
228. Eftekhari, A. Low Voltage Anode Materials for Lithium-Ion Batteries. *Energy Storage Mater.* **2017**, *7*, 157–180. [CrossRef]
229. Qian, X.; Xu, Q.; Hang, T.; Shanmugam, S.; Li, M. Electrochemical Deposition of Fe_3O_4 Nanoparticles and Flower-like Hierarchical Porous Nanoflakes on 3D Cu-Cone Arrays for Rechargeable Lithium Battery Anodes. *Mater. Des.* **2017**, *121*, 321–334. [CrossRef]
230. Davis, K.; Demopoulos, G.P. Effective Upcycling of NMC 111 to NMC 622 Cathodes by Hydrothermal Relithiation and Ni-Enriching Annealing. *Next Energy* **2024**, *4*, 100122. [CrossRef]
231. Evro, S.; Ajumobi, A.; Mayon, D.; Tomomewo, O.S. Navigating Battery Choices: A Comparative Study of Lithium Iron Phosphate and Nickel Manganese Cobalt Battery Technologies. *Future Batter.* **2024**, *4*, 100007. [CrossRef]
232. Celadon, A.; Sun, H.; Sun, S.; Zhang, G. Review Batteries for Electric Vehicles: Technical Advancements, Environmental Challenges, and Market Perspectives. *Susmat* **2024**, *4*, e234. [CrossRef]
233. EVDB Which Cars Have LFP Batteries? Available online: https://evdb.nz/ev-battery (accessed on 14 February 2025).
234. Zeekr The New Zeekr 5.5C EV Batteries Can Charge at the Fastest Rate in the World. Available online: https://www.zeekrlife.com/global/posts/the-new-zeekr-5-5c-ev-batteries-can-charge-at-the-fastest-rate-in-the-world (accessed on 14 February 2025).
235. Mark Andrews General Motors to Produce Ultra-Fast 6C Charging EV in China. Available online: https://carnewschina.com/2024/09/26/general-motors-to-produce-ultra-fast-6c-charging-ev-in-china/ (accessed on 14 February 2025).
236. Mark Andrews Geely's Aegis Short Blade LFP Battery—Full Details Including Extreme Safety. Available online: https://carnewschina.com/2024/06/28/geelys-aegis-short-blade-lfp-battery-full-details-including-extreme-safety/ (accessed on 14 February 2025).
237. Green Car Congress, "Gotion High-Tech Launches New L600 LMFP Astroinno Battery; Single-Cell Energy Density of 240Wh/kg". Available online: https://www.greencarcongress.com/2023/05/20230524-astroinno.html (accessed on 20 February 2025).
238. Hasan, S.; Islam, M.S.; Bashar, S.A.; Tamzid, A.A.N.; Hossain, R.B.; Haque, M.A.; Faishal, R. Beyond Lithium-Ion: The Promise and Pitfalls of BYD's Blade Batteries for Electric Vehicles. *E3S Web Conf.* **2023**, *469*, 00005. [CrossRef]
239. Hasselwander, S.; Meyer, M.; Österle, I. Techno-Economic Analysis of Different Battery Cell Chemistries for the Passenger Vehicle Market. *Batteries* **2023**, *9*, 379. [CrossRef]
240. Waseem, M.; Lakshmi, G.S.; Ahmad, M.; Suhaib, M. Energy Storage Technology and Its Impact in Electric Vehicle: Current Progress and Future Outlook. *Next Energy* **2025**, *6*, 100202. [CrossRef]
241. Almadani, M.M.M.; Emmanuel Oni, O.; Mary Longe, O.; Olatomiwa, L. Comparison of Battery Chemistries for Electric Vehicle Applications. In Proceedings of the International Conference on Industrial Engineering and Operations Management, Pretoria, South Africa, 23 April 2024; IEOM Society International: Johannesburg, South Africa; Pretoria, South Africa, 2024.
242. Sharma, S.; Panwar, A.K.; Tripathi, M.M. Storage Technologies for Electric Vehicles. *J. Traffic Transp. Eng. (Engl. Ed.)* **2020**, *7*, 340–361. [CrossRef]
243. Takami, N.; Inagaki, H.; Tatebayashi, Y.; Saruwatari, H.; Honda, K.; Egusa, S. High-Power and Long-Life Lithium-Ion Batteries Using Lithium Titanium Oxide Anode for Automotive and Stationary Power Applications. *J. Power Sources* **2013**, *244*, 469–475. [CrossRef]
244. Vishnumurthy, K.A.; Girish, K.H. A Comprehensive Review of Battery Technology for E-Mobility. *J. Indian Chem. Soc.* **2021**, *98*, 100173. [CrossRef]
245. Zhang, H.; Zhang, J. An Overview of Modification Strategies to Improve $LiNi_{0.8}Co_{0.1}Mn_{0.1}O_2$ (NCM811) Cathode Performance for Automotive Lithium-Ion Batteries. *eTransportation* **2021**, *7*, 100105. [CrossRef]
246. Dasari, H.A.; Rammohan, A. Evaluating Fault Detection Strategies for Lithium-Ion Batteries in Electric Vehicles. *Eng. Res. Express* **2024**, *6*, 032302. [CrossRef]
247. Karmakar, A.; Zhou, H.; Vishnugopi, B.S.; Mukherjee, P.P. Li-Ion Battery Safety: A Perspective on Hierarchy of Scales. *Annu. Rev. Heat Transf.* **2023**, *26*, 11–68. [CrossRef]
248. Jia, Z.; Jin, K.; Mei, W.; Qin, P.; Jinhua, S.; Wang, Q. Advances and Perspectives in Fire Safety of Lithium-Ion Battery Energy Storage Systems. *eTransportation* **2025**, *24*, 100390. [CrossRef]
249. Comanescu, C. Ensuring Safety and Reliability: An Overview of Lithium-Ion Battery Service Assessment. *Batteries* **2025**, *11*, 6. [CrossRef]
250. Menale, C.; Mancino, A.N.; Vitiello, F.; Sglavo, V.; Vellucci, F.; Caiazzo, L.; Bubbico, R. Analysis of the Thermal Runaway Mitigation Performances of Dielectric Fluids During Overcharge Abuse Tests of Lithium-Ion Cells with Lithium Titanate Oxide Anodes. *WEVJ* **2024**, *15*, 554. [CrossRef]
251. Shahid, S.; Agelin-Chaab, M. A Review of Thermal Runaway Prevention and Mitigation Strategies for Lithium-Ion Batteries. *Energy Convers. Manag. X* **2022**, *16*, 100310. [CrossRef]
252. Wang, H.; Xu, H.; Zhang, Z.; Wang, Q.; Jin, C.; Wu, C.; Xu, C.; Hao, J.; Sun, L.; Du, Z.; et al. Fire and Explosion Characteristics of Vent Gas from Lithium-Ion Batteries after Thermal Runaway: A Comparative Study. *eTransportation* **2022**, *13*, 100190. [CrossRef]

253. Shan, T.; Zhang, P.; Wang, Z.; Zhu, X. Insights into Extreme Thermal Runaway Scenarios of Lithium-Ion Batteries Fire and Explosion: A Critical Review. *J. Energy Storage* **2024**, *88*, 111532. [CrossRef]
254. Shi, P.; Zhu, H.; Dong, X.; Hai, B. Research Progress on Thermal Runaway Warning Methods and Fire Extinguishing Technologies for Lithium-Ion Batteries. *WEVJ* **2025**, *16*, 81. [CrossRef]
255. Liu, Y.; Pan, Y.; Wang, H.; Li, H.; Bao, H.; Zhao, Z.; Liu, B. Mechanical Issues of Lithium-Ion Batteries in Road Traffic Conditions: A Review. *Thin-Walled Struct.* **2024**, *201*, 111985. [CrossRef]
256. Zhao, J.; Feng, X.; Tran, M.-K.; Fowler, M.; Ouyang, M.; Burke, A.F. Battery Safety: Fault Diagnosis from Laboratory to Real World. *J. Power Sources* **2024**, *598*, 234111. [CrossRef]
257. Xiao, Y.; Yang, F.; Gao, Z.; Liu, M.; Wang, J.; Kou, Z.; Lin, Y.; Li, Y.; Gao, L.; Chen, Y.; et al. Review of Mechanical Abuse Related Thermal Runaway Models of Lithium-Ion Batteries at Different Scales. *J. Energy Storage* **2023**, *64*, 107145. [CrossRef]
258. Wang, Z.; Zhao, Q.; Wang, S.; Song, Y.; Shi, B.; He, J. Aging and Post-Aging Thermal Safety of Lithium-Ion Batteries under Complex Operating Conditions: A Comprehensive Review. *J. Power Sources* **2024**, *623*, 235453. [CrossRef]
259. Cui, Z.; Manthiram, A. Thermal Stability and Outgassing Behaviors of High-Nickel Cathodes in Lithium-Ion Batteries. *Angew. Chem. Int. Ed.* **2023**, *62*, e202307243. [CrossRef]
260. Liang, T.; Cheng, D.; Chen, J.; Wu, X.; Xiong, H.; Yu, S.; Zhang, Z.; Liu, H.; Liu, S.; Song, X. Evolution from Passive to Active Components in Lithium Metal and Lithium-Ion Batteries Separators. *Mater. Today Energy* **2024**, *45*, 101684. [CrossRef]
261. Ha, J.; Lee, J.; Kim, Y.-T.; Choi, J. Effective Approaches to Preventing Dendrite Growth in Lithium Metal Anodes: A Review. *Appl. Chem. Eng.* **2023**, *34*, 365–382. [CrossRef]
262. Mishra, R.; Anne, M.; Das, S.; Chavva, T.; Shelke, M.V.; Pol, V.G. Glory of Fire Retardants in Li-Ion Batteries: Could They Be Intrinsically Safer? *Adv. Sustain. Syst.* **2024**, *8*, 2400273. [CrossRef]
263. Yun, S.; Liang, X.; Xi, J.; Liao, L.; Cui, S.; Chen, L.; Li, S.; Hu, Q. Electrolytes for High-Safety Lithium-Ion Batteries at Low Temperature: A Review. *Polymers* **2024**, *16*, 2661. [CrossRef] [PubMed]
264. Cui, Z.; Liu, C.; Wang, F.; Manthiram, A. Navigating Thermal Stability Intricacies of High-Nickel Cathodes for High-Energy Lithium Batteries. *Nat. Energy* **2025**, *10*, 490–501. [CrossRef]
265. Yerkinbekova, Y.; Kumarov, A.; Tatykayev, B.; Mentbayeva, A.; Repo, E.; Laakso, E. Ni-Rich Cathode Materials with Concentration Gradients for High-Energy and Safe Lithium-Ion Batteries: A Comprehensive Review. *J. Power Sources* **2025**, *626*, 235686. [CrossRef]
266. Tang, Z.; Feng, D.; Xu, Y.; Chen, L.; Zhang, X.; Ma, Q. Safety Issues of Layered Nickel-Based Cathode Materials for Lithium-Ion Batteries: Origin, Strategies and Prospects. *Batteries* **2023**, *9*, 156. [CrossRef]
267. Bin Abu Sofian, A.D.A.; Imaduddin, I.S.; Majid, S.R.; Kurniawan, T.A.; Chew, K.W.; Lay, C.-H.; Show, P.L. Nickel-Rich Nickel–Cobalt–Manganese and Nickel–Cobalt–Aluminum Cathodes in Lithium-Ion Batteries: Pathways for Performance Optimization. *J. Clean. Prod.* **2024**, *435*, 140324. [CrossRef]
268. Wu, Z.; Zhang, C.; Yuan, F.; Lyu, M.; Yang, P.; Zhang, L.; Zhou, M.; Wang, L.; Zhang, S.; Wang, L. Ni-Rich Cathode Materials for Stable High-Energy Lithium-Ion Batteries. *Nano. Energy* **2024**, *126*, 109620. [CrossRef]
269. Tran, M.-K.; Mevawalla, A.; Aziz, A.; Panchal, S.; Xie, Y.; Fowler, M. A Review of Lithium-Ion Battery Thermal Runaway Modeling and Diagnosis Approaches. *Processes* **2022**, *10*, 1192. [CrossRef]
270. Wang, Q.; Mao, B.; Stoliarov, S.I.; Sun, J. A Review of Lithium Ion Battery Failure Mechanisms and Fire Prevention Strategies. *Prog. Energy Combust. Sci.* **2019**, *73*, 95–131. [CrossRef]
271. Feng, X.; Ouyang, M.; Liu, X.; Lu, L.; Xia, Y.; He, X. Thermal Runaway Mechanism of Lithium Ion Battery for Electric Vehicles: A Review. *Energy Storage Mater.* **2018**, *10*, 246–267. [CrossRef]
272. Dai, Y.; Panahi, A. Thermal Runaway Process in Lithium-Ion Batteries: A Review. *Next Energy* **2025**, *6*, 100186. [CrossRef]
273. Shen, R.; Quan, Y.; McIntosh, J.D.; Salem, A.; Wang, Q. Fire Safety of Battery Electric Vehicles: Hazard Identification, Detection, and Mitigation. *SAE Int. J. Electrified Veh.* **2024**, *13*, 279–294. [CrossRef]
274. Choudhary, A.; Fatima, S.; Panigrahi, B.K. State-of-the-Art Technologies in Fault Diagnosis of Electric Vehicles: A Component-Based Review. *IEEE Trans. Transp. Electrif.* **2023**, *9*, 2324–2347. [CrossRef]
275. He, H.; Sun, F.; Wang, Z.; Lin, C.; Zhang, C.; Xiong, R.; Deng, J.; Zhu, X.; Xie, P.; Zhang, S.; et al. China's Battery Electric Vehicles Lead the World: Achievements in Technology System Architecture and Technological Breakthroughs. *Green Energy Intell. Transp.* **2022**, *1*, 100020. [CrossRef]
276. European Parliament. *Directive 2006/66/EC of the European Parliament and of the Council of 6 September 2006 on Batteries and Accumulators and Waste Batteries and Accumulators and Repealing Directive 91/157/EEC*; European Parliament and the Council: Brussels, Belgium, 2018.
277. Ahn, H.; Kim, D.; Lee, M.; Nam, K.W. Challenges and Possibilities for Aqueous Battery Systems. *Commun. Mater.* **2023**, *4*, 37. [CrossRef]
278. Guan, S.; Peng, Q.; Guo, X.; Zheng, Y.; Liao, E.; Sun, S.; Shin, K.; Liu, B.; Zhou, X.; Zou, C.; et al. A Review of the Advances and Prospects of Aqueous Dual-Ion Batteries. *Chem. Eng. J.* **2024**, *493*, 152864. [CrossRef]

279. Khan, T.; Garg, A.K.; Gupta, A.; Madan, A.K.; Jain, P.K. Comprehensive Review on Latest Advances on Rechargeable Batteries. *J. Energy Storage* **2023**, *57*, 106204. [CrossRef]
280. Heo, J.J.; Ryu, J. A Critical Review of Ultrafast Charging Dual-Ion Batteries. *Korean J. Chem. Eng.* 2024. [CrossRef]
281. Nzereogu, P.U.; Oyesanya, A.; Ogba, S.N.; Ayanwunmi, S.O.; Sobajo, M.S.; Chimsunum, V.C.; Ayanwunmi, V.O.; Amoo, M.O.; Adefemi, O.T.; Chukwudi, C.C. Solid-State Lithium-Ion Battery Electrolytes: Revolutionizing Energy Density and Safety. *Hybrid Adv.* **2025**, *8*, 100339. [CrossRef]
282. Koech, A.K.; Mwandila, G.; Mulolani, F. A Review of Improvements on Electric Vehicle Battery. *Heliyon* **2024**, *10*, e34806. [CrossRef] [PubMed]
283. Gandolfo, M.; Versaci, D.; Francia, C.; Bodoardo, S.; Amici, J. Enhancing the Safety and Stability of Lithium Metal Batteries through the Use of Composite Ionogels. *Electrochim. Acta* **2023**, *463*, 142857. [CrossRef]
284. Umair, M.; Zhou, S.; Li, W.; Rana, H.T.H.; Yang, J.; Cheng, L.; Li, M.; Yu, S.; Wei, J. Oxide Solid Electrolytes in Solid-State Batteries. *Batter. Amp. Supercaps* **2024**, e202400667. [CrossRef]
285. Rufino Júnior, C.A.; Sanseverino, E.R.; Gallo, P.; Amaral, M.M.; Koch, D.; Kotak, Y.; Diel, S.; Walter, G.; Schweiger, H.-G.; Zanin, H. Unraveling the Degradation Mechanisms of Lithium-Ion Batteries. *Energies* **2024**, *17*, 3372. [CrossRef]
286. Ai, S.; Wu, X.; Wang, J.; Li, X.; Hao, X.; Meng, Y. Research Progress on Solid-State Electrolytes in Solid-State Lithium Batteries: Classification, Ionic Conductive Mechanism, Interfacial Challenges. *Nanomaterials* **2024**, *14*, 1773. [CrossRef]
287. Zhao, Z.; Liang, W.; Su, S.; Jiang, X.; Bando, Y.; Zhang, B.; Ma, Z.; Wang, X. Advances of Solid Polymer Electrolytes with High-Voltage Stability. *Next Mater.* **2025**, *7*, 100364. [CrossRef]
288. Joshi, A.; Mishra, D.K.; Singh, R.; Zhang, J.; Ding, Y. A Comprehensive Review of Solid-State Batteries. *Appl. Energy* **2025**, *386*, 125546. [CrossRef]
289. Hu, Y.; Li, W.; Zhu, J.; Hao, S.-M.; Qin, X.; Fan, L.-Z.; Zhang, L.; Zhou, W. Multi-Layered Electrolytes for Solid-State Lithium Batteries. *Next Energy* **2023**, *1*, 100042. [CrossRef]
290. Cheng, Z.; Liu, T.; Zhao, B.; Shen, F.; Jin, H.; Han, X. Recent Advances in Organic-Inorganic Composite Solid Electrolytes for All-Solid-State Lithium Batteries. *Energy Storage Mater.* **2021**, *34*, 388–416. [CrossRef]
291. Sashmitha, K.; Rani, M.U. A Comprehensive Review of Polymer Electrolyte for Lithium-Ion Battery. *Polym. Bull.* **2023**, *80*, 89–135. [CrossRef]
292. Liu, Y.; Zhao, C.; Du, J.; Zhang, X.; Chen, A.; Zhang, Q. Research Progresses of Liquid Electrolytes in Lithium-Ion Batteries. *Small* **2023**, *19*, 2205315. [CrossRef] [PubMed]
293. Li, Z.; Wang, L.; Huang, X.; He, X. Lithium Bis(Trifluoromethanesulfonyl)Imide (LiTFSI): A Prominent Lithium Salt in Lithium-Ion Battery Electrolytes—Fundamentals, Progress, and Future Perspectives. *Adv. Funct. Mater.* **2024**, *34*, 2408319. [CrossRef]
294. Gong, Y.; Wang, C.; Xin, M.; Chen, S.; Xu, P.; Li, D.; Liu, J.; Wang, Y.; Xie, H.; Sun, X.; et al. Ultra-Thin and High-Voltage-Stable Bi-Phasic Solid Polymer Electrolytes for High-Energy-Density Li Metal Batteries. *Nano Energy* **2024**, *119*, 109054. [CrossRef]
295. Pereira, J.; Souza, R.; Moreira, A.; Moita, A. An Overview of the Ionic Liquids and Their Hybrids Operating in Electrochemical Cells and Capacitors. *Ionics* **2024**, *30*, 4343–4385. [CrossRef]
296. Inada, R.; Yasuda, S.; Hosokawa, H.; Saito, M.; Tojo, T.; Sakurai, Y. Formation and Stability of Interface between Garnet-Type Ta-Doped Li7La3Zr2O12 Solid Electrolyte and Lithium Metal Electrode. *Batteries* **2018**, *4*, 26. [CrossRef]
297. Ahmad, H.; Kubra, K.T.; Butt, A.; Nisar, U.; Iftikhar, F.J.; Ali, G. Recent Progress, Challenges, and Perspectives in the Development of Solid-State Electrolytes for Sodium Batteries. *J. Power Sources* **2023**, *581*, 233518. [CrossRef]
298. Chen, J.; Wu, J.; Wang, X.; Zhou, A.; Yang, Z. Research Progress and Application Prospect of Solid-State Electrolytes in Commercial Lithium-Ion Power Batteries. *Energy Storage Mater.* **2021**, *35*, 70–87. [CrossRef]
299. Lannelongue, P.; Lindberg, S.; Gonzalo, E.; Golov, A.; Bonilla, F.; Lopez Del Amo, J.M.; Marchandier, T.; Tron, A.; Carrasco, J.; López-Aranguren, P. Stable Cycling of Halide Solid State Electrolyte Enabled by a Dynamic Layered Solid Electrolyte Interphase between Li Metal and $Li_3YCl_4Br_2$. *Energy Storage Mater.* **2024**, *72*, 103733. [CrossRef]
300. He, B.; Zhang, F.; Xin, Y.; Xu, C.; Hu, X.; Wu, X.; Yang, Y.; Tian, H. Halogen Chemistry of Solid Electrolytes in All-Solid-State Batteries. *Nat. Rev. Chem.* **2023**, *7*, 826–842. [CrossRef]
301. Nikodimos, Y.; Su, W.; Hwang, B.J. Halide Solid-State Electrolytes: Stability and Application for High Voltage All-Solid-State Li Batteries. *Adv. Energy Mater.* **2023**, *13*, 2202854. [CrossRef]
302. Zhang, X.; Cheng, S.; Fu, C.; Yin, G.; Wang, L.; Wu, Y.; Huo, H. Advancements and Challenges in Organic–Inorganic Composite Solid Electrolytes for All-Solid-State Lithium Batteries. *Nano-Micro Lett.* **2025**, *17*, 2. [CrossRef] [PubMed]
303. Yang, T.; Wang, C.; Zhang, W.; Xia, Y.; Huang, H.; Gan, Y.; He, X.; Xia, X.; Tao, X.; Zhang, J. A Critical Review on Composite Solid Electrolytes for Lithium Batteries: Design Strategies and Interface Engineering. *J. Energy Chem.* **2023**, *84*, 189–209. [CrossRef]
304. *Polymer Electrolytes and Their Composites for Energy Storage/Conversion Devices*; Emerging materials and technologies; Sharma, A.L., Arya, A., Gaur, A., Eds.; CRC Press: Boca Raton, FL, USA, 2023; ISBN 978-1-003-20866-2.
305. Boaretto, N.; Garbayo, I.; Valiyaveettil-SobhanRaj, S.; Quintela, A.; Li, C.; Casas-Cabanas, M.; Aguesse, F. Lithium Solid-State Batteries: State-of-the-Art and Challenges for Materials, Interfaces and Processing. *J. Power Sources* **2021**, *502*, 229919. [CrossRef]

306. Sun, Y.-Y.; Zhang, Q.; Yan, L.; Wang, T.-B.; Hou, P.-Y. A Review of Interfaces within Solid-State Electrolytes: Fundamentals, Issues and Advancements. *Chem. Eng. J.* **2022**, *437*, 135179. [CrossRef]
307. Hun, Q.; Lan, L.; Lu, X.; Hu, Q.; Liang, X.; Guo, Y.; Wang, Y. Bilayer Heterostructure Electrolytes Were Prepared by a UV-Curing Process for High Temperature Lithium-Ion Batteries. *Polymers* **2024**, *16*, 2972. [CrossRef]
308. Scheller, M.; Durdel, A.; Frank, A.; Jossen, A. Model-Based Performance Evaluation of Hybrid Solid-State Batteries: Impact of Laser-Ablated Geometrical Structures. *Batteries* **2024**, *10*, 392. [CrossRef]
309. Burazer, S.; Popović, J. Mechanochemical Synthesis of Solid-State Electrolytes. *Inorganics* **2024**, *12*, 54. [CrossRef]
310. Van Der Ven, A.; McMeeking, R.M.; Clément, R.J.; Garikipati, K. Ferroelastic Toughening: Can It Solve the Mechanics Challenges of Solid Electrolytes? *Curr. Opin. Solid State Mater. Sci.* **2023**, *27*, 101056. [CrossRef]
311. Ning, Z.; Li, G.; Melvin, D.L.R.; Chen, Y.; Bu, J.; Spencer-Jolly, D.; Liu, J.; Hu, B.; Gao, X.; Perera, J.; et al. Dendrite Initiation and Propagation in Lithium Metal Solid-State Batteries. *Nature* **2023**, *618*, 287–293. [CrossRef]
312. John Carey Solid-State Batteries Could Revolutionize EVs and More—If They Can Surmount Technical and Financial Hurdles. Available online: https://www.pnas.org/doi/10.1073/pnas.2425219121 (accessed on 5 February 2025).
313. Biba, Jacob 7 New Battery Technologies to Watch. Available online: https://builtin.com/hardware/new-battery-technologies (accessed on 27 February 2025).
314. Tan, A.K.X.; Paul, S. Beyond Lithium: Future Battery Technologies for Sustainable Energy Storage. *Energies* **2024**, *17*, 5768. [CrossRef]
315. Singh, A.N.; Islam, M.; Meena, A.; Faizan, M.; Han, D.; Bathula, C.; Hajibabaei, A.; Anand, R.; Nam, K.-W. Unleashing the Potential of Sodium-Ion Batteries: Current State and Future Directions for Sustainable Energy Storage. *Adv. Funct. Mater.* **2023**, *33*, 2304617. [CrossRef]
316. Biemolt, J.; Jungbacker, P.; Van Teijlingen, T.; Yan, N.; Rothenberg, G. Beyond Lithium-Based Batteries. *Materials* **2020**, *13*, 425. [CrossRef]
317. Qi, T.; Xiong, K.; Zhang, X. Research Progress of Carbon Materials in the Anodes of Sodium-Ion Batteries. *J. Power Sources* **2025**, *626*, 235721. [CrossRef]
318. Yin, J.; Zhang, Y.S.; Liang, H.; Zhang, W.; Zhu, Y. Synthesis Strategies of Hard Carbon Anodes for Sodium-Ion Batteries. *Mater. Rep. Energy* **2024**, *4*, 100268. [CrossRef]
319. Chen, F.; Li, H.; Qiao, X.; Wang, R.; Hu, C.; Chen, T.; Niu, Y.; Zhong, B.; Wu, Z.; Guo, X. The Chance of Sodium Titanate Anode for the Practical Sodium-Ion Batteries. *Chin. J. Chem. Eng.* **2024**, *72*, 226–244. [CrossRef]
320. Wei, C.; Qu, D.; Li, Q.; Sun, Z.; Song, Z.; Guan, H.; Niu, L. Recent Advances on Transition Metal Chalcogenide for Sodium-Ion Batteries. *Batteries* **2023**, *9*, 467. [CrossRef]
321. Shkrob, I.A.; Wishart, J.F.; Abraham, D.P. What Makes Fluoroethylene Carbonate Different? *J. Phys. Chem. C* **2015**, *119*, 14954–14964. [CrossRef]
322. Zhao, Y.; Zhang, Z.; Zheng, Y.; Luo, Y.; Jiang, X.; Wang, Y.; Wang, Z.; Wu, Y.; Zhang, Y.; Liu, X.; et al. Sodium-Ion Battery at Low Temperature: Challenges and Strategies. *Nanomaterials* **2024**, *14*, 1604. [CrossRef]
323. Yu, Z.; Gu, Y.; Sun, Q.; Zheng, Y.; Zhang, Y.; Zhang, M.; Zhang, D.; Zhang, Z.; Jiang, Y. Research Progress of Modified Metal Current Collectors in Sodium Metal Anodes. *Chin. Chem. Lett.* **2024**, *36*, 109997. [CrossRef]
324. Chen, C.; Yao, W.; Tang, Y. Emerging Solutions to Enable the Efficient Use of Sodium Metal Anodes: Progress and Perspectives. *Adv Funct Mater.* **2024**, *34*, 2310833. [CrossRef]
325. Li, Y.; Zhou, Q.; Weng, S.; Ding, F.; Qi, X.; Lu, J.; Li, Y.; Zhang, X.; Rong, X.; Lu, Y.; et al. Interfacial Engineering to Achieve an Energy Density of over 200 Wh Kg^{-1} in Sodium Batteries. *Nat Energy* **2022**, *7*, 511–519. [CrossRef]
326. Zhang, L.; Wang, J.; Ji, W.; Tapia-Ruiz, N.; Winter, M.; Li, J. Structural Evolution Mechanisms and Design Strategies of Layered Cathodes for Sodium-Ion Batteries. *Next Energy* **2025**, *7*, 100241. [CrossRef]
327. Wang, C.; Liu, L.; Zhao, S.; Liu, Y.; Yang, Y.; Yu, H.; Lee, S.; Lee, G.-H.; Kang, Y.-M.; Liu, R.; et al. Tuning Local Chemistry of P2 Layered-Oxide Cathode for High Energy and Long Cycles of Sodium-Ion Battery. *Nat Commun* **2021**, *12*, 2256. [CrossRef]
328. Sathiya, M.; Jacquet, Q.; Doublet, M.; Karakulina, O.M.; Hadermann, J.; Tarascon, J. A Chemical Approach to Raise Cell Voltage and Suppress Phase Transition in O_3 Sodium Layered Oxide Electrodes. *Adv. Energy Mater.* **2018**, *8*, 1702599. [CrossRef]
329. Zheng, B.; Qian, H.; Cheng, G.; Yuan, C.; Cheng, Y.; Wang, M.-S.; Liu, X.; Xiang, Y. Ultrafast Lattice Engineering for High Energy Density and High-Rate Sodium-Ion Layered Oxide Cathodes. *Energy Storage Mater.* **2025**, *74*, 103868. [CrossRef]
330. Rajagopalan, R.; Zhang, Z.; Tang, Y.; Jia, C.; Ji, X.; Wang, H. Understanding Crystal Structures, Ion Diffusion Mechanisms and Sodium Storage Behaviors of NASICON Materials. *Energy Storage Mater.* **2021**, *34*, 171–193. [CrossRef]
331. Zhou, Q.; Wang, L.; Li, W.; Zhao, K.; Liu, M.; Wu, Q.; Yang, Y.; He, G.; Parkin, I.P.; Shearing, P.R.; et al. Sodium Superionic Conductors (NASICONs) as Cathode Materials for Sodium-Ion Batteries. *Electrochem. Energy Rev.* **2021**, *4*, 793–823. [CrossRef]
332. Deng, L.; Yu, F.-D.; Xia, Y.; Jiang, Y.-S.; Sui, X.-L.; Zhao, L.; Meng, X.-H.; Que, L.-F.; Wang, Z.-B. Stabilizing Fluorine to Achieve High-Voltage and Ultra-Stable Na3V2(PO4)2F3 Cathode for Sodium Ion Batteries. *Nano Energy* **2021**, *82*, 105659. [CrossRef]

333. Zhang, J.; Liu, Y.; Zhao, X.; He, L.; Liu, H.; Song, Y.; Sun, S.; Li, Q.; Xing, X.; Chen, J. A Novel NASICON-Type Na_4 $MnCr(PO_4)_3$ Demonstrating the Energy Density Record of Phosphate Cathodes for Sodium-Ion Batteries. *Adv. Mater.* **2020**, *32*, 1906348. [CrossRef]
334. Tang, Y.; Li, W.; Feng, P.; Zhou, M.; Wang, K.; Wang, Y.; Zaghib, K.; Jiang, K. High-Performance Manganese Hexacyanoferrate with Cubic Structure as Superior Cathode Material for Sodium-Ion Batteries. *Adv Funct Mater.* **2020**, *30*, 1908754. [CrossRef]
335. Bauer, A.; Song, J.; Vail, S.; Pan, W.; Barker, J.; Lu, Y. The Scale-up and Commercialization of Nonaqueous Na-Ion Battery Technologies. *Adv. Energy Mater.* **2018**, *8*, 1702869. [CrossRef]
336. Zhang, H.; Gao, Y.; Liu, X.; Yang, Z.; He, X.; Li, L.; Qiao, Y.; Chen, W.; Zeng, R.; Wang, Y.; et al. Organic Cathode Materials for Sodium-Ion Batteries: From Fundamental Research to Potential Commercial Application. *Adv Funct Mater.* **2022**, *32*, 2107718. [CrossRef]
337. Wang, S.; Wang, L.; Zhu, Z.; Hu, Z.; Zhao, Q.; Chen, J. All Organic Sodium-Ion Batteries with $Na_4C_8H_2O_6$. *Angew. Chem.* **2014**, *126*, 6002–6006. [CrossRef]
338. Chu, S.; Guo, S.; Zhou, H. Advanced Cobalt-Free Cathode Materials for Sodium-Ion Batteries. *Chem. Soc. Rev.* **2021**, *50*, 13189–13235. [CrossRef] [PubMed]
339. Zhang, W.; Lu, J.; Guo, Z. Challenges and Future Perspectives on Sodium and Potassium Ion Batteries for Grid-Scale Energy Storage. *Mater. Today* **2021**, *50*, 400–417. [CrossRef]
340. Meng, J.; Jia, G.; Yang, H.; Wang, M. Recent Advances for SEI of Hard Carbon Anode in Sodium-Ion Batteries: A Mini Review. *Front. Chem.* **2022**, *10*, 986541. [CrossRef]
341. Huang, Z.-X.; Zhang, X.-L.; Zhao, X.-X.; Zhao, Y.-Y.; Aravindan, V.; Liu, Y.-H.; Geng, H.; Wu, X.-L. Electrode/Electrolyte Additives for Practical Sodium-Ion Batteries: A Mini Review. *Inorg. Chem. Front.* **2023**, *10*, 37–48. [CrossRef]
342. Yue, Y.; Jia, Z.; Li, Y.; Wen, Y.; Lei, Q.; Duan, Q.; Sun, J.; Wang, Q. Thermal Runaway Hazards Comparison between Sodium-Ion and Lithium-Ion Batteries Using Accelerating Rate Calorimetry. *Process Saf. Environ. Prot.* **2024**, *189*, 61–70. [CrossRef]
343. Bhutia, P.T.; Grugeon, S.; El Mejdoubi, A.; Laruelle, S.; Marlair, G. Safety Aspects of Sodium-Ion Batteries: Prospective Analysis from First Generation Towards More Advanced Systems. *Batteries* **2024**, *10*, 370. [CrossRef]
344. Zhou, Y.; Xu, G.; Lin, J.; Zhu, J.; Pan, J.; Fang, G.; Liang, S.; Cao, X. A Multicationic-Substituted Configurational Entropy-Enabled NASICON Cathode for High-Power Sodium-Ion Batteries. *Nano Energy* **2024**, *128*, 109812. [CrossRef]
345. Liao, J.; Yuan, Z.; Hu, Q.; Sheng, X.; Song, L.; Xu, Y.; Du, Y.; Zhou, X. Heat-Resistant Carbon-Coated Potassium Magnesium Hexacyanoferrate Nanoplates for High-Performance Potassium-Ion Batteries. *Angew Chem Int Ed* **2024**, *63*, e202409145. [CrossRef] [PubMed]
346. Canepa, P.; Sai Gautam, G.; Hannah, D.C.; Malik, R.; Liu, M.; Gallagher, K.G.; Persson, K.A.; Ceder, G. Odyssey of Multivalent Cathode Materials: Open Questions and Future Challenges. *Chem. Rev.* **2017**, *117*, 4287–4341. [CrossRef]
347. Zhang, W.; Yin, J.; Sun, M.; Wang, W.; Chen, C.; Altunkaya, M.; Emwas, A.; Han, Y.; Schwingenschlögl, U.; Alshareef, H.N. Direct Pyrolysis of Supermolecules: An Ultrahigh Edge-Nitrogen Doping Strategy of Carbon Anodes for Potassium-Ion Batteries. *Adv. Mater.* **2020**, *32*, 2000732. [CrossRef] [PubMed]
348. Wang, L.; Li, Y.; Wang, B.; Jing, Z.; Chen, M.; Zhai, Y.; Kong, Z.; Iqbal, S.; Zeng, S.; Sun, X.; et al. Controllable Growth of Crystal Facets Enables Superb Cycling Stability of Anode Material for Potassium Ion Batteries. *Adv Funct Mater.* **2024**, *34*, 2406988. [CrossRef]
349. Ji, B.; Yao, W.; Zheng, Y.; Kidkhunthod, P.; Zhou, X.; Tunmee, S.; Sattayaporn, S.; Cheng, H.-M.; He, H.; Tang, Y. A Fluoroxalate Cathode Material for Potassium-Ion Batteries with Ultra-Long Cyclability. *Nat Commun* **2020**, *11*, 1225. [CrossRef] [PubMed]
350. Liang, Y.; Dong, H.; Aurbach, D.; Yao, Y. Current Status and Future Directions of Multivalent Metal-Ion Batteries. *Nat. Energy* **2020**, *5*, 646–656. [CrossRef]
351. Shah, R.; Mittal, V.; Matsil, E.; Rosenkranz, A. Magnesium-Ion Batteries for Electric Vehicles: Current Trends and Future Perspectives. *Adv. Mech. Eng.* **2021**, *13*, 16878140211003398. [CrossRef]
352. Dominko, R.; Bitenc, J.; Berthelot, R.; Gauthier, M.; Pagot, G.; Di Noto, V. Magnesium Batteries: Current Picture and Missing Pieces of the Puzzle. *J. Power Sources* **2020**, *478*, 229027. [CrossRef]
353. Mizrahi, O.; Amir, N.; Pollak, E.; Chusid, O.; Marks, V.; Gottlieb, H.; Larush, L.; Zinigrad, E.; Aurbach, D. Electrolyte Solutions with a Wide Electrochemical Window for Rechargeable Magnesium Batteries. *J. Electrochem. Soc.* **2008**, *155*, A103. [CrossRef]
354. Doe, R.E.; Han, R.; Hwang, J.; Gmitter, A.J.; Shterenberg, I.; Yoo, H.D.; Pour, N.; Aurbach, D. Novel, Electrolyte Solutions Comprising Fully Inorganic Salts with High Anodic Stability for Rechargeable Magnesium Batteries. *Chem. Commun.* **2014**, *50*, 243–245. [CrossRef]
355. Liu, F.; Guoqin, C.; Jinjin, B.; Honghong, L.; Yan, Z.; Guosheng, S.; Aiguo, Z.; Li, Z.F.; Junhua, H. Recent Advances Based on Mg Anodes and Their Interfacial Modulation in Mg Batteries. *J. of Magnesium and Alloys* **2022**, *10*, 2699–2716. [CrossRef]
356. Ortiz-Rodríguez, J.C.; Velázquez, J.M. Structure–Reactivity Relationships in Chevrel Phase Electrocatalysts for Small-Molecule Reduction Reactions. *Curr. Opin. Electrochem.* **2022**, *34*, 101002. [CrossRef]

357. Liu, Y.; Qu, B.; Li, S.; Lian, X.; Luo, Y.; Shen, X.; Xu, C.; Wang, J.; Pan, F. Recent Advances and Prospects of Chalcogenide Cathodes for Rechargeable Magnesium Batteries. *Adv Funct Mater.* **2024**, *34*, 2405586. [CrossRef]
358. Shin, C.-H.; Lee, H.-Y.; Gyan-Barimah, C.; Yu, J.-H.; Yu, J.-S. Magnesium: Properties and Rich Chemistry for New Material Synthesis and Energy Applications. *Chem. Soc. Rev.* **2023**, *52*, 2145–2192. [CrossRef] [PubMed]
359. Yang, Y.; Xiong, X.; Chen, J.; Peng, X.; Chen, D.; Pan, F. Research Advances of Magnesium and Magnesium Alloys Worldwide in 2022. *J. Magnes. Alloys* **2023**, *11*, 2611–2654. [CrossRef]
360. Wang, J.; Tan, S.; Zhang, G.; Jiang, Y.; Yin, Y.; Xiong, F.; Li, Q.; Huang, D.; Zhang, Q.; Gu, L.; et al. Fast and Stable Mg2+ Intercalation in a High Voltage NaV2O2(PO4)2F/rGO Cathode Material for Magnesium-Ion Batteries. *Sci. China Mater.* **2020**, *63*, 1651–1662. [CrossRef]
361. Chen, Z.; Yang, Q.; Wang, D.; Chen, A.; Li, X.; Huang, Z.; Liang, G.; Wang, Y.; Zhi, C. Tellurium: A High-Performance Cathode for Magnesium Ion Batteries Based on a Conversion Mechanism. *ACS Nano.* **2022**, *16*, 5349–5357. [CrossRef]
362. Ma, X.; Zhao, B.; Liu, H.; Tan, J.; Li, H.; Zhang, X.; Diao, J.; Yue, J.; Huang, G.; Wang, J.; et al. H_2 O-Mg^{2+} Waltz-Like Shuttle Enables High-Capacity and Ultralong-Life Magnesium-Ion Batteries. *Adv. Sci.* **2024**, *11*, 2401005. [CrossRef]
363. Han, W.; Li, Z.; Wang, R. Progress Study on the Development of Lithium-Sulfur Batteries. *IOP Conf. Ser. Earth Environ. Sci.* **2022**, *1011*, 012004. [CrossRef]
364. Dörfler, S.; Althues, H.; Härtel, P.; Abendroth, T.; Schumm, B.; Kaskel, S. Challenges and Key Parameters of Lithium-Sulfur Batteries on Pouch Cell Level. *Joule* **2020**, *4*, 539–554. [CrossRef]
365. Raza, H.; Bai, S.; Cheng, J.; Majumder, S.; Zhu, H.; Liu, Q.; Zheng, G.; Li, X.; Chen, G. Li-S Batteries: Challenges, Achievements and Opportunities. *Electrochem. Energy Rev.* **2023**, *6*, 29. [CrossRef]
366. Vedhanarayanan, B.; Seetha Lakshmi, K.C. Beyond Lithium-Ion: Emerging Frontiers in next-Generation Battery Technologies. *Front. Batter. Electrochem.* **2024**, *3*, 1377192. [CrossRef]
367. Wang, J.; Wang, H.; Jia, S.; Zhao, Q.; Zheng, Q.; Ma, Y.; Ma, T.; Li, X. Recent Advances in Inhibiting Shuttle Effect of Polysulfide in Lithium-Sulfur Batteries. *J. Energy Storage* **2023**, *72*, 108372. [CrossRef]
368. Sun, J.; Wang, T.; Gao, Y.; Pan, Z.; Hu, R.; Wang, J. Will lithium-sulfur Batteries Be the next beyond-lithium Ion Batteries and Even Much Better? *InfoMat* **2022**, *4*, e12359. [CrossRef]
369. GELION Lithium-Sulfur Cell High Energy Density Milestone Achieved. Available online: https://gelion.com/news/lithium-sulfur-cell-high-energy-density-milestone-achieved/ (accessed on 5 March 2025).
370. Ni, W. Perspectives on Advanced Lithium–Sulfur Batteries for Electric Vehicles and Grid-Scale Energy Storage. *Nanomaterials* **2024**, *14*, 990. [CrossRef]
371. Oh, B.; Yoon, B.; Ahn, S.; Jang, J.; Lim, D.; Seo, I. High-Energy-Density Lithium–Sulfur Battery Based on a Lithium Polysulfide Catholyte and Carbon Nanofiber Cathode. *Energies* **2024**, *17*, 5258. [CrossRef]
372. Leckie, T.J.; Robertson, S.D.; Brightman, E. Recent Advances in in Situ/Operando Characterization of Lithium–Sulfur Batteries. *Energy Adv.* **2024**, *3*, 2479–2502. [CrossRef]
373. Sun, Y.-K. Emerging All-Solid-State Lithium–Sulfur Batteries: Holy Grails for Future Secondary Batteries. *ACS Energy Lett.* **2024**, *9*, 5092–5095. [CrossRef]
374. Zhang, X.; Ma, B.; Huang, S.; Han, D. Review on Multi-Functional Separator for Li-S Batteries. *Sustain. Polym. Energy* **2023**, *1*, 10003. [CrossRef]
375. Ovc-Okene, D.; Shankar, L.S.; Vizintin, A.; Kun, R. Revitalizing Li–S Batteries: The Power of Electrolyte Additives. *RSC Adv.* **2025**, *15*, 5381–5404. [CrossRef]
376. Bi, C.-X.; Hou, L.-P.; Li, Z.; Zhao, M.; Zhang, X.-Q.; Li, B.-Q.; Zhang, Q.; Huang, J.-Q. Protecting Lithium Metal Anodes in Lithium–Sulfur Batteries: A Review. *Energy Mater Adv.* **2023**, *4*, 0010. [CrossRef]
377. Jan, W.; Khan, A.D.; Iftikhar, F.J.; Ali, G. Recent Advancements and Challenges in Deploying Lithium Sulfur Batteries as Economical Energy Storage Devices. *J. Energy Storage* **2023**, *72*, 108559. [CrossRef]
378. Tiwari, S.; Yadav, V.; Poonia, A.K.; Pal, D. Exploring Advances in Sulfur Composite Cathodes for Lithium-Sulfur Batteries: A Comprehensive Review. *J. Energy Storage* **2024**, *94*, 112347. [CrossRef]
379. Cai, D.; Zheng, F.; Li, Y.; Zhang, C.; Qin, Z.; Li, W.; Liu, Y.; Li, A.; Zhang, J. Design of Coatings for Sulfur-Based Cathode Materials in Lithium-Sulfur Batteries: A Review. *Chem. Asian J.* **2024**, *19*, e202400099. [CrossRef]
380. Sandeep Rao, K.; Dutta Pathak, D.; Mandal, B.P.; Debnath, A.K.; Tyagi, A.K. Lithium–Sulfur Battery Performance Enhancement by Optimal Loading of Graphene Oxide on Separator with MXene/Reduced Graphene Oxide/Sulfur Cathode. *Mater. Today Sustain.* **2023**, *23*, 100414. [CrossRef]
381. Innocenti, A.; Bresser, D.; Garche, J.; Passerini, S. A Critical Discussion of the Current Availability of Lithium and Zinc for Use in Batteries. *Nat. Commun* **2024**, *15*, 4068. [CrossRef]
382. Mageto, T.; Bhoyate, S.D.; Mensah-Darkwa, K.; Kumar, A.; Gupta, R.K. Development of High-Performance Zinc-Ion Batteries: Issues, Mitigation Strategies, and Perspectives. *J. Energy Storage* **2023**, *70*, 108081. [CrossRef]

383. Nazir, G.; Rehman, A.; Lee, J.-H.; Kim, C.-H.; Gautam, J.; Heo, K.; Hussain, S.; Ikram, M.; AlObaid, A.A.; Lee, S.-Y.; et al. A Review of Rechargeable Zinc–Air Batteries: Recent Progress and Future Perspectives. *Nano-Micro Lett.* **2024**, *16*, 138. [CrossRef]
384. Iqbal, A.; El-Kadri, O.M.; Hamdan, N.M. Insights into Rechargeable Zn-Air Batteries for Future Advancements in Energy Storing Technology. *J. Energy Storage* **2023**, *62*, 106926. [CrossRef]
385. Tang, L.; Peng, H.; Kang, J.; Chen, H.; Zhang, M.; Liu, Y.; Kim, D.H.; Liu, Y.; Lin, Z. Zn-Based Batteries for Sustainable Energy Storage: Strategies and Mechanisms. *Chem. Soc. Rev.* **2024**, *53*, 4877–4925. [CrossRef]
386. Lu, Q.; Zou, X.; Bu, Y.; An, L.; Wang, Y.; Shao, Z. What Matters in Engineering Next-Generation Rechargeable Zn-Air Batteries? *Next Energy* **2023**, *1*, 100025. [CrossRef]
387. Hong, S.; Choi, Z.; Hwang, B.; Matic, A. Research Trends and Future Perspectives on Zn-Ion Batteries Using Ga-Based Liquid Metal Coatings on Zn Anodes. *ACS Energy Lett.* **2024**, *9*, 5421–5433. [CrossRef]
388. Huang, R.-B.; Wang, M.-Y.; Xiong, J.-F.; Zhang, H.; Tian, J.-H.; Li, J.-F. Anode Optimization Strategies for Zinc–Air Batteries. *eScience* **2024**, 100309. [CrossRef]
389. Abbasi, A.; Hosseini, S.; Somwangthanaroj, A.; Cheacharoen, R.; Olaru, S.; Kheawhom, S. Discharge Profile of a Zinc-Air Flow Battery at Various Electrolyte Flow Rates and Discharge Currents. *Sci. Data* **2020**, *7*, 196. [CrossRef]
390. Gourley, S.W.D.; Brown, R.; Adams, B.D.; Higgins, D. Zinc-Ion Batteries for Stationary Energy Storage. *Joule* **2023**, *7*, 1415–1436. [CrossRef]
391. Domenico Frattini Electrically Rechargeable Zinc-Air Batteries: One of the Most Sustainable Technologies of the Post-Lithium Era. Available online: https://cicenergigune.com/en/blog/electrically-rechargeable-zinc-air-batteries-the-most-sustainable-technologies (accessed on 3 February 2025).
392. *ECOS—European Environmental Citizens' Organisation for Standardisation Europe Needs an Ambitious Regulatory Framework to Guarantee Sustainability of Batteries*; ECOS: Brussels, Belgium, 2019.
393. Rufino Júnior, C.A.; Riva Sanseverino, E.; Gallo, P.; Koch, D.; Diel, S.; Walter, G.; Trilla, L.; Ferreira, V.J.; Pérez, G.B.; Kotak, Y.; et al. Towards to Battery Digital Passport: Reviewing Regulations and Standards for Second-Life Batteries. *Batteries* **2024**, *10*, 115. [CrossRef]
394. Vasileios, R.; Patricia, U. *Implementing the EU Digital Battery Passport*; CEPS: Brussels, Belgium, 2024.
395. Bassi, A. *Rules for the Calculation of the Carbon Footprint of Electric Vehicle Batteries (CFB-EV)*; Publications Office of the European Union: Luxembourg, 2023.
396. European Commission. *Joint Research Centre. Clean Energy Technology Observatory, Batteries for Energy Storage in the European Union: Status Report on Technology Development, Trends, Value Chains and Markets: 2022*; Publications Office: Luxembourg, 2022.
397. JRC RMIS—Raw Materials Information System. Available online: https://rmis.jrc.ec.europa.eu/ (accessed on 7 March 2025).
398. Battery Pass Consortium. Unlocking the Value of the EU Battery Passport; Systemiq, Amsterdam, Berlin, Jakarta, London, Munich, Paris and São Paulo. 2024. Available online: https://www.systemiq.earth/unlocking-the-value-of-the-eu-battery-passport/ (accessed on 31 March 2025).
399. Murdock, B.E.; Toghill, K.E.; Tapia-Ruiz, N. A Perspective on the Sustainability of Cathode Materials Used in Lithium-Ion Batteries. *Adv. Energy Mater.* **2021**, *11*, 2102028. [CrossRef]
400. Tan, J.; Keiding, J.K. The Cobalt and Lithium Global Supply Chains: Status, Risks and Recommendations. In *MiMa Rapport*; GEUS: Copenhagen, Denmark, 2023; Volume 2023. [CrossRef]
401. Accardo, A.; Dotelli, G.; Musa, M.L.; Spessa, E. Life Cycle Assessment of an NMC Battery for Application to Electric Light-Duty Commercial Vehicles and Comparison with a Sodium-Nickel-Chloride Battery. *Appl. Sci.* **2021**, *11*, 1160. [CrossRef]
402. Yudhistira, R.; Khatiwada, D.; Sanchez, F. A Comparative Life Cycle Assessment of Lithium-Ion and Lead-Acid Batteries for Grid Energy Storage. *J. Clean. Prod.* **2022**, *358*, 131999. [CrossRef]
403. Peters, J.F.; Baumann, M.; Zimmermann, B.; Braun, J.; Weil, M. The Environmental Impact of Li-Ion Batteries and the Role of Key Parameters—A Review. *Renew. Sustain. Energy Rev.* **2017**, *67*, 491–506. [CrossRef]
404. Porzio, J.; Scown, C.D. Life-Cycle Assessment Considerations for Batteries and Battery Materials. *Adv. Energy Mater.* **2021**, *11*, 202100771. [CrossRef]
405. Quan, J.; Zhao, S.; Song, D.; Wang, T.; He, W.; Li, G. Comparative Life Cycle Assessment of LFP and NCM Batteries Including the Secondary Use and Different Recycling Technologies. *Sci. Total Environ.* **2022**, *819*, 153105. [CrossRef] [PubMed]
406. Meegoda, J.N.; Malladi, S.; Zayas, I.C. End-of-Life Management of Electric Vehicle Lithium-Ion Batteries in the United States. *Clean Technol.* **2022**, *4*, 1162–1174. [CrossRef]
407. Reinhart, L.; Vrucak, D.; Woeste, R.; Lucas, H.; Rombach, E.; Friedrich, B.; Letmathe, P. Pyrometallurgical Recycling of Different Lithium-Ion Battery Cell Systems: Economic and Technical Analysis. *J. Clean. Prod.* **2023**, *416*, 137834. [CrossRef]
408. Thomas, F.; Mahdi, L.; Lemaire, J.; Santos, D.M.F. Technological Advances and Market Developments of Solid-State Batteries: A Review. *Materials* **2024**, *17*, 239. [CrossRef]

409. GLOBE NEWSWIRE Solid State Battery Market Size and Forecast 2020–2030: Market Poised for Strong Growth with Projected CAGR of Over 30% from 2023 to 2030. Available online: https://www.globenewswire.com/news-release/2024/12/10/2994815/28124/en/Solid-State-Battery-Market-Size-and-Forecast-2020-2030-Market-Poised-for-Strong-Growth-with-Projected-CAGR-of-Over-30-from-2023-to-2030.html (accessed on 17 April 2025).
410. Nekahi, A.; Dorri, M.; Rezaei, M.; Bouguern, M.D.; Madikere Raghunatha Reddy, A.K.; Li, X.; Deng, S.; Zaghib, K. Comparative Issues of Metal-Ion Batteries toward Sustainable Energy Storage: Lithium vs. *Sodium. Batteries* **2024**, *10*, 279. [CrossRef]
411. Fraunhofer Institute for Manufacturing Technology and Advanced Materials IFAM BMBF Project Launched for Fast Industrial Implementation of Sodium-Ion Technology. Available online: https://www.ifam.fraunhofer.de/en/Press_Releases/bmbf-project-sodium-ion-technology.html (accessed on 17 April 2025).
412. Endo, C.; Kaufmann, T.; Schmuch, R.; Thielmann, A. *Unav Benchmarking International Battery Policies*; Fraunhofer-Gesellschaft: München, Germany, 2024.
413. Transport & Environment. *An Industrial Blueprint for Batteries in Europe*; Transport & Environment: Brussels, Belgium, 2024.

MDPI AG
Grosspeteranlage 5
4052 Basel
Switzerland
Tel.: +41 61 683 77 34

Batteries Editorial Office
E-mail: batteries@mdpi.com
www.mdpi.com/journal/batteries

www.ingramcontent.com/pod-product-compliance
Lightning Source LLC
Chambersburg PA
CBHW041414210926
51850CB00013B/3271
* 9 7 8 3 7 2 5 8 4 4 2 3 4 *